Calculus Refresher

By A. ALBERT KLAF

DOVER PUBLICATIONS, INC.
NEW YORK

This Dover edition, first published in 1956, is an un-
abridged and unaltered republication of the work originally
published in 1944 by Whittlesey House, a division of the
McGraw-Hill Book Company, under the title *Calculus Refresher for Technical Men*.

Standard Book Number: 486-20370-0
Library of Congress Catalog Card Number: 57-13502

Manufactured in the United States of America
Dover Publications, Inc., 31 East 2nd Street, Mineola, N.Y. 11501

PREFACE

Today the watchword is *speed*—for the student, the engineer, the technician, the designer, the statistician—in short, for all who desire rapid integration into a mechanized world.

The primary purpose of this book, therefore, is to make available, for ready and rapid use, a "refresher" on the fundamental concepts, methods, and practical applications of simple calculus. It is designed, chiefly, for those who have once studied the subject in the usual lengthy volumes and have found themselves swamped by details, not easily understood and, hence, more readily forgotten.

The subject matter is divided into the two customary general sections, one on differential calculus and one on integral calculus. These are followed by a third section on applications of calculus to various fields of technology. The book omits as many nonessentials as possible and, at the same time, retains enough details to clarify fundamental concepts.

For ease of comprehension, the entire subject is offered in the modern question and answer form—a unique presentation of calculus. Typical examples and problems are worked out in detail to illustrate the laws and principles involved in their solution. This gives the student a sense of quick mastery of the usable fundamentals of the subject.

It is hoped that this refresher will prove adaptable as a ready reference book, especially where rapidity of attainment is essential.

A. ALBERT KLAF.

CONTENTS

SECTION I SIMPLE DIFFERENTIAL CALCULUS

SECTION I

SIMPLE DIFFERENTIAL CALCULUS

Chapter I

CONSTANTS—VARIABLES—FUNCTIONS—
INCREMENTS

1. What is a constant?

A quantity whose value is fixed.

2. What is a numerical or absolute constant?

A constant that always has the same value, as 1, 2, 7, $\sqrt{3}$, π, ϵ, etc.

3. What is an arbitrary constant?

A constant that continues to have the same value throughout one problem but may have another value in a different problem. It may be represented by a letter from the beginning of the alphabet, as a, b, c, d, etc.

4. What is a variable?

A quantity that may assume an indefinite number of values in the same problem and which may be represented usually by a letter from the end of the alphabet, as s, t, u, v, w, x, y, or z.

5. What is meant by an interval of a variable?

The variable is considered confined to take on all values lying only between two numbers, as $[a, b]$, a being less than b.

6. When is a variable said to vary continuously through an interval [a, b]?

When the variable assumes in succession all possible intermediate values from a to b.

7. What is a function?

A function is a relationship between variables.

$\left.\begin{array}{l} \dfrac{y}{x} = \tan 60° \\[2mm] x^2 + y^2 = r^2 \end{array}\right\}$ are functions of x and y, for they indicate a relationship between x and y.

If $s = \dfrac{gt^2}{2}$, where s is the distance a body will fall in t sec. and g is the constant acceleration due to gravity, s is said to be a function of t.

8. What is a dependent variable?

A variable whose value depends upon the value of another variable.

If $y = x^2 \tan 30°$, then y is the dependent variable. It depends on the value of the variable x.

If $s = \dfrac{gt^2}{2}$, then s is the dependent variable. It depends on the value of the variable t.

9. What is an independent variable?

A variable whose value determines the value of the related dependent variable.

If $y = x^2 \tan 30°$, then x is the independent variable. It determines the value of y.

If $s = \dfrac{gt^2}{2}$, then t is the independent variable. It determines the value of s.

10. What is the general relationship of the dependent and independent variables?

The dependent variable is a *function* of the independent variable. In the above, y is a function of x; s is a function of t.

11. May the dependent variable ever be taken as the independent variable?

Frequently, when two variables are related, either may be taken as the independent variable and the other as the dependent.

Example

In a circle whose radius is r and whose area is A, r may be assumed as depending upon A or A as depending upon r. A change in either variable will cause a corresponding change in the other.

12. What is an implicit function?

It is a function expressing an *unsolved* relationship between the variables.

$$\left.\begin{array}{l} \dfrac{y}{x} = \tan 60° \\[2mm] x^2 + y^2 = a^2 \\[2mm] z + y^2 \tan x = a \end{array}\right\} \begin{array}{l} \text{are functions expressing an unsolved} \\ \text{relationship between variables.} \end{array}$$

13. What is an explicit function?

A function expressing a *solved* relationship between the variables. One variable is solved in terms of the other.

$y = x \tan 60°$; y is an explicit function of x.

$x = \dfrac{y}{\tan 60°}$; x is an explicit function of y.

$x = \pm \sqrt{a^2 - y^2}$; x is an explicit function of y.

$z = a - y^2 \tan x$; z is an explicit function of y and x.

The dependent variable is therefore the value of the explicit function.

14. What are the usual symbols for expressing a function in a general way?

$y = f(x)$ is read "*y* is a function of *x*" and means "*y* depends on the value of *x*."

$s = F(t)$ is read "*s* is a function of *t*" and means "*s* depends on the value of *t*."

$u = \phi(v)$ is read "*u* is a function of *v*" and means "*u* depends on the value of *v*."

Other letters may also be used to express a function. Different letters are used to represent the functions when different relations exist between the variables. The same letter is used when the same relation exists though the variables are different.

15. What are the usual symbols for expressing an implicit function in a general way?

$F(x, y, z); f(x, y, z); \phi(x, y, z)$.   Each expression indicates an implicit function in terms of x, y, z.

16. What are the usual symbols for expressing an explicit function in a general way?

$x = F(y, z); x = f(y, z); x = \phi(y, z)$.   Each expression denotes x as an explicit function of y and z.

17. How may the general symbol, as $f(x)$, for a function be used to indicate substitutions for the variable in the function?

If $f(x) = x^2 + x - 3$
then $f(a) = a^2 + a - 3$
and $f(2) = (2)^2 + 2 - 3 = 3$

18. When is a function said to be single valued for $x = a$?

When only one value of the function corresponds to $x = a$. If $y = 3x + 2$, then y is single valued for every value of x.

19. When is a function multiple valued for $x = a$?

When two or more values of the function correspond to $x = a$. If $x^2 + y^2 = 9$ or $y = \pm \sqrt{9 - x^2}$, then, for every value of x numerically less than 3, there correspond two real values of y.

20. What is meant by continuity and discontinuity of a function?

The function $y = 2x^2$ is continuous for all values of x because, if x varies continuously from any value $x = a$ to $x = b$, then y will vary continuously from $y = 2a^2$ to $y = 2b^2$, and any point $P(x, y)$ will move continuously along the graph. Here the function is defined for all values of the independent variable.

The function $y = \frac{2}{x}$ is not defined for $x = 0$, as $\frac{2}{0}$ is meaningless. There is no point on the graph for $x = 0$, and the function is discontinuous for $x = 0$. But if x increases continuously through any interval $[a, b]$ that does not include $x = 0$, then y will decrease continuously from $\frac{2}{a}$ to $\frac{2}{b}$ and the point $P(x, y)$ will move continuously along the graph within that interval. For a definition, see the chapter on Limits.

21. What is an increment?

An increment is any change or growth of a variable or of a function. It is the difference found by subtracting the first value from the changed, or second, value of the variable or function. The change or growth may be positive or negative according as the variable or function increases or decreases when changing.

Example

The area of a circle $A = \pi r^2$. The area A is a function of the radius r (A depends on r); any change in r is called an increment of r, and a corresponding change in A is called an increment of A.

22. What symbol is used to denote an increment?

The Greek letter Δ (delta) denotes an increment.

$$\Delta x = \text{an increment of } x$$

In $A = \pi r^2$, $\Delta r =$ an increment of the radius

$$\Delta A = \text{a corresponding increment of the function}$$

PROBLEMS

1. How would you express the circumference of a circle as a function of its area?

2. How would you express the area of a square as a function of its diagonal?

3. What is the relation of the surface of a sphere to its volume?

4. How would you express the intensity of stress on the outer fiber of a beam as a function of the bending moment and section modulus?

5. What is the expression for the relation of the pressure head to the pressure and weight of a liquid?

6. How would you express the horsepower of an electric power machine as a function of the electromotive force (e.m.f.) and the current?

7. If $f(x) = \sin x$, what is (a) $f(0°)$; (b) $f(30°)$; (c) $f(\sin^{-1} 1)$; (d) $f(\cos^{-1} - \frac{1}{2})$; (e) $f(\cot^{-1} - 1)$?

8. If $F(x, y) = 5x^3 y + 3x^2 y - 6y^2$, what is (a) $F(x, -y)$; (b) $F(-x, y)$; (c) $F(-x, -y)$?

9. If $F(x) = 2^x$, what is (a) $F(y)$; (b) $F(x + y)$?

10. If $y = \cos x$, what is the explicit expression of x in terms of y?

11. If $x = 4^y$, what is the explicit expression of y in terms of x?

12. If $x^2 + y^2 = 8$, how would you express each variable as an explicit function of the other?

Chapter II

LIMITS

23. What is meant by the limit of a variable?

If a variable x approaches more and more closely a constant value c, so that $c - x$ eventually becomes and remains less, in absolute value, than any preassigned positive number, however small, the constant c is said to be the limit of x.

Example

a. If a variable assumes the successive values of $\frac{1}{2}$, $\frac{1}{4}$, $\frac{1}{8}$, $\frac{1}{16}$, $\cdots$, $\frac{1}{2^n}$, the variable approaches zero as a limit because the difference between zero and the variable eventually becomes and remains less than any preassigned number, however small. For instance, let the assigned number be $\frac{1}{100,000}$. The sixteenth value of the variable is $\frac{1}{65,536}$, and the seventeenth value of the variable is $\frac{1}{131,072}$; the latter differs numerically from zero by less than $\frac{1}{100,000}$ as do all the subsequent values.

b. Let the values of x be $3 + 1$, $3 + \frac{1}{2}$, $3 + \frac{1}{4}$, $3 + \frac{1}{8}$, $\cdots$, $3 + \frac{1}{2^n}$ without end. Then the limit of x is 3.

24. How may we illustrate that a variable x is approaching 1 as a limit?

Let line $OA = 1$.

If a point x starts from O and during the first second moves half the length to x_1, during the next second half the

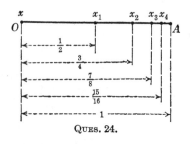

QUES. 24.

remaining distance to x_2, continuing in this way to move half the remaining distance during each successive second, then the distance that the point x is from O is a variable of which $OA = 1$ is the limit because the difference between $OA = 1$ and the variable ultimately becomes and remains less than any preassigned number, however small.

25. Should we be concerned with the smallness of the fraction by which the variable differs from its limit?

The smallness of the fraction by which the variable approximates the limit is of no primary importance. All we are interested in is the limit, which is fixed.

26. What symbols are used to indicate that a variable or a function is approaching a limit?

$x \to a$ means "x approaches a as a limit."
$\lim\limits_{x \to a} [f(x)] = c$ means "the limit of $f(x)$, as x approaches a as a limit, is c."

27. When is a variable said to increase without limit, or to become infinite?

When the variable changes in such a manner that it becomes and remains greater than any assigned positive number, however great.

$x \to \infty$ means "x increases without limit, or becomes infinite."

$\lim_{x \to \infty} [f(x)] = c$ means "the limit of $f(x)$, as x becomes infinite, is c."

28. When is a variable said to decrease without limit, or to become infinite negatively?

When the variable changes in such a manner that it becomes and remains less than any assigned negative number, however great in absolute value.

$x \to -\infty$ means "x decreases without limit, or becomes infinite negatively."

29. What are the elementary theorems of limits?

T1. When two variables, each approaching a limit, are equal for all their successive values, their limits are equal.

T2. When a constant is added to a variable that approaches a limit, then the limit of their sum is the sum of the constant and the limit of the variable.

T3. When a variable that approaches a limit is multiplied by a constant, then the limit of their product is the product of the constant and the limit of the variable.

T4. The limit of a sum of a number of variables, each of which is approaching a limit, is the sum of their respective limits.

T5. The limit of a product of a number of variables, each of which is approaching a limit, is the product of their respective limits.

T6. The limit of the quotient of two variables, each of which is approaching a limit, is the quotient of their limits, except when the limit of the divisor is zero. If the limit of the divisor is zero, the *limit of the quotient* may have a definite finite value or the quotient may become infinite, but it is not determined by finding the quotient of the limits of the two variables.

30. If $y = \dfrac{10}{x+3}$ and x is a variable approaching 2 as a limit, what is the limit of y?

By T2, $\qquad \lim\limits_{x \to 2} (x + 3) = 5$

$\qquad\qquad\qquad \therefore y = \tfrac{10}{5} = 2$

and y is a variable approaching 2 as a limit.

31. What is the $\lim\limits_{x \to 2} (x^2 + 3x)$?

$\qquad \lim\limits_{x \to 2} (x^2) = 4, \quad$ since $x^2 = x \cdot x$, by T5

$\qquad\qquad \lim\limits_{x \to 2} (3x) = 3 \cdot \lim\limits_{x \to 2} x = 6 \qquad$ by T3

$\qquad \therefore \lim\limits_{x \to 2} (x^2 + 3x) = 4 + 6 = 10 \qquad$ by T4

32. What is the $\lim\limits_{x \to 2} \left(\dfrac{x^2 - 5}{x + 3}\right)$?

$\qquad \lim\limits_{x \to 2} (x^2 - 5) = 4 - 5 = -1 \quad$ by T5 and T2

$\qquad \lim\limits_{x \to 2} (x + 3) = 5 \qquad\qquad$ by T2

$\qquad \therefore \lim\limits_{x \to 2} \left(\dfrac{x^2 - 5}{x + 3}\right) = -\dfrac{1}{5} \qquad$ by T6

33. When is a function $f(x)$ said to be continuous and when discontinuous for $x = a$?

A function $f(x)$ is continuous for $x = a$ when

$$\lim_{x \to a} f(x) = f(a)$$

This means that, if the limit of the function, as x approaches a as a limit, is obtained by substituting a for x, then the function is continuous for $x = a$. The function is discontinuous for $x = a$ if this condition is not satisfied.

Example

$\lim\limits_{x \to 2} (x^2 + 3x) = 10$. It is seen that the limit of the function as $x \to 2$ is exactly the value of the function when 2 is substituted for x. Therefore, the function $x^2 + 3x$ is said to be continuous for $x = 2$.

34. Is $y = \dfrac{x^2 - 9}{x - 3}$ **continuous for** $x = 2$?

For $x = 2$, $y = f(x) = f(2) = \dfrac{(2)^2 - 9}{2 - 3} = 5$

Also, $\lim\limits_{x \to 2} \dfrac{x^2 - 9}{x - 3} = 5$

Therefore, the function is continuous for $x = 2$.

35. When a function is not already defined for $x = a$, **is it sometimes possible to give the function such a value for** $x = a$ **as to satisfy the condition of continuity?**

Example

$y = \dfrac{x^2 - 9}{x - 3}$ is not defined for $x = 3$ because there would be division by zero.   But for any other value of x,

$$y = \dfrac{(x + 3)(x - 3)}{x - 3} = x + 3$$

and $\lim\limits_{x \to 3} (x + 3) = 6$

$$\therefore \lim\limits_{x \to 3} \dfrac{x^2 - 9}{x - 3} = 6$$

Now if we arbitrarily let the value of the function be 6 for $x = 3$, the function becomes continuous for this value.

36. When is a function continuous in an interval?

When it is continuous for all values of x in this interval.

37. When is a function discontinuous for $x = a$?

If $f(x)$ becomes infinite as x approaches a as a limit, then $f(x)$ is discontinuous for $x = a$, that is, when

$$\lim\limits_{x \to a} f(x) = \infty$$

38. May a function have a limiting value when the independent variable becomes infinite?

It certainly may; for $\lim\limits_{x \to \infty} \dfrac{1}{x} = 0$; $\lim\limits_{x \to \infty} \dfrac{c}{x} = 0$. And in general, if $f(x)$ approaches the constant value c as a limit when $x \to \infty$, $\lim\limits_{x \to \infty} f(x) = c$.

Example

$f(x) = 3 + \dfrac{1}{(x-1)^2}$. As x becomes infinite in either sense, the fraction tends to zero and $f(x)$ approaches 3 as a limit. That is, $\lim\limits_{x \to \infty} f(x) = 3$.

39. What are some special limits that occur frequently?

(a) $\lim\limits_{x \to 0} \dfrac{a}{x} = \infty$ (b) $\lim\limits_{x \to \infty} ax = \infty$

(c) $\lim\limits_{x \to \infty} \dfrac{x}{a} = \infty$ (d) $\lim\limits_{x \to \infty} \dfrac{a}{x} = 0$

The constant a is not zero.

40. What is the limit $\lim\limits_{x \to 3} \dfrac{5x}{6 - 2x}$?

$$\lim\limits_{x \to 3} [6 - 2x] = 0$$

$$\therefore \lim\limits_{x \to 3} \frac{5x}{6 - 2x} = \infty$$

41. What is the limit $\lim\limits_{x \to \infty} \left[\dfrac{x^4 + x^3 + x^2 + 2}{3x^4 + x^2 - 1} \right]$?

Divide the numerator and denominator by x^4.

Then $\dfrac{x^4 + x^3 + x^2 + 2}{3x^4 + x^2 - 1} = \dfrac{1 + \dfrac{1}{x} + \dfrac{1}{x^2} + \dfrac{2}{x^4}}{3 + \dfrac{1}{x^2} - \dfrac{1}{x^4}}$

As x becomes very large, the fractions $\frac{1}{x}$, $\frac{1}{x^2}$, $\frac{2}{x^4}$, and $\frac{1}{x^4}$ become very small and approach 0 as a limit, and the value of the fraction approaches $\frac{1}{3}$ as a limit.

$$\therefore \lim_{x \to \infty} \left[\frac{x^4 + x^3 + x^2 + 2}{3x^4 + x^2 - 1} \right] = \frac{1}{3}$$

42. What is the limit [tan θ]?
$$\theta \to \frac{\pi}{2}$$

As θ approaches $\frac{\pi}{2}$ from values of θ smaller than $\frac{\pi}{2}$, tan θ increases without limit.

But when θ approaches $\frac{\pi}{2}$ from larger values, then tan θ decreases without limit.

Therefore, tan $\frac{\pi}{2}$ has no numerical value.

43. What is the limit $\left[\dfrac{\sin x}{x} \right]$ where x is in radians?
$$x \to 0$$

For values of x near 0, $\dfrac{\sin x}{x}$ approaches 1 as a limit.

For $x = 0$, $\sin x = 0$ and $\dfrac{\sin x}{x} = \dfrac{0}{0}$, which is indeterminate.

PROBLEMS
Prove that the limits of the following are as given:

1. $\lim\limits_{x \to 0} \left[\dfrac{5 + 3x}{7 - 2x} \right] = \dfrac{5}{7}$

2. $\lim\limits_{x \to 1} \left[\dfrac{4x^2 - 3x + 1}{x^2 - 3x + 2} \right] = \infty$

3. $\lim\limits_{x \to \infty} \left[\dfrac{5x^2 - 3x + 1}{x^2 + x - 1} \right] = 5$

4. $\lim\limits_{x \to \frac{\pi}{2}} [\sec x] = \infty$

5. $\lim\limits_{x \to \infty} \left[\dfrac{x^4 - 1}{2x^3 + 1} \right] = \infty$

6. $\lim\limits_{x \to 0} \left[\dfrac{\cos x}{x} \right] = \infty$

7. $\lim\limits_{x \to \infty} \left[\dfrac{4 - 3x^2}{2x + 4x^2} \right] = -\dfrac{3}{4}$

8. $\lim\limits_{x \to 0} \left[\dfrac{\sin x}{x + 1} \right] = 0$

9. $\lim\limits_{x \to 2} \left[\dfrac{x^2 + x - 12}{x^2 - 9} \right] = \dfrac{6}{5}$

10. $\lim\limits_{x \to \frac{\pi}{2}} \left[\dfrac{x}{\tan x} \right] = 0$

Chapter III

DERIVATIVES

44. What is the primary concern of differential calculus?

The primary concern is the determination of the rates of change or growth of related variables, *i.e.*, the amount of change or growth in the function (dependent variable) per unit change or growth in the value of the independent variable. Growth or change may be positive or negative.

45. How are independent and dependent variables represented in plotting a function?

Abscissas represent the changing values of the independent variable, and ordinates represent the corresponding changed values of the function (dependent variable).

46. What is meant by the average rate of change or growth of a function in any interval?

It is the amount of growth during the interval divided by the number of units in the interval, or the amount of growth in the value of the function (ordinate) during the interval divided by the amount of growth in the value of the independent variable during the interval.

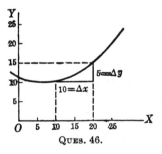

Ques. 46.

In the figure the increment or growth in the function (ordinate) is $\Delta y = 5$. The growth of the independent variable (abscissa) is from 10 to 20 or $\Delta x = 10$.

$$\therefore \text{ Average rate of growth} = \frac{\Delta y}{\Delta x} = \frac{5}{10} = 0.5$$

14

In general, if $y = f(x)$ and we give x the increment or growth Δx, then y becomes

$$y + \Delta y = f(x + \Delta x)$$

Subtracting

$$y \quad\quad = f(x)$$

to get the growth of the function

we get

$$\Delta y = f(x + \Delta x) - f(x)$$

Now divide by Δx to get the average rate of growth.

We get $\dfrac{\Delta y}{\Delta x} = \dfrac{f(x + \Delta x) - f(x)}{\Delta x}$ = the average rate of growth of y with respect to x in the interval from x to $x + \Delta x$.

47. If a ball is thrown into the air and $h = 100t - 16t^2$ expresses the relation between the height h in feet and the time t in seconds, what is the average rate of change (speed in this case) throughout the interval from $t = 2$ seconds to $t = 2.01$ seconds?

When $t = 2$,

$$h = 100 \cdot 2 - 16 \cdot (2)^2 = 136 \text{ ft.}$$

When $t = 2.01$,

$$h = 100 \cdot 2.01 - 16 \cdot (2.01)^2 = 136.3584 \text{ ft.}$$

$$\therefore \frac{\Delta h}{\Delta t} = \frac{0.3584}{0.01} = 35.84 \text{ ft./sec.} = \text{ the average rate of}$$
change (speed) throughout the interval

48. When is the average rate of growth of two related variables uniform or constant?

Only when the function is described by a straight line.

Example

If $y = mx + b$ and x takes on an increment,

then $y + \Delta y = m(x + \Delta x) + b = mx + m \cdot \Delta x + b$

Subtract $y \qquad\qquad = \qquad\qquad mx \qquad\qquad + b$

$$\Delta y = m \cdot \Delta x$$

and $\dfrac{\Delta y}{\Delta x} = m = \text{slope} = \text{constant}$

Either variable changes uniformly with respect to the other.

49. When are two related variables changing nonuniformly with respect to one another?

When the function is described by a curve, then the ratio of the corresponding increments is variable.

QUES. 48.

Example

If $s = \dfrac{gt^2}{2}$ and t takes on an increment,

then $s + \Delta s = \dfrac{g}{2}(t + \Delta t)^2 = \dfrac{g}{2}t^2 + gt \cdot \Delta t + \dfrac{g}{2}(\Delta t)^2$

Subtract s $\qquad\qquad\qquad = \dfrac{gt^2}{2}$

getting $\Delta s = gt \cdot \Delta t + \dfrac{g}{2}(\Delta t)^2$

Dividing by Δt,

$$\dfrac{\Delta s}{\Delta t} = gt + \dfrac{g}{2} \cdot \Delta t$$

Therefore, $\dfrac{\Delta s}{\Delta t}$ varies with t, and the change is nonuniform, or not constant. Different values of t give different values of $\dfrac{\Delta s}{\Delta t}$.

50. How may we obtain the instantaneous direction of a curve at any point?

By drawing a tangent line to the curve at the point.

51. What is meant by the instantaneous rate of change or growth of a function?

Since the slope of a straight line is a measure of the rate of change of a function represented by the line, then the slope of a tangent line is a measure of the instantaneous rate of growth of the function at the point in question.

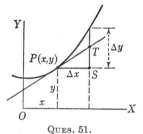

QUES. 51.

The tangent line indicates the change or growth during an interval, if the growth continues throughout the interval as at the beginning of the interval.

ST indicates the change or growth during the interval Δx, if the growth continues as at the beginning of the interval. It is the instantaneous increment for point $P(x, y)$.

52. Why is it necessary to examine the average rates of change or growth of the variables for very small intervals?

Because "instantaneous" may be misleading owing to the fact that any change, however small, requires an interval, however minute.

Example

The speed of an automobile at a particular instant cannot be taken as the number of miles traveled during an hour containing the instant or even as the number of feet covered during a second containing the instant. The smaller the interval, however, the closer the average rate for the interval approaches the instantaneous rate at the beginning of the interval.

53. What is the specific relationship of the instantaneous rate to that of the approaching average rate as the interval continues to become smaller and smaller?

The rate at any instant is the *limit* that the average rate is approaching as the interval is made smaller and smaller or is made to approach zero.

54. Does this mean that we are dealing with an approxi-mate quantity and that the instantaneous rate is approximated?

Not at all.   To be sure, we are not concerned with the variable difference, no matter how small, between the average rate and the instantaneous rate as the interval approaches zero.   We are interested only in the limit (instantaneous rate) which the variable is approaching and which is a precise quantity.

55. Can you show that an expression with very large coefficients may approach some small finite number as a limit as the interval Δx approaches zero?

Let us consider $y = 1 + 1,000,000\,\Delta x + 1,000,000,000(\Delta x)^2$

If $\Delta x = 0.1$,
$y = 1 + 100,000 + 10,000,000 \qquad = 10,100,001$
If $\Delta x = 0.01$,
$y = 1 + 10,000 + 100,000 \qquad = 110,001$
If $\Delta x = 0.001$,
$y = 1 + 1,000 + 1,000 \qquad = 2,001$
If $\Delta x = 0.0001$,
$y = 1 + 100 + 10 \qquad = 111$
If $\Delta x = 0.00001$,
$y = 1 + 10 + 0.1 \qquad = 11.1$
If $\Delta x = 0.000001$,
$y = 1 + 1 + 0.001 \qquad = 2.001$
If $\Delta x = 0.000000001$,
$y = 1 + 0.001 + 0.000000001 \qquad = 1.001000001$
If $\Delta x = 0.00000000001$,
$y = 1 + 0.00001 + 0.0000000000001 = 1.0000100000001$

As we take Δx smaller, making it approach zero as a limit, we make y come close at will to 1, which is the limit of the above function. The limit is a definite, precise quantity that we always desire to get and that is what we actually do get.

56. How would you show graphically that the average rate of increase is approaching the instantaneous rate as a limit when the interval approaches zero?

$\dfrac{\Delta y}{\Delta x}$ is the slope of the secant through points A, A_1 and represents the average rate of increase of the function y for the interval Δx.

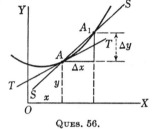

As Δx approaches the limiting value zero, point A_1 approaches the limiting position A and the secant SS revolves about A to the limiting position TT, which is the tangent to the curve at A.

Therefore, TT is the exact limiting position of the secant as Δx approaches zero.

QUES. 56.

The slope of the tangent line TT is a measure of the exact limit that the ratio $\dfrac{\Delta y}{\Delta x}$ approaches as Δx approaches zero.

The slope of the tangent line TT is the instantaneous rate at the point A.

57. If a ball is thrown into the air and $h = 100t - 16t^2$ expresses the relation between the height h in feet and the time t in seconds, what limit is the speed approaching when t approaches 2 seconds?

Start with an interval of time beginning at 2 seconds, and let 0.01 be the duration of the interval.

When $t = 2$,

$$h = 100 \cdot 2 - 16 \cdot (2)^2 = 136 \text{ ft.}$$

When $t = 2.01$,

$$\Delta t = 0.01 \quad \text{and} \quad h = 100 \cdot 2.01 - 16 \cdot (2.01)^2$$
$$= 136.3584 \text{ ft.}$$

obtaining $\Delta h = 0.3584$ and $\dfrac{\Delta h}{\Delta t} = \dfrac{0.3584}{0.01} = 35.84$, which is the average rate of growth (speed) during the interval.

Continue to follow through the calculations, making the interval smaller and smaller.

Δt, sec.	Δh, ft.	$\dfrac{\Delta h}{\Delta t}$, ft./sec.
0.01	0.3584	35.84
0.001	0.035984	35.984
0.0001	0.00359984	35.9984
0.00001	0.0003599984	35.99984
0.000001	0.000035999984	35.999984

As the interval Δt becomes smaller and smaller, approaching zero as a limit, the average rate of growth $\dfrac{\Delta h}{\Delta t}$ comes closer and closer to some limiting number, possibly 36.

NOTE.—Although both Δh and Δt approach zero as a limit, $\dfrac{\Delta h}{\Delta t}$ does not do so.

58. How can we find the exact value of this limiting number for $t = 2$ seconds?

We are given

$$h = 100t - 16t^2 \tag{1}$$

Now, if t grows to $t + \Delta t$, then h grows to $h + \Delta h$. And Eq. (1) becomes

$$h + \Delta h = 100(t + \Delta t) - 16(t + \Delta t)^2$$

or $\qquad h + \Delta h = 100t + 100\,\Delta t - 16t^2 - 32t \cdot \Delta t$
$$- 16(\Delta t)^2$$

Subtract $h \qquad = 100t \qquad\qquad -16t^2$ to get the growth
of the function

obtaining $\qquad \Delta h = \qquad\qquad 100\,\Delta t \qquad -32t \cdot \Delta t$
$$- 16(\Delta t)^2$$

Dividing by Δt to get the average rate of growth for the interval Δt,

$$\frac{\Delta h}{\Delta t} = 100 - 32t - 16 \cdot \Delta t = \text{average rate of increase, or}$$

growth, for any interval Δt

Now let Δt approach zero as a limit. Then, by the theorems of limits, $\dfrac{\Delta h}{\Delta t}$ approaches $100 - 32t$, which is then the limiting value approached by the average speed when Δt approaches zero. This limit is a precise quantity representing the exact rate of increase at the instant t.

Now, for $t = 2$,

$$100 - 32t = 100 - 32 \cdot 2 = 36 \text{ ft./sec.}$$

which is therefore the limit for the preceding question.

59. What is the rate of change or growth of function y with respect to x at the point $x = 5$ when given $2y = x^2 + 3x - 1$?

Let x grow to $x + \Delta x$; then y becomes $y + \Delta y$, and the given function becomes

$$2(y + \Delta y) = (x + \Delta x)^2 + 3(x + \Delta x) - 1$$

or $\qquad 2y + 2\,\Delta y = x^2 + 2x \cdot \Delta x + (\Delta x)^2 + 3x$
$$+ 3 \cdot \Delta x - 1$$

Subtract $2y \qquad = x^2 \qquad\qquad\qquad + 3x \qquad - 1$

getting $\qquad 2\,\Delta y = \qquad 2x \cdot \Delta x + (\Delta x)^2 + 3 \cdot \Delta x$

Dividing by Δx,

$$\frac{2\,\Delta y}{\Delta x} = \frac{2x \cdot \Delta x}{\Delta x} + \frac{(\Delta x)^2}{\Delta x} + \frac{3\,\Delta x}{\Delta x} = 2x + \Delta x + 3$$

and $\qquad \dfrac{\Delta y}{\Delta x} = x + \dfrac{\Delta x}{2} + \dfrac{3}{2}$

Now let Δx become smaller and smaller and approach zero. Then $\dfrac{\Delta y}{\Delta x}$ approaches $x + \dfrac{3}{2}$, which is the precise limit.

Now, for $x = 5$,

$$\lim_{\Delta x \to 0} \frac{\Delta y}{\Delta x} = 5 + \frac{3}{2} = 6\frac{1}{2}$$

This means that the instantaneous rate of change or growth of the function represented by the curve at the point $x = 5$ is $6\frac{1}{2}$.

The function, it is seen, changes $6\frac{1}{2}$ times as fast as the independent variable x at $x = 5$.

The slope of the tangent at $x = 5$ is $6\frac{1}{2}$.

60. What is a derivative?

A derivative $= \lim\limits_{\Delta x \to 0} \dfrac{\Delta y}{\Delta x}$, which means the *limiting* value of the ratio $\dfrac{\Delta y}{\Delta x}$ ($=$ average rate of growth) as the interval Δx of the independent variable approaches the limit zero.

It is the *instantaneous rate of growth* or change of a function at the point in question.

61. What symbols are used to express a derivative of y with respect to x?

$\dfrac{dy}{dx}$ is the usual symbol for the derivative of y with respect to x.

Other symbols to denote a derivative are:

If $y = f(x)$,

then $\qquad y' = \dfrac{d}{dx}\, y = \dfrac{d}{dx}\, f(x) = f'(x)$

Thus $\dfrac{d}{dx}\, f(x)$ means the derivative of $f(x)$ with respect to x.

62. What is differentiation?

The process of finding the derivative $\dfrac{dy}{dx}$, or $\lim\limits_{\Delta x \to 0} \dfrac{\Delta y}{\Delta x}$.

63. How is the derivative generally determined?

By the general process indicated in

$$\frac{d}{dx}[f(x)] = \lim_{\Delta x \to 0}\left[\frac{f(x + \Delta x) - f(x)}{\Delta x}\right]$$

or

$$\frac{dy}{dx} = \lim_{\Delta x \to 0}\frac{\Delta y}{\Delta x}$$

Example

If $y = 3x^2 + 2x = f(x)$,

then $\dfrac{d}{dx}(3x^2 + 2x)$

$$= \lim_{\Delta x \to 0}\left\{\frac{[3(x + \Delta x)^2 + 2(x + \Delta x)] - [3x^2 + 2x]}{\Delta x}\right\}$$

64. Is the symbol $\dfrac{dy}{dx}$ to be considered as a fraction?

The expression $\dfrac{\Delta y}{\Delta x}$ is actually a fraction since Δy and Δx are always finite quantities having definite values, but the symbol $\dfrac{dy}{dx}$, in the sense that it is a symbol for a derivative, is to be considered as a *limiting value of a fraction*. This symbol does possess fractional properties, however, and will be considered in the chapter on Differentials.

65. What is an infinitesimal?

An infinitesimal is a *variable* that approaches zero as a limit. A constant, however small numerically, is not an infinitesimal.

66. Is an infinitesimal restricted to values in the immediate neighborhood of zero?

It need not be so restricted. For instance, in the definition of the derivative $\frac{dy}{dx} = \lim_{\Delta x \to 0} \frac{\Delta y}{\Delta x}$, Δx and Δy are infinitesimals, but they may initially have large absolute values. Only because these increments approach zero as a limit are they infinitesimals. They ultimately become and remain less in numerical value than any preassigned constant, however small.

67. What is the instantaneous rate of growth of the function $y = \dfrac{2x}{x+1}$ for any value of x and for $x = 2$?

$$y + \Delta y = \frac{2(x + \Delta x)}{(x + \Delta x) + 1}$$

$$\Delta y = \frac{2x + 2 \cdot \Delta x}{x + \Delta x + 1} - \frac{2x}{x + 1}$$

$$= \frac{(2x + 2 \cdot \Delta x)(x + 1) - 2x(x + \Delta x + 1)}{(x + \Delta x + 1)(x + 1)}$$

$$= \frac{2x^2 + 2x \cdot \Delta x + 2x + 2 \cdot \Delta x - 2x^2 - 2x \cdot \Delta x - 2x}{(x + \Delta x + 1)(x + 1)}$$

$$= \frac{2 \cdot \Delta x}{(x + \Delta x + 1)(x + 1)}$$

$$\frac{\Delta y}{\Delta x} = \frac{2}{(x + \Delta x + 1)(x + 1)}$$

$$\frac{dy}{dx} = \lim_{\Delta x \to 0} \frac{\Delta y}{\Delta x} = \frac{2}{(x + 1)^2} = \text{the instantaneous rate of growth for any value of } x$$

For $x = 2$, $\qquad \dfrac{dy}{dx} = \dfrac{2}{(2 + 1)^2} = \dfrac{2}{9}$

68. What is the instantaneous rate of growth of the volume V of a right circular cone with respect to the radius of its base r, when $r = 10$ inches and the altitude of

the cone always remains equal to the diameter of the base?

Volume of a cone $= \frac{1}{3}h\pi r^2$, but $h = 2r$ (in this case).

$$\therefore V = \frac{2}{3}\pi r^3$$

where V is the function and r is the independent variable.

$$V + \Delta V = \frac{2}{3}\pi(r + \Delta r)^3$$

$$= \frac{2\pi}{3}[r^3 + 3r^2 \cdot \Delta r + 3r \cdot (\Delta r)^2 + (\Delta r)^3]$$

$$V + \Delta V = \frac{2\pi r^3}{3} + 2\pi r^2 \cdot \Delta r + 2\pi r (\Delta r)^2$$

$$+ \frac{2}{3}\pi (\Delta r^3)$$

Subtract $V \qquad = \frac{2\pi r^3}{3}$

getting $\qquad \overline{\Delta V = \qquad 2\pi r^2 \cdot \Delta r + 2\pi r (\Delta r)^2}$

$$+ \frac{2}{3}\pi (\Delta r)^3$$

and $\qquad \dfrac{\Delta V}{\Delta r} = 2\pi r^2 + 2\pi r \cdot \Delta r + \dfrac{2}{3}\pi (\Delta r)^2$

$$\frac{dV}{dr} = \lim_{\Delta r \to 0} \frac{\Delta V}{\Delta r} = 2\pi r^2$$

The rate of change of volume with respect to the radius is shown to be 2π times the square of the instantaneous value of the radius.

Therefore, for $r = 10$, $\dfrac{dV}{dr} = 2\pi \cdot (10)^2 = 200\pi$

The volume is therefore growing instantaneously at the rate of 200π cu. in. per inch increase of the radius.

69. Can the derivative $\dfrac{dy}{dx}$ of a polynomial algebraic expression always be found by the process indicated in

$$\frac{d}{dx}[f(x)] = \lim_{\Delta x \to 0}\left[\frac{f(x + \Delta x) - f(x)}{\Delta x}\right]?$$

This method can always be used but is laborious. Rules have been formulated to cover all the elementary types of expressions, algebraic, trigonometric, and logarithmic functions.

PROBLEMS

1. If $y = 8x$, **(a)** what is Δy when x increases by Δx? **(b)** What is Δy when $x = 3$ and $\Delta x = 2$? **(c)** What is Δy when $x = 5$ and $\Delta x = 2$? **(d)** Does Δy change as x changes?

2. If $y = 3x^2 + 5$, what is Δy when $x = 3$ and **(a)** when $\Delta x = 0.2$; **(b)** when $\Delta x = 0.02$; **(c)** when $\Delta x = 0.002$?

3. If $s = 16t^2$ and $t = 3$, what is $\dfrac{\Delta s}{\Delta t}$ when **(a)** $\Delta t = 1$; **(b)** when $\Delta t = 0.1$; **(c)** when $\Delta t = 0.01$; **(d)** when $\Delta t = 0.001$; **(e)** when $\Delta t = 0.0001$? **(f)** What is $\dfrac{ds}{dt}$?

4. If $y = \sin x$, what is Δy when $x = 36°20'$ and **(a)** when $\Delta x = 1'$; **(b)** when $\Delta x = 20''$?

5. If A is the area of a rectangle of base x and constant height 6 in., **(a)** what is ΔA for $\Delta x = \frac{1}{4}$ in.? **(b)** Does ΔA vary with x?

6. If $y = \sin x$, what is Δy when $x = \dfrac{\pi}{4}$ and $\Delta x = \dfrac{\pi}{8}$?

7. If $A = x^2$, what is ΔA for $x = 4$ and $\Delta x = \frac{1}{2}$?

8. If $A = \pi r^2$, what is ΔA for $r = 5$ and $\Delta r = 0.1$?

9. If $y = \log_{10} x$, what is Δy when $x = 100$ and $\Delta x = 0.01$?

10. If $y = \dfrac{1}{x + 1}$, what is Δy when x becomes $x + \Delta x$?

11. If a train travels s miles in t hr., what is the meaning of **(a)** $\dfrac{s}{t}$; **(b)** $\dfrac{\Delta s}{\Delta t}$; **(c)** $\dfrac{ds}{dt}$?

12. If $y = x^3$, what is $\dfrac{dy}{dx}$ when $x = 1$?

13. If $y = 3x^2 - 4$, what is $\dfrac{dy}{dx}$, **(a)** when $x = 5$; **(b)** when $x = \sqrt{5}$; **(c)** when $x = -\sqrt{5}$?

14. Using the general process indicated by $\lim\limits_{\Delta x \to 0} \left[\dfrac{f(x + \Delta x) - f(x)}{\Delta x} \right]$, differentiate **(a)** $y = 4x^3$; **(b)** $y = \dfrac{x^3 + 1}{x}$; **(c)** $s = 2t - t^2$.

CHAPTER IV

DIFFERENTIATION
ELEMENTARY RULES—ALGEBRAIC

70. What is the rule for differentiating a power when it is a whole number?

Multiply by the power, and reduce the exponent by 1.

Example

a. If $y = x^n$,

then $\qquad \dfrac{dy}{dx} = n \cdot x^{n-1} = $ the derivative

b. If $y = x^4$,

then $\qquad \dfrac{dy}{dx} = 4 \cdot x^{4-1} = 4x^3 = $ the derivative

c. If $y = x^{-n}$,

then $\qquad \dfrac{dy}{dx} = -n \cdot x^{-n-1} = $ the derivative

d. If $y = x^{-4}$,

then $\qquad \dfrac{dy}{dx} = -4 \cdot x^{-4-1} = -4x^{-5} = $ the derivative

71. What is the rule for differentiating a power when it is a fraction?

The same as for a whole number: Multiply by the power and reduce the exponent by 1.

Example

a. If $y = x^{\pm\frac{1}{n}}$,

then $\qquad \dfrac{dy}{dx} = \pm \dfrac{1}{n} \cdot x^{\pm\frac{1}{n}-1}$ = the derivative

b. If $y = x^{\frac{1}{2}}$,

then $\qquad \dfrac{dy}{dx} = \dfrac{1}{2} \cdot x^{\frac{1}{2}-1} = \dfrac{1}{2} x^{-\frac{1}{2}}$ = the derivative

72. What is the derivative of a constant?

The derivative of a constant is zero. $\dfrac{d}{dx}(c) = 0$, since c does not change while the variable grows.

Example

If $y = 6$, then $\dfrac{dy}{dx} = 0$.

73. What is the derivative of a variable with respect to itself?

The derivative of a variable with respect to itself is unity.

Example

If $y = x$, then $dy = dx$, and the growths are equal.

$$\therefore \frac{dy}{dx} = \frac{dx}{dx} = 1$$

74. What is the rule for differentiating when there is an added constant in the function?

The constant disappears in the derivative.

Example

If $y = x^4 + 8$,

then $\qquad\qquad \dfrac{dy}{dx} = 4x^3$ = the derivative

75. What is the rule for differentiating when there is a constant as a multiplier?

The constant remains a multiplier in the derivative.

Example

If $y = ⑨x^2$,

then $\dfrac{dy}{dx} = ⑨ \cdot 3x^2 = 27x^2 =$ the derivative

76. What is the rule for differentiating when the function is divided by a constant?

The constant appears unchanged in the derivative as with a multiplier constant; for, instead of multiplying by c, we multiply by $\dfrac{1}{c}$.

Example

If $y = \dfrac{x^3}{⑨}$,

then $\dfrac{dy}{dx} = \dfrac{1}{⑨} \cdot 3x^2 = \dfrac{x^2}{3} =$ the derivative

77. What is the rule for differentiating a sum or difference of two or more terms?

Differentiate one term after the other in succession.

Example

If $y = (x^3 + 8) \pm (7x^6 + 10)$,

then $\dfrac{dy}{dx} = 3x^2 \pm 42x^5 =$ the derivative

78. What is the rule for differentiating a product of two terms?

a. Add the products of each term by the derivative of the other term.

b. Or multiply out, and then differentiate each term separately.

Example

If $y = (x^3 + 5)(6x^5 + 7)$,

then $\dfrac{dy}{dx} = (x^3 + 5) \cdot \dfrac{d}{dx}(6x^5 + 7) + (6x^5 + 7) \cdot \dfrac{d}{dx}(x^3 + 5)$

$\qquad = (x^3 + 5) \cdot 30x^4 + (6x^5 + 7) \cdot 3x^2$

$\qquad = 30x^7 + 150x^4 + 18x^7 + 21x^2$

Finally, $\dfrac{dy}{dx} = 48x^7 + 150x^4 + 21x^2 = $ the derivative

Or first multiplying out, we get

$$y = 6x^8 + 30x^5 + 7x^3 + 35.$$

Then $\dfrac{dy}{dx} = 6 \cdot 8x^{8-1} + 30 \cdot 5x^{5-1} + 7 \cdot 3x^{3-1}$

$\qquad = 48x^7 + 150x^4 + 21x^2 = $ the derivative

79. What is the rule for differentiating a quotient of two terms of the independent variable?

(a) Multiply the denominator by the derivative of the numerator.

(b) Multiply the numerator by the derivative of the denominator.

(c) Subtract (b) from (a) and divide by the square of the denominator.

Example

If $y = \dfrac{3x^4 + 5}{x^3 + 6}$, then

(a) $(x^3 + 6) \cdot \dfrac{d}{dx}(3x^4 + 5) = (x^3 + 6) \cdot 12x^3 = 12x^6$

$\qquad\qquad + 72x^3$

(b) $(3x^4 + 5) \cdot \dfrac{d}{dx}(x^3 + 6) = (3x^4 + 5) \cdot 3x^2 = 9x^6 + 15x^2$

(c) $\dfrac{(12x^6 + 72x^3) - (9x^6 + 15x^2)}{(x^3 + 6)^2} = \dfrac{3x^6 + 72x^3 - 15x^2}{(x^3 + 6)^2}$

$\qquad\qquad\qquad = $ the derivative

80. If $y = x^{3b}$, what is the derivative?

$$\frac{dy}{dx} = 3b \cdot x^{3b-1}$$

81. If $y = \sqrt[a]{\dfrac{1}{x^c}}$, what is the derivative?

$$y = x^{-\frac{c}{a}}$$

$$\frac{dy}{dx} = -\frac{c}{a} \cdot x^{-\frac{c}{a}-1} = -\frac{c}{a} x^{-\frac{(c+a)}{a}} = -\frac{c}{a} \sqrt[a]{\frac{1}{x^{c+a}}}$$

82. If $v = \sqrt[4]{\dfrac{1}{y^7}}$, what is the derivative?

$$v = y^{-\frac{7}{4}}$$

$$\frac{dv}{dy} = -\frac{7}{4} \cdot y^{-\frac{7}{4}-1} = -\frac{7}{4} \cdot y^{-\frac{11}{4}} = -\frac{7}{4} \sqrt[4]{\frac{1}{y^{11}}}$$

83. If $y = \dfrac{ax^n - 2}{b}$, what is the derivative?

$$y = \frac{a}{b} x^n - \frac{2}{b}$$

$$\frac{dy}{dx} = \frac{a}{b} \cdot n \cdot x^{n-1} - 0$$

or

$$\frac{dy}{dx} = \frac{na}{b} \cdot x^{n-1}$$

84. If $y = \dfrac{x^6}{8} + \dfrac{4}{9}$, what is the derivative?

$$\frac{dy}{dx} = \frac{1}{8} \cdot 6 \cdot x^{6-1} + 0 = \frac{3}{4} x^5$$

85. What is the derivative of y^n with respect to y^7?

$$\frac{d}{d(y^7)}(y^n) = \frac{ny^{n-1}}{7y^{7-1}} = \frac{n}{7} \cdot y^{n-1-6} = \frac{n}{7} \cdot y^{n-7}$$

86. If $y = \dfrac{ax^4 + b}{x^3 + c}$**, what is the derivative?**

$$\frac{dy}{dx} = \frac{(x^3 + c) \cdot \dfrac{d}{dx}(ax^4 + b) - (ax^4 + b) \cdot \dfrac{d}{dx}(x^3 + c)}{(x^3 + c)^2}$$

$$= \frac{(x^3 + c) \cdot a \cdot 4x^{4-1} - (ax^4 + b) \cdot 3 \cdot x^{3-1}}{(x^3 + c)^2}$$

$$= \frac{4ax^3(x^3 + c) - 3x^2(ax^4 + b)}{(x^3 + c)^2}$$

$$= \frac{4ax^6 + 4acx^3 - 3ax^6 - 3bx^2}{(x^3 + c)^2} = \frac{ax^6 + 4acx^3 - 3bx^2}{(x^3 + c)^2}$$

87. If $y = (x^2 - 3x)(4x + 5)$**, what is the derivative?**

$$\frac{dy}{dx} = (x^2 - 3x) \cdot \frac{d}{dx}(4x + 5)$$
$$+ (4x + 5) \cdot \frac{d}{dx}(x^2 - 3x)$$

or $\dfrac{dy}{dx} = (x^2 - 3x) \cdot 4 \cdot 1 \cdot x^{1-1}$
$$+ (4x + 5)(2x^{2-1} - 3 \cdot 1 \cdot x^{1-1})$$

$$\frac{dy}{dx} = 4(x^2 - 3x) + (4x + 5)(2x - 3)$$
$$= 4x^2 - 12x + 8x^2 - 2x - 15 = 12x^2 - 14x - 15$$

88. What is the relative rate of increase of x and y of the function $y = 2x^2$ **when** $x = 1, 3,$ **and 5, respectively?**

$$\frac{dy}{dx} = 2 \cdot 2 \cdot x^{2-1} = 4x$$

When $x = 1$,

$$\frac{dy}{dx} = 4 \cdot 1 = 4 \quad \text{or} \quad dy = 4\,dx$$

When $x = 3$,

$$\frac{dy}{dx} = 4 \cdot 3 = 12 \quad \text{or} \quad dy = 12\,dx$$

When $x = 5$,

$$\frac{dy}{dx} = 4 \cdot 5 = 20 \qquad \text{or} \qquad dy = 20\,dx$$

89. **What is the relative rate of increase of the heat with respect to the temperature when the temperature is 20°C., if the heat H required to raise a unit weight of water from zero degrees centigrade to a temperature T is given by the relation**

$$H = T + 0.000023T^2 + 0.00000032T^3?$$

$$\frac{dH}{dT} = 1 \cdot T^{1-1} + 2 \cdot 0.000023 \cdot T^{2-1} + 3 \cdot 0.00000032T^{3-1}$$

$$= 1 + 0.000046T + 0.00000096T^2$$

Now, when the temperature is 20°C.,

$$\frac{dH}{dT} = 1 + 0.000046 \times 20 + 0.00000096 \cdot (20)^2$$

$$= 1 + 0.00092 + 0.000384 = 1.001304$$

or $\qquad dH = 1.001304 \cdot dT$

90. **How would you compare the sensitiveness of a pyrometer at 900, 1100, and 1300°C., if its scale reads 40 ($= \phi_1$), when the actual temperature is 1000°C. ($= T_1$), and if it is related to the temperature T of the observed body according to the formula**

$$\frac{\phi}{\phi_1} = \left(\frac{T}{T_1}\right)^3?$$

The sensitiveness is the rate of variation of the scale reading with regard to the actual temperature or $\dfrac{d\phi}{dT}$.

From the given relation,

$$\phi = \frac{\phi_1}{T_1^3} \cdot T^3$$

But when $\phi_1 = 40$,

$$T_1 = 1,000°$$

Then

$$\phi = \frac{40T^3}{(1,000)^3}$$

and $\dfrac{d\phi}{dT} = \dfrac{1}{(1,000)^3} \cdot 40 \cdot 3 \cdot T^{3-1} = \dfrac{120T^2}{(1,000)^3} = \dfrac{3T^2}{25,000,000}$

Now when $T = 900°C$.,

$$\frac{d\phi}{dT} = \frac{3 \cdot (900)^2}{25,000,000} = 0.0972$$

When $T = 1100°C$.,

$$\frac{d\phi}{dT} = \frac{3 \cdot (1,100)^2}{25,000,000} = 0.1452$$

When $T = 1300°C$.,

$$\frac{d\phi}{dT} = \frac{3 \cdot (1,300)^2}{25,000,000} = 0.2028$$

Therefore the sensitiveness is increased about 50 per cent from 900 to 1100°C. and almost 50 per cent again from 1100 to 1300°C.

91. If the relation between the electric resistance R of a wire at the temperature $T°C$. and the resistance R_0 of the same wire at zero degrees centigrade is found to be $\dfrac{R}{R_0} = (1 + aT + b\sqrt[3]{T})$ where a and b are constants, what is the rate of variation of R with regard to the temperature?

$$R = R_0(1 + aT + b\sqrt[3]{T}) = R_0 + R_0aT + R_0bT^{\frac{1}{3}}$$

$$\frac{dR}{dT} = 0 + aR_0 + R_0 \cdot b \cdot \frac{1}{3} \cdot T^{\frac{1}{3}-1} = aR_0 + \frac{bR_0}{3\sqrt[3]{T^2}}$$

or $\dfrac{dR}{dT} = R_0\left(a + \dfrac{b}{3\sqrt[3]{T^2}}\right)$

92. What is the rule for finding the derivative of an implicit function of x and y?

Differentiate the terms of the equation as given, by applying the usual rules for finding the derivative, and then solve for $\dfrac{dy}{dx}$.

Example

If $ax^5 + 3x^2y - y^5x = 16 =$ an implicit function in x and y,

then $\quad \dfrac{d}{dx}(ax^5) + \dfrac{d}{dx}(3x^2y) - \dfrac{d}{dx}(y^5x) = \dfrac{d}{dx}(16)$

or $a \cdot 5 \cdot x^{5-1} + 3x^2 \cdot \dfrac{dy}{dx} + y \cdot 3 \cdot 2 \cdot x^{2-1}$

$$- y^5 \cdot 1 - x \cdot 5 \cdot y^{5-1} \cdot \dfrac{dy}{dx} = 0$$

or $\quad 5ax^4 + 3x^2 \cdot \dfrac{dy}{dx} + 6xy - y^5 - 5xy^4 \cdot \dfrac{dy}{dx} = 0$

$$(3x^2 - 5xy^4)\dfrac{dy}{dx} = y^5 - 5ax^4 - 6xy$$

$$\therefore \dfrac{dy}{dx} = \dfrac{y^5 - 5ax^4 - 6xy}{3x^2 - 5xy^4}$$

93. If $x^2 + y^2 = 16$, how would you find $\dfrac{dy}{dx}$ as an implicit function of x and y?

It is seen that y is a function of x, and the left-hand member is therefore the sum of two functions of x, which we now differentiate.

$$2x + 2y \cdot \dfrac{dy}{dx} = 0 \quad \text{or} \quad 2y\dfrac{dy}{dx} = -2x$$

$$\therefore \dfrac{dy}{dx} = -\dfrac{x}{y}$$

94. When would you differentiate as an implicit function?

When it is either not convenient or not possible to express one variable as an explicit function of the other.

Example

In the above question, if y were expressed as an explicit function of x before differentiating, the work would not be so simple. Solving for y, $y = \pm \sqrt{16 - x^2}$ and the $\dfrac{dy}{dx}$ is not so simple to find. The method of such a solution will be considered later (page 42).

95. If $x^3y^2 = 16$, how would you find (a) $\dfrac{dy}{dx}$ and (b) $\dfrac{dx}{dy}$ each as an implicit function of x and y?

a. For $\dfrac{dy}{dx}$,

$$x^3 \cdot \frac{d}{dx}(y^2) + y^2 \cdot \frac{d}{dx}(x^3) = \frac{d}{dx}(16)$$

or $\qquad x^3 \cdot 2 \cdot y^{2-1} \cdot \dfrac{dy}{dx} + y^2 \cdot 3 \cdot x^{3-1} = 0$

$$2yx^3 \cdot \frac{dy}{dx} + 3x^2y^2 = 0$$

Solving for $\dfrac{dy}{dx}$,

$$2yx^3 \cdot \frac{dy}{dx} = -3x^2y^2$$

or $\qquad \dfrac{dy}{dx} = -\dfrac{3x^2y^2}{2x^3y} = -\dfrac{3y}{2x}$

b. For $\dfrac{dx}{dy}$,

$$x^3 \cdot \frac{d}{dy}(y^2) + y^2 \cdot \frac{d}{dy}(x^3) = \frac{d}{dy}(16)$$

or $\qquad x^3 \cdot 2 \cdot y^{2-1} + y^2 \cdot 3x^{3-1} \cdot \dfrac{dx}{dy} = 0$

$$2x^3y + 3y^2 \cdot x^2 \cdot \frac{dx}{dy} = 0$$

Solving for $\dfrac{dx}{dy}$,

$$\frac{dx}{dy} = -\frac{2x^3y}{3x^2y^2} = -\frac{2x}{3y}$$

NOTE.—This latter result for (b) could have been obtained directly as the reciprocal of the first result for (a).

96. How would you successively differentiate $y = x^6$?

First derivative $= 6x^{6-1} = 6x^5$
Second derivative $= 5 \cdot 6x^{5-1} = 30x^4$
Third derivative $= 4 \cdot 30x^{4-1} = 120x^3$
Fourth derivative $= 3 \cdot 120x^{3-1} = 360x^2$
Fifth derivative $= 2 \cdot 360x^{2-1} = 720x^1 = 720x$
Sixth derivative $= 1 \cdot 720x^{1-1} = 720x^0 = 720$
Seventh derivative $= 0 \cdot 720x^{0-1} = 0$

97. What are the symbols for successive differentiation?

If $y = x^n = f(x) = \phi(x)$,
First derivative $= nx^{n-1} = f'(x) = \phi'(x)$
Second derivative $= n(n-1)x^{n-2} = f''(x) = \phi''(x)$
Third derivative $= n(n-1)(n-2)x^{n-3} = f'''(x)$
$$= \phi'''(x)$$

98. What are the alternative symbols for successive differentiation?

If $y = f(x)$,

First derivative $= f'(x) = \dfrac{dy}{dx}$

Second derivative $= f''(x) = \dfrac{d\left(\dfrac{dy}{dx}\right)}{dx} = \dfrac{d^2y}{(dx)^2} = \dfrac{d^2y}{dx^2}$ (this is the most convenient symbol)

Third derivative $= f'''(x) = \dfrac{d^3y}{dx^3}$

99. How would you differentiate
$$y = 5x^4 - 4x^3 + 2x^2 - x + 4$$
successively?

$$\frac{dy}{dx} = f'(x) = 4 \cdot 5x^{4-1} - 3 \cdot 4x^{3-1} + 2 \cdot 2x^{2-1}$$
$$- 1 \cdot x^{1-1} + 0$$

or $f'(x) = 20x^3 - 12x^2 + 4x - 1$

$$\frac{d^2y}{dx^2} = f''(x) = 3 \cdot 20x^{3-1} - 2 \cdot 12x^{2-1} + 1 \cdot 4x^{1-1} - 0$$
$$= 60x^2 - 24x + 4$$

$$\frac{d^3y}{dx^3} = f'''(x) = 2 \cdot 60x^{2-1} - 1 \cdot 24x^{1-1} + 0 = 120x - 24$$

$$\frac{d^4y}{dx^4} = f''''(x) = 1 \cdot 120x^{1-1} - 0 = 120$$

$$\frac{d^5y}{dx^5} = f'''''(x) = 0$$

100. How would you differentiate $y = F(x) = 4x(x^3 - 5)$ successively?

$$F'(x) = \frac{dy}{dx} = 4x \cdot \frac{d}{dx}(x^3 - 5) + (x^3 - 5) \cdot \frac{d}{dx}(4x)$$

or $F'(x) = 4x \cdot 3x^2 + (x^3 - 5)4 = 12x^3 + 4x^3 - 20$
$$= 16x^3 - 20$$

$$F''(x) = \frac{d^2y}{dx^2} = 48x^2$$

$$F'''(x) = \frac{d^3y}{dx^3} = 96x$$

$$F''''(x) = \frac{d^4y}{dx^4} = 96$$

$$F'''''(x) = \frac{d^5y}{dx^5} = 0$$

101. How would you find the $\frac{d^2y}{dx^2}$ of the equation for the hyperbola $b^2x^2 - a^2y^2 = a^2b^2$, which is an implicit function?

Differentiate with respect to x implicitly.

$$b^2 \cdot 2x - a^2 \cdot 2y \cdot \frac{dy}{dx} = 0 \quad \text{or} \quad 2b^2x = 2a^2y \frac{dy}{dx}$$

Solving for $\frac{dy}{dx}$,

$$\frac{dy}{dx} = \frac{2b^2x}{2a^2y} = \frac{b^2x}{a^2y} \tag{1}$$

Differentiate again, and note that y is a function of x.

$$\frac{d^2y}{dx^2} = \frac{a^2y \cdot \frac{d}{dx}(b^2x) - b^2x \cdot \frac{d}{dx}(a^2y)}{(a^2y)^2}$$

$$= \frac{a^2y \cdot b^2 - b^2x \cdot a^2 \cdot \frac{dy}{dx}}{a^4y^2} = \frac{a^2b^2y - a^2b^2x \cdot \frac{dy}{dx}}{a^4y^2}$$

Now substitute for $\frac{dy}{dx}$ its value from Eq. (1).

Then $\quad \dfrac{d^2y}{dx^2} = \dfrac{a^2b^2y - a^2b^2x \cdot \dfrac{b^2x}{a^2y}}{a^4y^2} = \dfrac{a^2b^2y - \dfrac{b^4x^2}{y}}{a^4y^2}$

$$= -\frac{b^2(b^2x^2 - a^2y^2)}{a^4y^3}$$

But, from the original equation, $b^2x^2 - a^2y^2 = a^2b^2$.

$$\therefore \frac{d^2y}{dx^2} = -\frac{b^2 \cdot a^2b^2}{a^4y^3} = -\frac{a^2b^4}{a^4y^3} = -\frac{b^4}{a^2y^3}$$

PROBLEMS

Differentiate the following:

1. $y = x^{17}$

2. $y = \sqrt[5]{x^{-7}}$

3. $v = \sqrt[4]{u}$

4. $y = 4x^c$

5. $y = x^{-\frac{1}{4}}$

6. $y = \sqrt[a]{x^2}$

7. $y = x^{3b}$

8. $v = u^{3.8}$

9. $y = \dfrac{x^7}{9} - \dfrac{4}{7}$

10. $y = b\sqrt{y} - \frac{1}{4}\sqrt{b}$

11. $2ax + by = 3ay - 4bx + (x + 2y)\sqrt{a - b}$

12. $y = \frac{3}{8}x^4 + 7x^2 - 9x + 6$

13. $x = 3b\sqrt{ay^3} + \dfrac{6a\sqrt[4]{b}}{y} + 4\sqrt{ab}$

14. $x = 3.1\sqrt[4]{\dfrac{1}{\alpha^3}} + \dfrac{8.6}{\sqrt[3]{\alpha}} + 8$

15. $v = (4t^2 + 2.3t + 5)^2$

16. $x = 0.7y^4(y - 6)$

17. $y = \left(\alpha + \dfrac{3}{\alpha}\right)\left(\sqrt[3]{\alpha} + \dfrac{1}{\sqrt[3]{\alpha}}\right)$

18. $x = \dfrac{2b}{3 - b\sqrt{y} - b^3 x^2}$

19. $x = \dfrac{y^3}{y^4 + 3}$

20. $y = \dfrac{c + \sqrt[3]{x}}{c - \sqrt[3]{x}}$

21. $\theta = \dfrac{1 + b\sqrt{t^2}}{1 - b\sqrt[3]{t^2}}$

22. What is the rate of variation of the volume of a right circular cone with respect to the radius of its base when the radius is 10 in. and the height is 40 in.?

23. What would be the dimensions of the cone of Prob. 22 if when the radius is one-half its height a change of 1 in. in the radius causes a change of 1,200 cu. in. in the volume?

24. If the candle power of an incandescent electric lamp, c, is related to the voltage E, in the manner of $c = mE^n$ (where m and n are constants), what is the change of candle power per volt at 110 and 120 volts when $m = 0.6 \cdot 10^{-8}$ and $n = 5$?

25. What is the rate at which the pressure of the air is changing if, when the pressure is 50 lb./sq. in., the volume is 6 cu. ft. and it is increasing at the rate of 0.2 cu. ft./sec., assuming that the air expands according to the adiabatic law $c = pv^{1.41}$?

26. What is the speed with which a ball rolls down a slope when $t = 2$ sec. and when $t = 8$ sec., if the equation relating the distance in feet s and the time t in seconds is $s = 8t^2$?

27. What are the velocity and acceleration 5 sec. after a body begins to move and what are the corresponding values when the distance covered is 80 ft., if the distance x in feet that it travels from the start is given by $x = 0.4t^2 + 9.6$?

28. What is the angular velocity ω and the angular acceleration α of a wheel after 1 sec. and after it has turned one revolution, if the angle θ (in radians) turned through is given by $\theta = 4 + 3t - 0.2t^3$?

29. At what rate, in miles per hour, is a balloon rising at the end of 20 min. if $h =$ the miles it rises in t min. is given by $h = \dfrac{8t}{\sqrt{3,200 + t^2}}$?

30. At what rate does the area of a circular metal plate of diameter 4 in. increase if its radius increases at the rate of $\tfrac{1}{16}$ in./min.?

31. How fast is the distance of an airplane from a fort increasing 1 min. after it passes directly over the fort at a height of 10,000 ft. when it is flying horizontally at the rate of 200 ft./sec.?

32. What is the derivative as an implicit function of $x^4 + y^4 = 3a^2$?

Differentiate implicitly in the problems that follow:

33. Find $\dfrac{dy}{dx}$ of $\dfrac{x^2}{a^2} + \dfrac{y^2}{b^2} = 1$.

34. $x^6 + 3x^4y^3 + y^2 = 0$; find $\dfrac{dy}{dx}$ and $\dfrac{dx}{dy}$.

35. $(x + y)^{\frac{1}{2}} + (x - y)^{\frac{1}{2}} = a$; find $\dfrac{dy}{dx}$.

36. $\left(x + \dfrac{b}{s^3}\right)(s + c) = m$; find $\dfrac{dx}{ds}$ and $\dfrac{ds}{dx}$.

37. $y = \sqrt{x} + \sqrt[3]{x}$; find $\dfrac{dx}{dy}$.

38. $\sqrt{\dfrac{x}{y}} - \sqrt{\dfrac{y}{x}} = 10$; find $\dfrac{dy}{dx}$.

39. $x^4 + 4x^3y + 3y^2 = a^3$; find $\dfrac{dy}{dx}$.

40. $y^2 = 4ax$; find $\dfrac{dx}{dy}$.

41. Find $\dfrac{d^2y}{dx^2}$ of $y = 2x + 3x^2$.

42. Find $F''(x)$ of $y = \dfrac{x^2 + c}{x + c}$.

43. Find $f''(x)$ of $y = \dfrac{x^2}{a + x}$.

44. Find $\dfrac{d^2y}{dx^2}$ of $a^2x^2 + b^2y^2 = a^2b^2$.

45. Find $\phi''(x)$ of $x^3 + y^3 = 2$.

CHAPTER V

DIFFERENTIATION BY SUBSTITUTION

102. What is the rule for ordinary substitution?

$$\frac{dy}{dx} = \frac{dy}{du} \cdot \frac{du}{dx}$$

Function u is substituted to simplify the original function.

103. How would you differentiate $y = (x^3 + b^4)^{\frac{5}{2}}$ by substitution?

Let $(x^3 + b^4) = u$.

Then $\qquad\qquad\qquad y = u^{\frac{5}{2}} \qquad$ by substitution

and $\qquad\qquad \dfrac{dy}{du} = \dfrac{5}{2} u^{\frac{5}{2}-1} = \dfrac{5}{2} u^{\frac{3}{2}}$

Also, $\qquad\qquad \dfrac{du}{dx} = \dfrac{d}{dx}(x^3 + b^4) = 3x^2$

$\therefore \dfrac{dy}{dx} = \dfrac{dy}{du} \cdot \dfrac{du}{dx} = \dfrac{5}{2} u^{\frac{3}{2}} \cdot 3x^2 = \dfrac{5}{2}(x^3 + b^4)^{\frac{3}{2}} \cdot 3x^2$

$$= \frac{15}{2} x^2 (x^3 + b^4)^{\frac{3}{2}}$$

104. What is the derivative of $y = \sqrt{b + x}$ by substitution?

Let $(b + x) = u$.

Then $y = u^{\frac{1}{2}} \qquad\qquad$ by substitution

and $\dfrac{dy}{du} = \dfrac{1}{2} u^{\frac{1}{2}-1} = \dfrac{1}{2} u^{-\frac{1}{2}}$

Also, $\dfrac{du}{dx} = 1$

$\therefore \dfrac{dy}{dx} = \dfrac{dy}{du} \cdot \dfrac{du}{dx} = \dfrac{1}{2} u^{-\frac{1}{2}} \cdot 1 = \dfrac{1}{2}(b + x)^{-\frac{1}{2}} = \dfrac{1}{2\sqrt{b + x}}$

105. What is the derivative of $y = (x + \sqrt{x^2 + x + c})^4$ by substitution?

Let $u = x + \sqrt{x^2 + x + c}$.

Then $y = u^4$ by substitution

and $\dfrac{dy}{du} = 4u^3$

Also, $\dfrac{du}{dx} = 1 + \dfrac{d}{dx}(x^2 + x + c)^{\frac{1}{2}}$

$\therefore \dfrac{dy}{dx} = \dfrac{dy}{du} \cdot \dfrac{du}{dx} = 4u^3\left[1 + \dfrac{d}{dx}(x^2 + x + c)^{\frac{1}{2}}\right]$ (1)

Now let $v = (x^2 + x + c)^{\frac{1}{2}}$

and let $w = (x^2 + x + c)$

$\therefore v = w^{\frac{1}{2}}$ by substitution

Then $\dfrac{dv}{dw} = \dfrac{1}{2}w^{\frac{1}{2}-1} = \dfrac{1}{2}w^{-\frac{1}{2}}$

Also $\dfrac{dw}{dx} = 2x + 1$

$\dfrac{dv}{dx} = \dfrac{dv}{dw} \cdot \dfrac{dw}{dx} = \dfrac{1}{2}w^{-\frac{1}{2}}(2x + 1) = \dfrac{(2x + 1)}{2\sqrt{x^2 + x + c}}$

Substituting in Eq. (1),

$$\dfrac{dy}{dx} = \dfrac{dy}{du} \cdot \dfrac{du}{dx} = 4u^3\left[1 + \dfrac{(2x + 1)}{2\sqrt{x^2 + x + c}}\right]$$

$$= 4\left[x + \sqrt{x^2 + x + c}\right]^3\left[1 + \dfrac{(2x + 1)}{2\sqrt{x^2 + x + c}}\right]$$

106. How would you extend the process of ordinary substitution to three differential coefficients?

$$\dfrac{dy}{dx} = \dfrac{dy}{dv} \cdot \dfrac{dv}{dz} \cdot \dfrac{dz}{dx} \tag{1}$$

If $y = \sqrt{1 + v}$, and $v = \dfrac{6}{z^2}$, and $z = 2x^3$,

then $\dfrac{dy}{dv} = \dfrac{1}{2\sqrt{1 + v}}$ and $\dfrac{dv}{dz} = -\dfrac{12}{z^3}$

                                   and $\dfrac{dz}{dx} = 6x^2$

Now, substituting in Eq. (1), we get

$$\frac{dy}{dx} = \frac{1}{2\sqrt{1+v}} \cdot -\frac{12}{z^3} \cdot 6x^2 = -\frac{36x^2}{z^3\sqrt{1+v}}$$

$$= -\frac{36x^2}{8x^9\sqrt{1+\dfrac{6}{4x^6}}}$$

Finally, $\dfrac{dy}{dx} = -\dfrac{36x^2}{8x^9\sqrt{\dfrac{4x^6+6}{4x^6}}} = -\dfrac{9}{x^4\sqrt{4x^6+6}}$

107. How would you differentiate $x = \dfrac{1}{\sqrt{1+y^2}}$ by substitution?

Let $u = 1 + y^2$.

Then $\qquad \dfrac{du}{dy} = 2y \qquad$ and $\qquad x = u^{-\frac{1}{2}}$

Then $\qquad \dfrac{dx}{du} = -\dfrac{1}{2}u^{-\frac{3}{2}} = -\dfrac{1}{2}(1+y^2)^{-\frac{3}{2}}$

$\therefore \dfrac{dx}{dy} = \dfrac{dx}{du} \cdot \dfrac{du}{dy} = -\dfrac{1}{2}(1+y^2)^{-\frac{3}{2}} \cdot 2y = -\dfrac{y}{\sqrt{(1+y^2)^3}}$

108. How would you differentiate $x = \dfrac{1}{\sqrt{y^4-1}}$ by substitution?

Let $u = y^4 - 1$.

Then $\qquad \dfrac{du}{dy} = 4y^3 \qquad$ and $\qquad x = u^{-\frac{1}{2}}$

Then $\qquad \dfrac{dx}{du} = -\dfrac{1}{2}u^{-\frac{3}{2}} = -\dfrac{1}{2}(y^4-1)^{-\frac{3}{2}}$

$\therefore \dfrac{dx}{dy} = \dfrac{dx}{du} \cdot \dfrac{du}{dy} = -\dfrac{1}{2}(y^4-1)^{-\frac{3}{2}} \cdot 4y^3 = -\dfrac{2y^3}{\sqrt{(y^4-1)^3}}$

109. How would you differentiate $x = \dfrac{y}{a} \sqrt{2by - y^2}$?

$$\frac{dx}{dy} = \frac{y}{a} \cdot \frac{d}{dy}(2by - y^2)^{\frac{1}{2}} + (2by - y^2)^{\frac{1}{2}} \cdot \frac{d}{dy}\left(\frac{y}{a}\right)$$

$$= \frac{y}{a} \cdot \frac{d}{dy}(2by - y^2)^{\frac{1}{2}} + \frac{1}{a} \cdot (2by - y^2)^{\frac{1}{2}}$$

Let $w = (2by - y^2)$

Then $\dfrac{dw}{dy} = (2b - 2y)$

Now let $s = (2by - y^2)^{\frac{1}{2}}$

Then $s = w^{\frac{1}{2}}$

and $\dfrac{ds}{dw} = \dfrac{1}{2}w^{-\frac{1}{2}} = \dfrac{1}{2w^{\frac{1}{2}}} = \dfrac{1}{2(2by - y^2)^{\frac{1}{2}}}$

$$\therefore \frac{ds}{dy} = \frac{ds}{dw} \cdot \frac{dw}{dy} = \frac{1}{2(2by - y^2)^{\frac{1}{2}}} \cdot (2b - 2y)$$

$$= \frac{(b - y)}{(2by - y^2)^{\frac{1}{2}}}$$

Finally, $\dfrac{dx}{dy} = \dfrac{y}{a} \cdot \dfrac{(b - y)}{(2by - y^2)^{\frac{1}{2}}} + \dfrac{1}{a}(2by - y^2)^{\frac{1}{2}}$

or $\dfrac{dx}{dy} = \dfrac{y(b - y) + (2by - y^2)}{a(2by - y^2)^{\frac{1}{2}}}$

$$= \frac{by - y^2 + 2by - y^2}{a(2by - y^2)^{\frac{1}{2}}} = \frac{3by - 2y^2}{a(2by - y^2)^{\frac{1}{2}}}$$

110. If $u = \dfrac{1}{8\sqrt{s}}$, $y = u^4 + \dfrac{u^2}{3}$, $z = \dfrac{4y^3}{\sqrt[4]{y - 3}}$, what is $\dfrac{dz}{ds}$?

$$\frac{dz}{dy} = \frac{(y - 3)^{\frac{1}{4}} \cdot 12y^2 - 4y^3 \cdot \dfrac{d}{dy}(y - 3)^{\frac{1}{4}}}{(y - 3)^{\frac{1}{2}}}$$

In differentiating $\dfrac{d}{dy}(y - 3)^{\frac{1}{4}}$, let $w = y - 3$

Then $\dfrac{dw}{dy} = 1$

and $\dfrac{d}{dy}(y-3)^{\frac{1}{4}} = \dfrac{d}{dy}(w^{\frac{1}{4}}) = \dfrac{1}{4}w^{-\frac{3}{4}}\cdot\dfrac{dw}{dy} = \dfrac{1}{4(y-3)^{\frac{3}{4}}}$

Then $\dfrac{dz}{dy} = 12y^2(y-3)^{-\frac{1}{4}} - \dfrac{y^3}{(y-3)^{\frac{5}{4}}} = \dfrac{12y^2(y-3)-y^3}{(y-3)^{\frac{5}{4}}}$

$\dfrac{dz}{dy} = \dfrac{12y^3 - 36y^2 - y^3}{(y-3)^{\frac{5}{4}}} = \dfrac{11y^3 - 36y^2}{(y-3)^{\frac{5}{4}}}$

Now $\dfrac{dy}{du} = 4u^3 + \dfrac{2}{3}u = \dfrac{12u^3 + 2u}{3}$

and $\dfrac{du}{ds} = \dfrac{1}{8}\cdot - \dfrac{1}{2}s^{-\frac{3}{2}} = -\dfrac{1}{16s^{\frac{3}{2}}}$

Finally, $\dfrac{dz}{ds} = \dfrac{dz}{dy}\cdot\dfrac{dy}{du}\cdot\dfrac{du}{ds}$

or $\dfrac{dz}{ds} = \dfrac{11y^3 - 36y^2}{(y-3)^{\frac{5}{4}}}\cdot\left(\dfrac{12u^3 + 2u}{3}\right)\cdot - \dfrac{1}{16s^{\frac{3}{2}}}$

$= -\dfrac{(11y^3 - 36y^2)(12u^3 + 2u)}{48s^{\frac{3}{2}}(y-3)^{\frac{5}{4}}}$

The values of y and u must be substituted in this expression in order to get the value in terms of s.

PROBLEMS

Find the derivative of the following:

1. $x = \sqrt{y^2 + a^2}$

2. $x = \dfrac{1}{\sqrt{a+y}}$

3. $y = \dfrac{\sqrt{x^2 - 2a^2}}{3x^2}$

4. $x = \dfrac{3a^2 + 4y^2}{(a+y)^2}$

5. $y = \sqrt{2x^2 + 4a^2}$

6. $y = \dfrac{2a}{\sqrt{2a - 3x^2}}$

7. $y = \dfrac{\sqrt[4]{x^3 - a}}{\sqrt{x^2 + a}}$

8. $y = \dfrac{x}{(x^2 + a)^{\frac{1}{2}}}$

9. $y = (x^2 + 4)^5$

10. $y = \sqrt{a^2 - x^2}$

11. $y = \dfrac{a^2 + x^2}{\sqrt{a^2 - x^2}}$

12. If $x = \dfrac{2y^2}{\sqrt{5}}$, $z = (1 + 3x)^2$, and $u = \dfrac{1}{\sqrt{1 + 2z}}$, what is $\dfrac{du}{dy}$?

13. If $x = \dfrac{1}{3}u^2$, $v = 5(x^2 + 2x)$, and $w = \dfrac{3}{v^2}$, what is $\dfrac{dw}{du}$?

14. If $x = 5y^2 + \sqrt{5}$, $z = \sqrt{x+3}$, and $v = \dfrac{1}{\sqrt{5} - 3z}$, what is $\dfrac{dv}{dy}$?

Chapter VI

INVERSE FUNCTIONS

111. What is the inverse function of $y = 5x$?

Solve for x to get the inverse function.

Then $\qquad x = \dfrac{y}{5} = $ the inverse function

112. What is the rule for the derivative of the inverse function?

It is the *reciprocal of the derivative of the original function* expressed as,

$$\frac{dy}{dx} = \frac{1}{\frac{dx}{dy}} \qquad \text{or} \qquad \frac{dy}{dx} \cdot \frac{dx}{dy} = 1$$

Example

If $y = 5x = $ the original function,

then $\qquad\qquad \dfrac{dy}{dx} = 5$

Now, $x = \dfrac{y}{5} = $ inverse function $\qquad \therefore \dfrac{dy}{dx} = \dfrac{1}{\frac{dx}{dy}} = \dfrac{1}{\frac{1}{5}} = 5$

and the derivative of the inverse function is

$$\frac{dx}{dy} = \frac{1}{5}$$

113. What is the derivative of the inverse function of $y = 5x^3$?

$\dfrac{dy}{dx} = 15x^2 = $ derivative of the original function

47

Therefore, the derivative of the inverse function is

$$\frac{dx}{dy} = \frac{1}{15x^2} = \text{reciprocal of the derivative of the original}$$
$$\text{function}$$

114. What is the derivative of the inverse function of $y = 5x^3$, worked out?

Solve for x.

$$x = \left(\frac{y}{5}\right)^{\frac{1}{3}} = \frac{y^{\frac{1}{3}}}{5^{\frac{1}{3}}} = \text{the inverse function}$$

Then $\dfrac{dx}{dy} = \dfrac{5^{\frac{1}{3}} \cdot \frac{1}{3}y^{\frac{1}{3}-1} - 0}{5^{\frac{1}{3}}} = \dfrac{1}{3y^{\frac{1}{3}} \cdot 5^{\frac{1}{3}}} = \dfrac{1}{3 \cdot (5x^3)^{\frac{1}{3}} \cdot 5^{\frac{1}{3}}}$

$$= \frac{1}{3 \cdot 5^{\frac{1}{3}} \cdot x^2 \cdot 5^{\frac{1}{3}}} = \frac{1}{3 \cdot 5 \cdot x^2} = \frac{1}{15x^2}$$

NOTE.—By rule this is simplified, as $\dfrac{dy}{dx} = 15x^2 = $ derivative of the original function.

$$\therefore \frac{dx}{dy} = \frac{1}{15x^2} = \text{derivative of inverse function}$$

115. What is the $\dfrac{dx}{dy}$ of $y = \sqrt{\dfrac{5}{x} - 1}$?

Let $u = \left(\dfrac{5}{x} - 1\right)$.

Then $\dfrac{du}{dx} = -\dfrac{5}{x^2}$

and $\quad y = u^{\frac{1}{2}} \qquad\qquad$ by substitution

Now, $\dfrac{dy}{du} = \dfrac{1}{2}u^{-\frac{1}{2}} = \dfrac{1}{2\sqrt{\dfrac{5}{x} - 1}}$

Then $\dfrac{dy}{dx} = \dfrac{dy}{du} \cdot \dfrac{du}{dx} = \dfrac{1}{2\sqrt{\dfrac{5}{x} - 1}} \cdot -\dfrac{5}{x^2} = -\dfrac{5}{2x^2\sqrt{\dfrac{5}{x} - 1}}$

By rule,

$$\therefore \frac{dx}{dy} = -\frac{2x^2 \sqrt{\frac{5}{x} - 1}}{5} = \text{reciprocal of the derivative of the original function}$$

116. What is the derivative $\frac{dx}{dy}$ obtained directly from the inverse function of $y = \sqrt{\frac{5}{x} - 1}$?

Squaring,

$$y^2 = \frac{5}{x} - 1$$

$$\therefore x = \frac{5}{y^2 + 1} = \text{inverse function}$$

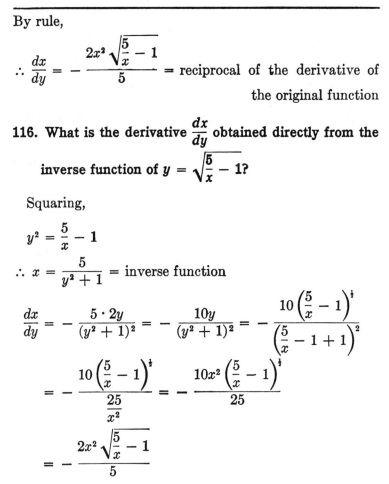

$$\frac{dx}{dy} = -\frac{5 \cdot 2y}{(y^2 + 1)^2} = -\frac{10y}{(y^2 + 1)^2} = -\frac{10\left(\frac{5}{x} - 1\right)^{\frac{1}{2}}}{\left(\frac{5}{x} - 1 + 1\right)^2}$$

$$= -\frac{10\left(\frac{5}{x} - 1\right)^{\frac{1}{2}}}{\frac{25}{x^2}} = -\frac{10x^2\left(\frac{5}{x} - 1\right)^{\frac{1}{2}}}{25}$$

$$= -\frac{2x^2 \sqrt{\frac{5}{x} - 1}}{5}$$

Compare by working out $\frac{dy}{dx}$ first and then getting $\frac{dx}{dy}$ as the reciprocal.   Which is simpler in this case?

117. What is the derivative $\frac{dy}{d\theta}$ of $y = \frac{1}{\sqrt[4]{\theta + 6}}$?

$$y = (\theta + 6)^{-\frac{1}{4}}, \qquad y^{-4} = \theta + 6, \qquad \theta = y^{-4} - 6 = \frac{1}{y^4} - 6$$

$$\theta = \frac{1}{y^4} - 6 = \text{inverse function}$$

$$\frac{d\theta}{dy} = -4y^{-5} = -4(\theta + 6)^{\frac{5}{4}} = -4\sqrt[4]{(\theta + 6)^5}$$

$$= \text{derivative of the inverse function}$$

$$\therefore \frac{dy}{d\theta} = -\frac{1}{4\sqrt[4]{(\theta + 6)^5}} = \text{derivative of the original function}$$

Obtain $\frac{dy}{d\theta}$ directly, and compare. Which method is simpler here?

PROBLEMS

1. What is the inverse function of (a) $x = y^2 + a^2$; (b) $x = a^y$; (c) $x = \sin y$?

2. Solve Probs. 2, 3, and 6 on page 39 and 1, 2, 3, 5, 8, and 10 on page 46 by inverse functions.

Chapter VII

DIFFERENTIATION
ELEMENTARY RULES—TRIGONOMETRIC

118. What is the derivative of $y = \sin \theta$?

$$\frac{d}{d\theta}(\sin \theta) = \cos \theta$$

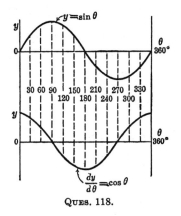

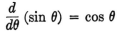

QUES. 118.

119. What is the derivative of $y = \cos \theta$?

$$\frac{d}{d\theta}(\cos \theta) = -\sin \theta$$

120. What is the derivative of $y = \tan \theta$?

$$\frac{d}{d\theta}(\tan \theta) = \sec^2 \theta$$

121. What is the derivative of $y = \cot \theta$?

$$\frac{d}{d\theta}(\cot \theta) = -\operatorname{cosec}^2 \theta$$

122. What is the second derivative of $y = \sin \theta$?

$$\frac{d^2}{d\theta^2} (\sin \theta) = F'' (\sin \theta) = - \sin \theta$$

123. What is the second derivative of $y = \cos \theta$?

$$\frac{d^2}{d\theta^2} (\cos \theta) = F'' (\cos \theta) = - \cos \theta$$

124. Only the functions $\left.\begin{array}{l} y = \sin \theta \\ y = \cos \theta \end{array}\right\}$ possess what unique characteristic?

It is unique with each of these functions that its second derivative = the original function with opposite sign.

125. What is the derivative of $y = \text{arc} \sin x$?

If $y = \text{arc} \sin x$,

then $\quad x = \sin y \quad$ or $\quad x^2 = \sin^2 y$

and $\quad \dfrac{dx}{dy} = \cos y$

Now $\quad \dfrac{dy}{dx} = \dfrac{1}{\cos y} =$ the derivative of the inverse function

But $\cos y = \sqrt{1 - \sin^2 y} = \sqrt{1 - x^2}$

$$\therefore \frac{dy}{dx} = \frac{1}{\sqrt{1 - x^2}}$$

QUES. 125.

126. What is the derivative of $y = \cos^3 \theta$?

If $y = (\cos \theta)^3$,

let $\quad\quad \cos \theta = u$

Then $\quad\quad \dfrac{du}{d\theta} = -\sin \theta$

and $\quad\quad y = u^3$

$\quad\quad\quad \dfrac{dy}{du} = 3u^2 = 3\cos^2 \theta$

Now $\quad \dfrac{dy}{d\theta} = \dfrac{dy}{du} \cdot \dfrac{du}{d\theta} = 3\cos^2 \theta \cdot (-\sin \theta)$

$\quad\quad \therefore \dfrac{dy}{d\theta} = -3\cos^2 \theta \cdot \sin \theta$

127. What is the derivative of $y = \tan 3\theta$?

If $y = \tan 3\theta$,

let $\quad\quad u = 3\theta$

Then $\quad\quad \dfrac{du}{d\theta} = 3$

and $\quad\quad y = \tan u \quad\quad$ by substitution

$\quad\quad\quad \dfrac{dy}{du} = \sec^2 u$

Finally, $\quad \dfrac{dy}{d\theta} = \dfrac{dy}{du} \cdot \dfrac{du}{d\theta} = \sec^2 u \cdot 3 = 3\sec^2 (3\theta)$

128. What is the derivative of $y = \sin (x + b)$?

If $y = \sin (x + b)$,

let $\quad\quad\quad\quad u = (x + b)$

Then $\quad\quad\quad\quad \dfrac{du}{dx} = 1$

and $y = \sin u \quad$ and $\quad \dfrac{dy}{du} = \cos u = \cos (x + b)$

Finally, $\dfrac{dy}{dx} = \dfrac{dy}{du} \cdot \dfrac{du}{dx} = \cos (x + b) \cdot 1 = \cos (x + b)$

129. What is the derivative of $y = \sin \theta \cdot \cos \theta$?

If $y = \sin \theta \cdot \cos \theta$,

then $\dfrac{dy}{d\theta} = \sin \theta \cdot \dfrac{d}{d\theta} (\cos \theta) + \cos \theta \cdot \dfrac{d}{d\theta} (\sin \theta)$

$$= \text{derivative of a product}$$

$\therefore \dfrac{dy}{d\theta} = -\sin^2 \theta + \cos^2 \theta = \cos^2 \theta - \sin^2 \theta$

130. What is the derivative of $y = \sqrt{1 + 3 \tan^2 \theta}$?

Let $u = 3 \tan^2 \theta$.

Then $\dfrac{du}{d\theta} = 3 \cdot 2 \tan \theta \cdot \dfrac{d}{d\theta}(\tan \theta) = 6 \tan \theta \sec^2 \theta$

and $y = (1 + u)^{\frac{1}{2}}$ and $\dfrac{dy}{du} = \dfrac{1}{2 \sqrt{1 + u}}$

$$\text{by substitution}$$

Finally, $\dfrac{dy}{d\theta} = \dfrac{dy}{du} \cdot \dfrac{du}{d\theta} = \dfrac{1}{2\sqrt{1+u}} \cdot 6 \tan \theta \sec^2 \theta$

$$= \dfrac{6 \tan \theta \sec^2 \theta}{2 \sqrt{1 + 3 \tan^2 \theta}}$$

131. What is the derivative of $y = \sec \theta$?

If $y = \sec \theta = \dfrac{1}{\cos \theta}$,

then $\dfrac{dy}{d\theta} = \dfrac{\sin \theta}{(\cos \theta)^2}$ (the derivative of a quotient)

or $\dfrac{dy}{d\theta} = \dfrac{\sin \theta}{\cos \theta} \cdot \dfrac{1}{\cos \theta} = \tan \theta \sec \theta = \text{the derivative}$

132. What is the derivative of $y = \tan \theta \sqrt{3 \sec \theta}$?

$\dfrac{dy}{d\theta} = \tan \theta \cdot \dfrac{d}{d\theta} (3 \sec \theta)^{\frac{1}{2}} + (3 \sec \theta)^{\frac{1}{2}} \cdot \dfrac{d}{d\theta} \cdot \tan \theta$

$$\text{(the derivative of a product)}$$

Now let $w = (3 \sec \theta)^{\frac{1}{2}}$ and let $u = 3 \sec \theta$.

Then $\dfrac{du}{d\theta} = 3 \sec \theta \cdot \tan \theta$

and $w = u^{\frac{1}{2}}$

Now $\dfrac{dw}{du} = \dfrac{1}{2} u^{-\frac{1}{2}} = \dfrac{1}{2 \sqrt{3 \sec \theta}}$

and $\dfrac{dw}{d\theta} = \dfrac{dw}{du} \cdot \dfrac{du}{d\theta} = \dfrac{1}{2 \sqrt{3 \sec \theta}} \cdot 3 \sec \theta \cdot \tan \theta$

Then $\dfrac{dy}{d\theta} = \dfrac{3 \sec \theta \cdot \tan \theta \cdot \tan \theta}{2 \sqrt{3 \sec \theta}} + \sqrt{3 \sec \theta} \cdot \sec^2 \theta$

$\dfrac{dy}{d\theta} = \dfrac{3 \tan^2 \theta \cdot \sec \theta}{2 \sqrt{3 \sec \theta}} + \sqrt{3 \sec \theta} \cdot \sec^2 \theta$

$= \dfrac{3 \tan^2 \theta \cdot \sec \theta + 2 \cdot 3 \sec \theta \cdot \sec^2 \theta}{2 \sqrt{3 \sec \theta}}$

$= \dfrac{3 \tan^2 \theta \sec \theta + 6 \sec^3 \theta}{2 \sqrt{3 \sec \theta}}$

$= \dfrac{3 \sec \theta (\tan^2 \theta + 2 \sec^2 \theta)}{2 \sqrt{3 \sec \theta}}$

$= \dfrac{(3 \sec \theta)^{\frac{1}{2}}(\tan^2 \theta + 2 \sec^2 \theta)}{2}$

Now $\tan^2 \theta = \sec^2 \theta - 1$

$\therefore \dfrac{dy}{d\theta} = \dfrac{(3 \sec \theta)^{\frac{1}{2}}(\sec^2 \theta - 1 + 2 \sec^2 \theta)}{2}$

$= \dfrac{\sqrt{3 \sec \theta} (3 \sec^2 \theta - 1)}{2}$

PROBLEMS

Differentiate the following:

1. $y = \sin^2 \theta$

2. $y = \sin^3 \theta$

3. $y = \sin 2\theta$

4. $y = \sin 3\theta$

5. $y = \arctan \theta$

6. $y = \arccos \theta$

7. $y = \text{arc sec } \theta$

8. $y = \sin 2\theta \cdot \cos 3\theta$

9. $y = \sin (\theta^2 + 2\theta - 3)$

10. $y = \sqrt{\dfrac{1 - \cos \theta}{1 + \cos \theta}}$

11. $y = \dfrac{1 - \cos \theta}{1 + \cos \theta}$

12. $y = \sin \frac{1}{2}\theta \cdot \cos 5\theta$

13. $y = a \sin \left(\theta - \dfrac{\pi}{2}\right)$

14. $y = a \cos (\omega t + \theta)$

15. $x = \dfrac{1}{2\pi} \cos 2\pi n t$

CHAPTER VIII

GEOMETRICAL MEANING OF DIFFERENTIATION

133. What is the geometrical meaning of $\frac{dy}{dx}$?

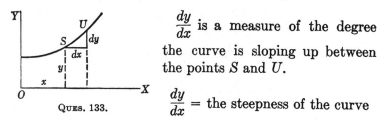

QUES. 133.

$\frac{dy}{dx}$ is a measure of the degree the curve is sloping up between the points S and U.

$\frac{dy}{dx}$ = the steepness of the curve

134. When is $\frac{dy}{dx}$ the slope of the curve along SU?

When SU = a straight line; *i.e.*, when S and U are very near each other.

135. When is the straight line SU tangent to the curve?

When SU is indefinitely small and S and U coincide.

136. What is the meaning of $\frac{dy}{dx}$ at any point on a curve?

It is the slope of the tangent at the point.

137. What is the value of $\frac{dy}{dx}$ at a point where the curve slopes up at 45 degrees?

$$\frac{dy}{dx} = \text{slope} = 1 \quad (\text{Here } \tan 45° = 1)$$
$$\therefore dy = dx$$

56

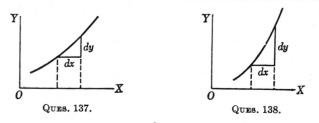

QUES. 137.   QUES. 138.

138. What is the value of $\dfrac{dy}{dx}$ at a point where the curve slopes up steeper than 45 degrees?

$$\frac{dy}{dx} > 1 \quad \text{(Here the tangent is greater than 1)}$$
$$\therefore dy > dx$$

139. What is the value of $\dfrac{dy}{dx}$ at a point where the curve slopes up at an angle less than 45 degrees?

$$\frac{dy}{dx} < 1 \quad \text{(Here the tangent is less than 1)}$$
$$\therefore dy < dx$$

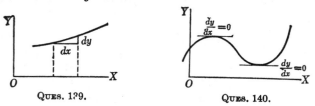

QUES. 139.   QUES. 140.

140. When is $\dfrac{dy}{dx}$ equal to zero?

When $dy = 0$ and therefore $\dfrac{dy}{dx} = 0$, which occurs at a horizontal place in a curve.

141. When is $\dfrac{dy}{dx}$ (a) a negative slope; (b) a positive slope?

a. When $dy =$ a step down and the curve is falling to the right.

b. When $dy =$ a step up and the curve is rising to the right.

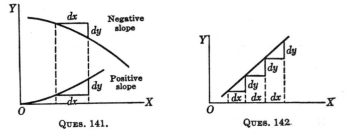

QUES. 141.   QUES. 142.

142. When is $\dfrac{dy}{dx}$ a constant?

When the curve is a straight line.   Here the slope $\dfrac{dy}{dx}$ obviously has no term in x.

143. When does the slope $\dfrac{dy}{dx}$ vary?

In a curve, when $\dfrac{dy}{dx}$ has a term in x.   This means that the slope varies with x.

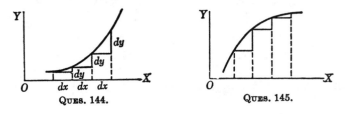

QUES. 144.   QUES. 145.

144. When is $\dfrac{dy}{dx}$ greater and greater to the right?

When the curve increases its upward bend to the right.

145. When is $\dfrac{dy}{dx}$ less and less to the right?

When the curve is flatter and flatter to the right.

146. When does the ordinate y pass through a minimum?

When $\frac{dy}{dx}$ is *first negative*, with diminishing values,

then $\frac{dy}{dx} = 0$ at the bottom of the trough (*with a minimum y*)

Then $\frac{dy}{dx}$ is *positive* with increasing values.

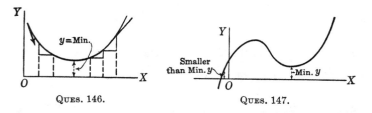

QUES. 146.   QUES. 147.

147. Is the minimum y the smallest y?

Not always. The minimum y is only the value of y corresponding to the bottom of the trough.

148. What is the characteristic of a minimum?

y must *increase* on either side of it.

149. How do you obtain the value of x for the minimum y?

Set $\frac{dy}{dx} = 0$. (This value is for the bottom of the trough.)

Solve for x, and then place this x in the original equation for the corresponding value of y.

150. When does y pass through a maximum?

When $\frac{dy}{dx}$ is first positive,

then $\qquad\qquad \frac{dy}{dx} = 0$ (with a maximum y)

Then $\frac{dy}{dx}$ is negative, and the curve slopes downward.

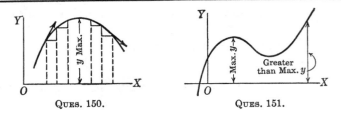

QUES. 150.   QUES. 151.

151. Is the maximum y the greatest y?

Not always.   The maximum y is only the value of y corresponding to the summit.

152. What is the characteristic of a maximum?

y must *decrease* on either side of it.

153. How do you obtain the value of x for the maximum y?

Set $\dfrac{dy}{dx} = 0$.   (This value is for the hilltop.)

Solve for x.   Then place this x in the original equation to get the corresponding value of y.

154. Can you show a curve in which $\dfrac{dy}{dx}$ is always positive and yet has a minimum $\dfrac{dy}{dx}$?

This is a special case.   The figure shows the least slope.

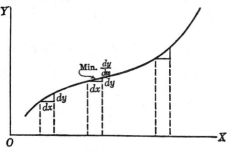

QUES. 154.

155. Can there be a maximum or minimum for a value of x that makes $\dfrac{dy}{dx} = \infty$?

Yes, when the function $f(x)$ is continuous and the characteristic that y decrease on either side of the value of x for a maximum and increase on either side for a minimum is satisfied.

There is a maximum at points A and C and a minimum at point B, but no maximum or minimum at D. Although at each of these points $\dfrac{dy}{dx} = \infty$, the characteristic for a maximum or minimum is satisfied at points A, B, and C and not at point D.

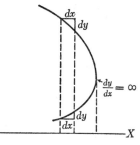

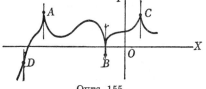

Ques. 155.   Ques. 156.

156. Can you show a curve having $\dfrac{dy}{dx}$ infinitely great?

$$-\frac{dy}{dx} \text{ in upper part}$$

$$+\frac{dy}{dx} \text{ in lower part}$$

$$\frac{dy}{dx} = \infty \text{ at nose}$$

This is a special case.

157. What is the slope $\dfrac{dy}{dx}$ at any point of the curve $y = x + a$?

$$\frac{dy}{dx} = 1 = \text{a constant slope}$$

The line ascends at a 45-deg. angle from a point where $y = a$ when $x = 0$.

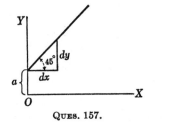

QUES. 157.

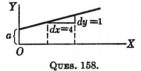

QUES. 158.

158. What is the slope $\dfrac{dy}{dx}$ at any point of the curve $y = bx + a$?

$$\frac{dy}{dx} = b = \text{a constant slope}$$

The line ascends at a constant angle whose tangent $= b$.

Example

Assume $b = \dfrac{1}{4}$. Then $\dfrac{dy}{dx} = \dfrac{1}{4}$ or $dx = 4 \cdot dy$

159. What is the slope $\dfrac{dy}{dx}$ at any point of the curve $y = bx^2 + a$?

$$\frac{dy}{dx} = 2bx \quad \text{(The steepness increases with } x\text{)}$$

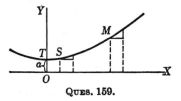

QUES. 159.

At the origin when $x = 0$,

$$\frac{dy}{dx} = 0 \quad \text{(The curve is horizontal at the origin)}$$

At $x = -1$,

$$\frac{dy}{dx} = -2b \text{ (The curve descends from left to right)}$$

160. What is the relation between the curve $y = \frac{1}{8}x^2 + 1$ and its derivative $\frac{dy}{dx} = \frac{1}{4}x$?

When	$x = 0$	1	2	3	4	5
then	$y = 1$	$1\frac{1}{8}$	$1\frac{1}{2}$	$2\frac{1}{8}$	3	$4\frac{1}{8}$
and	$\frac{dy}{dx} = 0$	$\frac{1}{4}$	$\frac{1}{2}$	$\frac{3}{4}$	1	$1\frac{1}{4}$

Observe that any ordinate of the $\frac{dy}{dx}$ curve = the corresponding slope of the original curve.

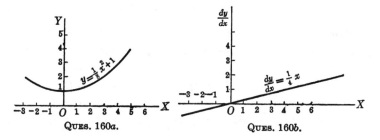

QUES. 160a.   QUES. 160b.

Therefore, ordinates of the $\frac{dy}{dx}$ curve are proportional to the slopes of the original curve.

161. What results when the slope $\frac{dy}{dx}$ of a curve abruptly changes from a positive to a negative value?

QUES. 161.

A sudden cusp is formed.

162. What is the slope of the tangent to the curve $y = \frac{1}{4x} + 5$ at $x = -2$?

$$\text{Slope} = \frac{dy}{dx} = -\frac{1}{4x^2} \quad \text{and at} \quad x = -2,$$

$$\frac{dy}{dx} = -\frac{1}{4(-2)^2} = -\frac{1}{16} = \text{slope of the tangent line and}$$

$$\text{curve at } x = -2$$

163. What is the equation of the above tangent line?

$y = ax + b =$ the equation of a straight line having a slope $= a$

$$\frac{dy}{dx} = a = \text{slope} = -\frac{1}{16}$$

From the equation of the curve $y = \dfrac{1}{4x} + 5$ and at $x = -2$,

$$y = \frac{1}{4(-2)} + 5 = 4\frac{7}{8}$$

These coordinates of the curve ($x = -2$, $y = 4\frac{7}{8}$) must also satisfy the equation of the tangent line at $x = -2$.
Then

$$y = ax + b = 4\tfrac{7}{8} = -\tfrac{1}{16}(-2) + b$$

from which $b = 4\frac{3}{4}$

$$\therefore y = -\tfrac{1}{16}x + 4\tfrac{3}{4} = \text{equation of tangent line}$$

164. What is the angle between the above tangent line and the curve $y = 3x^2 + 4\frac{3}{4}$?

Find the points of contact by solving simultaneously the equations of the tangent line and the curve

or

$$-\frac{x}{16} + \frac{19}{4} = 3x^2 + \frac{19}{4}$$

getting

$$3x^2 + \frac{x}{16} = 0, \qquad x\left(3x + \frac{1}{16}\right) = 0$$

The points of contact are at $x = 0$ and $x = -\tfrac{1}{48}$.
The slope at any point of the curve $y = 3x^2 + \tfrac{19}{4}$ is

$$\frac{dy}{dx} = 6x.$$

And at one point of contact, $x = 0$ and $\dfrac{dy}{dx} = 0$.   The slope of the tangent to the curve is horizontal.   At the other point of contact, $x = -\frac{1}{48}$ and

$$\frac{dy}{dx} = 6 \cdot \left(-\frac{1}{48}\right) = -\frac{1}{8}$$

and the tangent to the curve slopes **7.1 deg. downward** to the right.

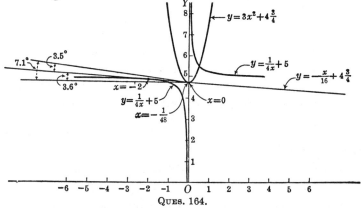

QUES. 164.

Now the slope of the tangent line itself $= -\frac{1}{16}$ (see above), which means a slope of 3.6 deg.

Therefore, at $x = 0$, the angle between the tangent line of the previous question and the tangent to the curve $= 3.6$ deg. and at $x = -\frac{1}{48}$ the angle between the tangent line of the previous question and the tangent to the curve $= (7.1 - 3.6) = 3.5$ deg.

165. **What are the coordinates of the point of contact of a straight line drawn through the point $x = 3$, $y = -2$ and tangent to the curve $y = x^2 - 3x + 2$?** •

$\dfrac{dy}{dx}$ of the curve $y = x^2 - 3x + 2 = 2x - 3$

which is the slope of the curve at any point, and which

must also be the slope of a straight line tangent to the curve.

Now $y = ax + b =$ the equation of a straight line having a slope $= a$

or $-2 = 3a + b =$ the equation of a straight line when passing through the given point $x = 3, y = -2$

Now the coordinates x and y of the point of contact must satisfy both the equation of the tangent and the equation of the curve.

$$y = x^2 - 3x + 2 \qquad \text{curve equation} \qquad (1)$$
$$y = ax + b \qquad \text{tangent equation} \qquad (2)$$
$$-2 = 3a + b \quad \text{tangent equation through } x = 3,$$
$$y = -2 \quad (3)$$
$$a = 2x - 3 = \text{slope of tangent} = \text{slope of curve} \qquad (4)$$

Now solve these equations simultaneously to get the point of contact.

$$x^2 - 3x + 2 = (ax + b) = (2x - 3)x + (-2 - 3a)$$
$$x^2 - 3x + 2 = 2x^2 - 3x - 2 - 3(2x - 3)$$
$$x^2 - 3x + 2 = 2x^2 - 3x - 2 - 6x + 9$$

which simplifies to

$$-x^2 + 6x - 5 = 0$$

which now becomes

$$x^2 - 6x + 5 = 0 \qquad \text{or} \qquad (x - 5)(x - 1) = 0,$$
$$x = 5 \text{ and } 1$$

Now put these values of x in Eq. (1) and get

When $\qquad x = 5, \qquad y = 25 - 15 + 2 = 12$ $\big\}$ the coordinates of the

and when $x = 1, \qquad y = 1 - 3 + 2 = 0$ $\big\}$ nates of the point of contact

PROBLEMS

1. What is the slope of the curve $y = x^3 - 2$ at $x = 3$?

2. What is the $\dfrac{dy}{dx}$ of $y = x^3 + 6x$ at $x = \frac{1}{2}$?

3. What is the value of x at which the slope $= 1$ in the equation $y^2 = 16 - x^2$?

4. What is the slope of the curve $\dfrac{x^2}{4} + \dfrac{y^2}{16} = 1$ when $x = 1$?

5. What is the equation of the tangent to the curve $y = 6 - 3x + x^3$ at $x = 3$?

6. At what angle do curves $y = 6x^2 + 4$ and $y = 2x^{2\cdot} - 3x + 8$ cut one another?

7. What are the coordinates of the point of contact of a straight line drawn through the point $(4, -3)$ and tangent to the curve $y = 4x^2 + 1$?

8. For what values of x is the curve $y = 2x^2$ rising, and for what values is it falling?

9. For what values of x is the function $y = 2x^3$ increasing, and for what values is it decreasing?

10. For what values of x is the curve $y = \dfrac{x^3}{2} - \dfrac{x^2}{4} - 8x$ rising and for what values falling, and for what value of x is y increasing four times as fast as x?

11. When is the derivative of the function of a circle with the center at the origin (a) positive; (b) negative; (c) zero?

12. How many times as rapidly as x is y increasing when $x = 6$ in $y = x^3 - 2x^2 - x$?

13. What are the maximum and minimum points of $y = x^3 - 2x^2 + 6$? Plot the curve.

14. What are the maximum and minimum points of $y = 4x - 2x^2 + 3$? Plot the curve.

CHAPTER IX

TANGENT—NORMAL—SUBTANGENT— SUBNORMAL

166. What is the inclination of a line?

The angle it makes with the positive part of the x axis.

167. What is the slope of a line?

The tangent of the inclination or

$$m = \text{slope} = \tan \theta$$

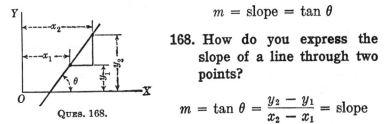

QUES. 168.

168. How do you express the slope of a line through two points?

$$m = \tan \theta = \frac{y_2 - y_1}{x_2 - x_1} = \text{slope}$$

169. What is the point-slope form of the equation of a straight line?

Let (x_1, y_1) be the coordinates of a definite point on a straight line.

Let (x, y) be the coordinates of any other point on the same line.

And let $m = $ the slope of this line.

Then $\dfrac{y - y_1}{x - x_1} = m = $ the slope of the line as of above

or $\quad y - y_1 = m(x - x_1) = $ point-slope form equation of a straight line

170. What is the equation of a line making an angle of 30 degrees with the x axis and passing through the point (2, 4)?

$$m = \tan 30° = \frac{\sqrt{3}}{3} \qquad x_1 = 2 \qquad y_1 = 4$$

$$\therefore y - 4 = \frac{\sqrt{3}}{3}(x - 2) \qquad \text{or}$$

$$\frac{\sqrt{3}}{3}x - y + 4 - \frac{2\sqrt{3}}{3} = 0 = \text{the equation of the line}$$

171. What is the equation of a tangent line to a curve $y = f(x)$ at a definite point (x_1, y_1)?

$$y - y_1 = m(x - x_1) = \text{point-slope form of a line}$$

But $m = \text{slope} = \dfrac{dy}{dx} = $ the derivative of the equation of

the curve at the point (x_1, y_1)

$$\therefore y - y_1 = \frac{dy}{dx}(x - x_1)$$

172. What is a normal to a curve at a definite point of the curve?

It is the perpendicular to the tangent to the curve at the point considered.

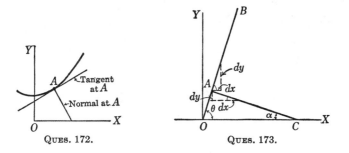

QUES. 172.   QUES. 173.

173. What is the relationship of the slopes of two lines when they are perpendicular to each other?

The slope of either line is the negative reciprocal of the slope of the other.

Example

If the slope of a line is 3, then the slope of a line perpendicular to it is $-\frac{1}{3}$.

$$\tan \theta = \frac{dy}{dx} = \text{slope of line } OB$$

$$-\frac{1}{\tan \theta} = -\tan \alpha = \text{slope of line } AC \perp \text{ to } OB$$

174. What is the equation of the normal to a curve $y = f(x)$ at a definite point (x_1, y_1)?

$$y - y_1 = \frac{dy}{dx}(x - x_1) = \text{the equation of the tangent to}$$

the curve at the point

$$\therefore \ y - y_1 = -\frac{1}{\dfrac{dy}{dx}}(x - x_1) = \text{the equation of the normal}$$

to the curve at the same point

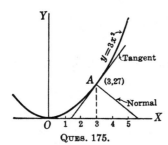

Qurs. 175.

175. What are the equations of the tangent and the normal to the curve $y = 3x^2$ at the point where $x = 3$?

$$\frac{dy}{dx} = 6x = 6 \cdot 3 = 18$$

Now in $y = 3x^2$, if $x_1 = 3$, $y_1 = 27$,

$$\therefore \ y - y_1 = \frac{dy}{dx}(x - x_1) = 18(x - 3) = \text{equation of the}$$

tangent

or $y - 27 = 18(x - 3)$

$18x - y + 27 - 54 = 0$

$18x - y - 27 = 0 = $ **equation of the tangent**

$y - 27 = -\frac{1}{18}(x - 3) = $ equation of the normal

$18y - 486 = -x + 3$

$x + 18y - 489 = 0 = $ **equation of the normal**

176. What is the length of the tangent at any point of a curve?

TP = the distance from the point of tangency to the intersection of the tangent line with the x axis = the length of the tangent.

177. What is the length of the normal?

PN = the distance from the tangent point to the intersection of the normal line with the x axis = the length of the normal.

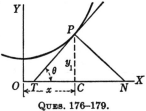

178. What is the length of the subtangent?

TC = the projection of the tangent on the x axis.

QUES. 176–179.

179. What is the length of the subnormal?

CN = the projection of the normal on the x axis.

180. What is the formula for the length of the tangent?

TP = length of the tangent = $PC \cdot$ cosec θ = y_1 cosec θ

Now $$\operatorname{cosec}^2 \theta = 1 + \cot^2 \theta$$

$$\operatorname{cosec} \theta = \sqrt{1 + \cot^2 \theta} = \sqrt{\left(1 + \frac{1}{\tan^2 \theta}\right) \frac{\tan^2 \theta}{\tan^2 \theta}}$$

$$\operatorname{cosec} \theta = \frac{\sqrt{\left(1 + \frac{1}{\tan^2 \theta}\right) \tan^2 \theta}}{\tan \theta} = \frac{\sqrt{1 + \tan^2 \theta}}{\tan \theta}$$

But $$\tan \theta = \frac{dy}{dx} = \text{slope}$$

$$\therefore\ TP = y_1 \operatorname{cosec} \theta = y_1 \frac{\sqrt{1 + \left(\frac{dy}{dx}\right)^2}}{\frac{dy}{dx}} = \text{length of the}$$

tangent

181. What is the formula for the length of the normal?

PN = length of the normal = $PC \sec \theta = y_1 \sec \theta$

Now $\sec \theta = \sqrt{1 + \tan^2 \theta}$

But $\dfrac{dy}{dx} = \tan \theta$

$$\therefore PN = y_1 \sec \theta = y_1 \sqrt{1 + \left(\dfrac{dy}{dx}\right)^2}$$

QUES. 180–183.

182. What is the formula for the length of the subtangent?

TC = length of the subtangent = $PC \cot \theta = \dfrac{y_1}{\tan \theta} = \dfrac{y_1}{\dfrac{dy}{dx}}$

183. What is the formula for the length of the subnormal?

CN = length of the subnormal = $PC \tan \theta = y_1 \dfrac{dy}{dx}$

184. What are the lengths of the tangent, normal, subtangent, and subnormal for the curve $y^2 = 2x$, at $x = 1$?

When $x = 1$, $y_1 = \sqrt{2}$.
Differentiate $y^2 = 2x$.

$2y \dfrac{dy}{dx} = 2$ or $\dfrac{dy}{dx} = \dfrac{1}{y}$ and $\dfrac{dy}{dx} = \dfrac{1}{\sqrt{2}}$

when $x = 1$

$$\text{Length of tangent} = y_1 \dfrac{\sqrt{1 + \left(\dfrac{dy}{dx}\right)^2}}{\dfrac{dy}{dx}} = \sqrt{2} \cdot \dfrac{\sqrt{1 + \left(\dfrac{1}{\sqrt{2}}\right)^2}}{\dfrac{1}{\sqrt{2}}}$$

$$= 2\sqrt{1.5} = 2.45$$

$$\text{Length of normal} = y_1 \sqrt{1 + \left(\dfrac{dy}{dx}\right)^2} = \sqrt{2} \cdot \sqrt{1 + \left(\dfrac{1}{\sqrt{2}}\right)^2}$$

$$= 1.73$$

$$\text{Subtangent} = \frac{y_1}{\dfrac{dy}{dx}} = \frac{\sqrt{2}}{\dfrac{1}{\sqrt{2}}} = \sqrt{2} \cdot \sqrt{2} = 2$$

$$\text{Subnormal} = y_1 \frac{dy}{dx} = \sqrt{2} \cdot \frac{1}{\sqrt{2}} = 1$$

185. What are parametric equations?

Equations of a curve with coordinates x and y of any point expressed as functions of (or depending upon) a third variable or parameter.

Example

$x = r \cos \theta$ ⎫ are parametric equations of a circle, θ
$y = r \sin \theta$ ⎭ being the parameter.

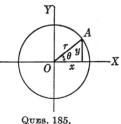

Each value of θ gives a value of x and of y that determines a point on the curve, and if θ varies from 0 to 2π the point A will describe a circle.

Eliminating the parameter θ by squaring and adding the parametric equations,

QUES. 185.

$$r^2 \cos^2 \theta + r^2 \sin^2 \theta = x^2 + y^2$$
$$r^2(\cos^2 \theta + \sin^2 \theta) = x^2 + y^2 \text{ (as } \cos^2 \theta + \sin^2 \theta = 1)$$

or $r^2 = x^2 + y^2 =$ the rectangular equation of the circle

186. How would you find the slope when given the parametric equations of the curve?

If $\qquad\qquad x = f(\theta)$
and $\qquad\qquad y = \phi(\theta)$

it is evident that θ is an inverse function of x.

Then $\qquad \dfrac{dy}{dx} = \dfrac{dy}{d\theta} \cdot \dfrac{d\theta}{dx} = \dfrac{\dfrac{dy}{d\theta}}{\dfrac{dx}{d\theta}} = \dfrac{\phi'(\theta)}{f'(\theta)}$

This is the slope at point (x, y) found when the parametric equations are given.

187. What are the equations of the tangent and normal and the lengths of the subtangent and subnormal of the ellipse $x = 3 \cos \theta$ and $y = 4 \sin \theta$ at the point where $\theta = 30°$?

The parameter is θ and $\dfrac{dx}{d\theta} = -3 \sin \theta$, and $\dfrac{dy}{d\theta} = 4 \cos \theta$.

Then $\dfrac{dy}{dx} = \dfrac{\dfrac{dy}{d\theta}}{\dfrac{dx}{d\theta}} = -\dfrac{4 \cos \theta}{3 \sin \theta} = -\dfrac{4}{3} \cot \theta =$ the slope at

$$\text{any point} = m$$

Now when $\theta = 30°$,

then $m_1 = -\frac{4}{3} \cot 30° = -\frac{4}{3} \cdot 1.7321 = -2.3095$

Substitute $\theta = 30°$ in the given equations to get

$$x_1 = 3 \cdot \cos 30° = 3 \cdot 0.866 = 2.598$$
$$y_1 = 4 \sin 30° = 4 \cdot \tfrac{1}{2} = 2.0$$
$$\therefore y - y_1 = m_1(x - x_1) = y - 2 = -2.31(x - 2.6)$$

or $\quad y = 2 + 6 - 2.31x = 8 - 2.31x =$ equation of tangent

and $y - y_1 = -\dfrac{1}{m_1}(x - x_1)$

or $\quad y - 2 = -\dfrac{1}{-2.31}(x - 2.6) = \dfrac{x}{2.31} - \dfrac{2.6}{2.31}$

$$= \dfrac{x}{2.31} - 1.125$$

$$y = 2 - 1.125 + \dfrac{x}{2.31} = 0.875 + \dfrac{x}{2.31}$$

$$= \dfrac{2.02 + x}{2.31}$$

$\therefore 2.31y = x + 2.02 =$ equation of normal

Length of subtangent $= \dfrac{y_1}{m_1} = -\dfrac{2}{2.31} = -0.87$

Length of subnormal $= m_1 y_1 = -2.31 \cdot 2 = -4.62$

188. How would you find the points of contact of the horizontal and vertical tangents to a curve given in parametric form?

Solve $\dfrac{dy}{d\theta} = 0$ for θ to get the horizontal tangents.

Solve $\dfrac{dx}{d\theta} = 0$ for θ to get the vertical tangents.

189. What are the points of contact of the horizontal and vertical tangents to $x = 2 - 3\sin\theta$, $y = 3 + 2\cos\theta$?

$\dfrac{dy}{d\theta} = -2\sin\theta = 0,$ and $\sin\theta = 0$ or

$$\theta = 0 \text{ or } \pi$$

$\dfrac{dx}{d\theta} = -3\cos\theta = 0,$ and $\cos\theta = 0,$ or

$$\theta = \frac{\pi}{2} \text{ or } \frac{3\pi}{2}$$

$\begin{cases} x = 2 - 3\sin(0) = 2 \\ y = 3 + 2\cos(0) = 5 \end{cases}$ $\begin{cases} x = 2 - 3\sin(\pi) = 2 \\ y = 3 + 2\cos(\pi) = 1 \end{cases}$

$\begin{cases} x = 2 - 3\sin\left(\dfrac{\pi}{2}\right) = -1 \\ y = 3 + 2\cos\left(\dfrac{\pi}{2}\right) = 3 \end{cases}$ $\begin{cases} x = 2 - 3\sin\left(\dfrac{3\pi}{2}\right) = 5 \\ y = 3 + 2\cos\left(\dfrac{3\pi}{2}\right) = 3 \end{cases}$

Therefore, the points $(2, 5)$ and $(2, 1)$ are the points of contact of the horizontal tangents; and $(-1, 3)$ and $(5, 3)$ are the points of contact of the vertical tangents.

190. How would you find the second derivative or rate of change of tangent to a curve given in parametric form?

The resulting first derivative $\dfrac{dy}{dx}$ (or y') was seen to be a function of the parameter. Therefore, to find the second derivative $\dfrac{d^2y}{dx^2}$ (or y''), merely use the formula again, replacing y by the first derivative (or y').

$$\therefore \frac{d^2y}{dx^2} = y'' = \frac{dy'}{dx} = \frac{\dfrac{dy'}{d\theta}}{\dfrac{dx}{d\theta}} = \frac{F'(\theta)}{f'(\theta)}$$

In Question 187, where $x = 3 \cos \theta$, $y = 4 \sin \theta$, we found

$$y' = \frac{dy}{dx} = -\frac{4}{3} \cot \theta$$

and
$$\frac{dx}{d\theta} = -3 \sin \theta$$

Now, differentiate y', getting

$$\frac{dy'}{d\theta} = \frac{4}{3} \operatorname{cosec}^2 \theta$$

$$\therefore y'' = \frac{\dfrac{dy'}{d\theta}}{\dfrac{dx}{d\theta}} = \frac{\frac{4}{3} \operatorname{cosec}^2 \theta}{-3 \sin \theta} = -\frac{4}{9} \cdot \frac{1}{\sin^2 \theta \cdot \sin \theta}$$

or
$$y'' = -\frac{4}{9 \sin^3 \theta}$$

PROBLEMS

1. What are the equations of the tangent and normal to the curve $y = 3x^2$ at the point where $x = 4$?

2. What is the equation of a line making an angle of $22\frac{1}{2}$ deg. with the x axis and passing through $(2, 4)$?

3. What are the equations of the tangent and normal to the ellipse $9x^2 + 16y^2 = 144$ at $x = 3$?

4. What is the equation of the tangent and normal to the curve $y = 4x^2 - 2$ at $x = 3$?

5. What is the equation of a line making an angle of 120 deg. with the x axis and passing through point $(-3, -1)$?

6. Write the equation of the lines (a) through point $(-2, 4)$, slope $= 3$; (b) point $(4, 6)$, inclination $= 45$ deg.; (c) point $(-3, -1)$, slope $= \frac{1}{4}$.

7. What are the lengths of the tangent, normal, subtangent, and subnormal for $y^2 = 8x$ at $x = 3$?

8. Find the lengths of the tangent and normal for $y = x^3 - 8x + 10$ at $x = 2$.

9. What are the lengths of the tangent, normal, subtangent, and subnormal for $2xy = 8$ at $x = 2$?

10. Find the equations of the tangent and normal and the lengths of subtangent and subnormal to $x = \cos \theta$, $y = \sin \theta$ at $\theta = \dfrac{\pi}{4}$.

11. Plot the curve and find the points of contact of the horizontal and vertical tangents of $x = 3 - 4 \sin \theta$, $y = 4 + 3 \cos \theta$.

12. Find $\dfrac{dy}{dx}$ and $\dfrac{d^2y}{dx^2}$ in terms of θ of $x = 5 \cos \theta$, $y = 2 \sin \theta$.

Chapter X

MAXIMA AND MINIMA

191. When does a quantity that varies continuously pass through a maximum value?

When the *values* immediately preceding and following are *smaller* than the maximum value.

192. When does a quantity that varies continuously pass through a minimum value?

When the values immediately preceding and following are *greater* than the minimum value.

193. What is the $\frac{dy}{dx}$ (slope of curve) at a maximum or a minimum?

$$\frac{dy}{dx} = 0$$

The slope is zero, or the tangent line is horizontal.

194. How do you obtain the value of x for a minimum or a maximum y?

Set $\frac{dy}{dx} = 0$, and solve for a real value of x.

Set this value of x in the original equation to find the corresponding y.

195. What is an equation of condition?

$\frac{dy}{dx} = 0$ is called an *equation of condition*.

196. What is the difference between an equation and an equation of condition?

An *equation* is true in itself.

An *equation of condition* is true only if a certain condition is fulfilled.

$\frac{dy}{dx}$ is not always equal to zero. It is only a condition, to learn how much x will be if $\frac{dy}{dx} = 0$ where the original curve is horizontal.

197. What is the meaning of $\frac{d^2y}{dx^2}$ (the second derivative) of a curve?

The rate at which the slope of the curve is changing (per unit of length x).

198. What is happening to the curve when $\frac{d}{dx}\left(\frac{dy}{dx}\right) = \frac{d^2y}{dx^2}$ is positive?

The slope itself is becoming *greater* upward to the right.

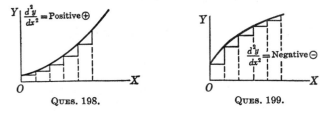

Ques. 198.   Ques. 199.

199. What is happening when $\frac{d^2y}{dx^2}$ is negative?

The slope is becoming *less* to the right.

200. What is the differential criterion for a minimum?

A minimum is indicated when

$$\frac{d^2y}{dx^2} = +, \text{positive,}$$

because to the left of M the slope is negative, and is getting less negative and to the right of M the slope is positive and is getting more positive.

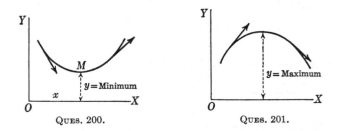

QUES. 200.   QUES. 201.

201. What is the differential criterion for a maximum?

A maximum is indicated when

$$\frac{d^2y}{dx^2} = -, \text{ negative,}$$

because to the left of the maximum the slope is positive and is getting less positive and to the right of the maximum the slope is negative and is getting more negative.

202. Where is the minimum or maximum of
$$y = 2x^2 - 8x + 6?$$

$$\frac{dy}{dx} = 4x - 8, \qquad \frac{d^2y}{dx^2} = 4 = +, \text{ positive (therefore a } \textit{mini-}$$
$$\textit{mum)}$$

$$\frac{dy}{dx} = 4x - 8 = 0, \qquad \text{and} \qquad x = 2$$

which is the value of x at the minimum.

Now substitute this in the original equation to get the corresponding y.

$$y = 2x^2 - 8x + 6 = 2 \cdot 2^2 - 8 \cdot 2 + 6 = 8 - 16 + 6$$
$$= -2$$

Therefore minimum is at $x = 2$, $y = -2$.

203. Where is the maximum or minimum of $y = 2x^2 - 3x$?

$\frac{dy}{dx} = 4x - 3,$ $\frac{d^2y}{dx^2} = 4 = +,$ positive, and indicates a minimum

Now set $\frac{dy}{dx}$ equal to zero, and solve to locate the minimum.

$$\frac{dy}{dx} = 4x - 3 = 0 \quad \therefore x = \frac{3}{4}$$

which we substitute in the original equation and get

$$y = 2(\tfrac{3}{4})^2 - 3 \cdot \tfrac{3}{4} = \tfrac{18}{16} - \tfrac{9}{4} = -\tfrac{9}{8}$$

Therefore, minimum is at $x = \tfrac{3}{4}$, $y = -\tfrac{9}{8}$.

204. When is there no maximum or minimum?

When $\frac{dy}{dx} = 0$ yields an impossible result.

Example

If $y = 3x^3 + 4x + 7$,

then $\frac{dy}{dx} = 9x^2 + 4 = 0,$ $x^2 = -\frac{4}{9},$ or

$$x = \pm\sqrt{-\frac{4}{9}}$$

Therefore y has no maximum or minimum as x is an imaginary quantity.

205. Where is the maximum or minimum of
$$y = \frac{x^3}{3} - \frac{5x^2}{2} + 6x + 4?$$

$\frac{dy}{dx} = x^2 - 5x + 6 = 0,$ $(x - 2)(x - 3) = 0,$

$$x = 3 \text{ and } 2$$

$$\frac{d^2y}{dx^2} = 2x - 5$$

For $x = 3$,

$$\frac{d^2y}{dx^2} = 2x - 5 = 2 \cdot 3 - 5 = \mathbf{+}, \text{ positive, and indicates a}$$
$$\text{minimum}$$

For $x = 2$,

$$\frac{d^2y}{dx^2} = 2x - 5 = 2 \cdot 2 - 5 = \mathbf{-}, \text{ negative, and indicates a}$$
$$\text{maximum}$$

For $x = 3$,

$$y = \frac{x^3}{3} - \frac{5x^2}{2} + 6x + 4 = \frac{3^3}{3} - \frac{5 \cdot 3^2}{2} + 6 \cdot 3 + 4$$
$$= 9 - \tfrac{45}{2} + 18 + 4 = 8\tfrac{1}{2}$$

QUES. 205.

For $x = 2$,

$$y = \frac{x^3}{3} - \frac{5x^2}{2} + 6x + 4 = \frac{2^3}{3} - \frac{5 \cdot 2^2}{2} + 6 \cdot 2 + 4$$
$$= \tfrac{8}{3} - 10 + 12 + 4 = 8\tfrac{2}{3}$$

Therefore, minimum is at $x = 3$, $y = 8\tfrac{1}{2}$, and maximum is at $x = 2$, $y = 8\tfrac{2}{3}$.

206. Where is the maximum or minimum of

$$y = \frac{x}{3 - x} + \frac{3 - x}{x}?$$

$$\frac{dy}{dx} = \frac{(3 - x) - [x \cdot (-1)]}{(3 - x)^2} + \frac{x(-1) - (3 - x) \cdot 1}{x^2} = 0$$

$$= \frac{3}{(3 - x)^2} - \frac{3}{x^2} = 0$$

or
$$\frac{1}{(3 - x)^2} - \frac{1}{x^2} = 0$$

which we simplify
$$x^2 - 9 + 6x - x^2 = 0$$
$$6x = 9, \qquad x = \tfrac{3}{2} = 1\tfrac{1}{2}$$

$$\frac{d^2y}{dx^2} = \frac{d}{dx}\left[\frac{3}{(3 - x)^2} - \frac{3}{x^2}\right] = \frac{-3\frac{d}{dx}(3 - x)^2}{(3 - x)^4} - \left(\frac{-6x}{x^4}\right)$$

$$\frac{d^2y}{dx^2} = \frac{-3(-6 + 2x)}{(3 - x)^4} + \frac{6x}{x^4} = \frac{18 - 6x}{(3 - x)^4} + \frac{6}{x^3}$$

Substitute $1\frac{1}{2}$, the value of x from $\frac{dy}{dx} = 0$, in $\frac{d^2y}{dx^2}$.

Then $\frac{d^2y}{dx^2} = \frac{18 - 6 \cdot \frac{3}{2}}{(3 - 1\frac{1}{2})^4} + \frac{6}{(\frac{3}{2})^3} = $ **+**, positive, *and indicates*

a minimum

Substitute $x = 1\frac{1}{2}$ in the original equation

$$y = \frac{x}{3 - x} + \frac{3 - x}{x},$$

getting, $y = \frac{\frac{3}{2}}{3 - \frac{3}{2}} + \frac{3 - \frac{3}{2}}{\frac{3}{2}} = 1 + 1 = 2$

Therefore, minimum is at $x = \frac{3}{2}$, $y = 2$

207. Where is the maximum or minimum of
$$y = \sqrt{2 + x} + \sqrt{2 - x}?$$

$$\frac{dy}{dx} = \frac{1}{2\sqrt{2 + x}} - \frac{1}{2\sqrt{2 - x}} = 0$$

$$\sqrt{2 + x} = \sqrt{2 - x}$$
$$2 + x = 2 - x$$
$$x = 0$$

Substitute this in the original equation.

For $x = 0$,

$$y = \sqrt{2 + x} + \sqrt{2 - x} = \sqrt{2} + \sqrt{2} = 2\sqrt{2}$$

$$\text{Now } \frac{d^2y}{dx^2} = \frac{1}{2} \cdot -\frac{1}{2}(2 + x)^{-\frac{3}{2}} \cdot 1 - \frac{1}{2} \cdot -\frac{1}{2}(2 - x)^{-\frac{3}{2}} \cdot$$

$$(-1)$$

$$\text{or} \quad \frac{d^2y}{dx^2} = -\frac{1}{4}(2 + x)^{-\frac{3}{2}} - \frac{1}{4}(2 - x)^{-\frac{3}{2}}$$

$$= -\frac{1}{4\sqrt{(2 + x)^3}} - \frac{1}{4\sqrt{(2 - x)^3}}$$

For $x = 0$,

$$\frac{d^2y}{dx^2} = -\frac{1}{4\sqrt{2^3}} - \frac{1}{4\sqrt{2^3}} = -\text{,} \quad \text{minus or negative, and}$$

indicates a maximum

Therefore, maximum is at $x = 0$, $y = 2\sqrt{2}$.

208. How would you split any number into two parts so that their product is a maximum?

Let $\quad\quad a =$ any number

and $\quad\quad x =$ one part

Then $\quad a - x =$ the other part

and $x(a - x) = xa - x^2 = y =$ the product of the two parts

For a maximum,

$$\frac{dy}{dx} = a - 2x = 0$$

Obtaining $x = \dfrac{a}{2}$, showing that the parts must be equal,

$$\frac{a}{2} \cdot \frac{a}{2} = \frac{a^2}{4} = \text{the value of the maximum product.}$$

209. If $a + b + c =$ a constant number, when is $a \cdot b \cdot c$ a maximum?

When $a = b = c$. This applies to any number of factors.

210. What are the maximum and minimum of the circle $(y - k)^2 + (x - l)^2 = r^2$?

$$(y - k)^2 = r^2 - (x - l)^2$$
$$y = k + \sqrt{r^2 - (x - l)^2}$$
$$\frac{dy}{dx} = \frac{1}{2}[r^2 - (x - l)^2]^{-\frac{1}{2}} \cdot -2(x - l) \cdot 1$$
$$= \frac{l - x}{[r^2 - (x - l)^2]^{\frac{1}{2}}} = 0$$

No value of x will make the denominator infinite. Therefore, the only condition to give zero is $x = l$.

Therefore, inserting this in the original equation of circle,

$$(y - k)^2 + (l - l)^2 = r^2$$
$$y - k = \pm r$$
$$\therefore y = k + r \text{ for an obvious maximum}$$
$$= k - r \text{ for an obvious minimum}$$

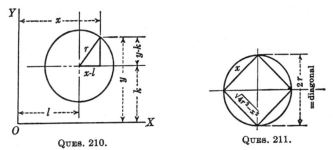

QUES. 210.   QUES. 211.

211. What rectangle of maximum area can be inscribed in a circle of radius r?

Let x = one side of the rectangle.

Then $\sqrt{(\text{Diagonal})^2 - x^2}$ = the other side = $\sqrt{(2r)^2 - x^2}$
$$= \sqrt{4r^2 - x^2}$$

$A = x\sqrt{4r^2 - x^2}$ = area of rectangle

$\dfrac{dA}{dx} = x \cdot \dfrac{d}{dx}(\sqrt{4r^2 - x^2}) + \sqrt{4r^2 - x^2} \cdot 1 = 0$ for a maximum or minimum

$$= x \cdot \tfrac{1}{2}(4r^2 - x^2)^{-\frac{1}{2}} \cdot (-2x) + \sqrt{4r^2 - x^2} = 0$$

or
$$-\frac{x^2}{(4r^2 - x^2)^{\frac{1}{2}}} + (4r^2 - x^2)^{\frac{1}{2}} = 0$$

which reduces to

$$\frac{-x^2 + 4r^2 - x^2}{(4r^2 - x^2)^{\frac{1}{2}}} = 0$$

or
$$\frac{4r^2 - 2x^2}{(4r^2 - x^2)^{\frac{1}{2}}} = 0$$

$$\therefore\ 4r^2 - 2x^2 = 0$$

and
$$x^2 = 2r^2$$

or
$$x = r\sqrt{2} = \text{one side}$$

and $(4r^2 - x^2)^{\frac{1}{2}} = (4r^2 - r^2 \cdot 2)^{\frac{1}{2}} = (2r^2)^{\frac{1}{2}} = r\sqrt{2} = \text{other side}$

Therefore, the figure is a square. Also, its side can be conceived as equal to the diagonal of another square whose own side = the radius of the circle.

212. What is the radius of the opening of a conical vessel for maximum capacity with a given slant height?

Let V = volume = capacity = $\frac{1}{3}$ height $\times$ area of base

Then
$$V = \frac{1}{3} \cdot h \cdot \pi r^2 = \frac{\pi r^2}{3}\sqrt{s^2 - r^2}$$

Now $\dfrac{dV}{dr} = \dfrac{\pi r^2}{3} \cdot \dfrac{d}{dr}(s^2 - r^2)^{\frac{1}{2}} + (s^2 - r^2)^{\frac{1}{2}} \cdot \dfrac{d}{dr} \cdot \dfrac{\pi r^2}{3} = 0$ for

a maximum or minimum

$$= \frac{\pi r^2}{3} \cdot -\frac{r}{(s^2 - r^2)^{\frac{1}{2}}} + \frac{2\pi r}{3} \cdot (s^2 - r^2)^{\frac{1}{2}} = 0$$

$$= -\frac{\pi r^3}{3(s^2 - r^2)^{\frac{1}{2}}} + \frac{2\pi r}{3} \cdot (s^2 - r^2)^{\frac{1}{2}} = 0$$

$$= \frac{-\pi r^3 + 2\pi r(s^2 - r^2)}{3(s^2 - r^2)^{\frac{1}{2}}} = 0 = \frac{2\pi r(s^2 - r^2) - \pi r^3}{3(s^2 - r^2)^{\frac{1}{2}}}$$

or
$$2\pi r(s^2 - r^2) - \pi r^3 = 0$$

$$2\pi r s^2 - 2\pi r^3 - \pi r^3 = 0$$

$$2\pi r s^2 - 3\pi r^3 = 0$$

$$2\pi s^2 = 3\pi r^2$$

$$2s^2 = 3r^2$$

$$r = \pm s \sqrt{\tfrac{2}{3}}$$

$$\therefore r = s \sqrt{\tfrac{2}{3}} \text{ for a maximum}$$

Obviously, the negative value cannot be used.

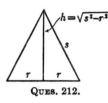

$h = \sqrt{s^2 - r^2}$

QUES. 212.

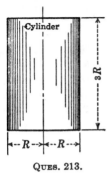

QUES. 213.

213. What is the volume increase per second, if the surface area increases at the rate of 32 square inches per second, of a cylinder where the height is always three times the radius?

$$\text{Surface area} = A = 2(\pi R^2) + 2\pi R \cdot 3R$$

or
$$A = 2\pi R^2 + 6\pi R^2 = 8\pi R^2$$

$$\text{Volume} = V = \pi R^2 \cdot 3R = 3\pi R^3$$

$$\frac{dA}{dt} = 16\pi R \frac{dR}{dt} = 32 \text{ sq. in. /sec.} \quad \text{(given)}$$

$$\therefore \frac{dR}{dt} = \frac{32}{16\pi R}$$

Now,
$$\frac{dV}{dt} = 9\pi R^2 \cdot \frac{dR}{dt}$$

and substituting the value of $\dfrac{dR}{dt}$

$$\frac{dV}{dt} = 9\pi R^2 \cdot \frac{32}{16\pi R} = 18R \text{ cu. in./sec.} = \text{the rate of volume}$$

increase per second

PROBLEMS

Find the maximum and minimum values of the following curves:

1. $y = 2x^5 - 3x^4$

2. $y = 2x^2 - x^4$

3 $y = \dfrac{1}{2x^2 + x - 5}$

4. $y = (1 + x)^{\frac{1}{2}}(2 - x)^{\frac{1}{2}}$

5. What is the altitude of a right circular cone of maximum volume that can be inscribed in a sphere of radius R?

6. What is the altitude of a right cylinder of maximum volume that can be inscribed in a right circular cone?

7. What are the most economical dimensions of a right circular cylindrical tank made of steel of uniform thickness and of fixed volume = 6,000 cu. ft.?

8. What is the least amount of canvas required to make a conical army tent 12 ft. high for maximum capacity?

9. What is the least cost of building a fence to enclose a rectangle containing 10 acres, at $1.50 for a 10-ft. length of fence?

10. What are the dimensions and volume of the largest cylindrical package that can be sent by parcel post if the regulations limit the size so that the length plus the girth equals 5 ft.?

11. For what current will a generator deliver the maximum power if the power delivered when the current is I amp. is $40I - 3I^2$?

12. Find the maximum and minimum values of $y = 1 - \dfrac{x^2}{3} + \dfrac{x^3}{12}$.

13. Find the maximum and minimum values of $y = 2x + 3 + \dfrac{6}{x^2}$.

14. Find the maximum and minimum values of $y = \dfrac{6x}{3 + x^2}$.

15. What are the dimensions of the strongest rectangular beam that can be cut from a round log 30 in. in diameter if the strength of the beam is proportional to the width and to the square of the depth?

16. How many telephone subscribers would result in a maximum net profit to the company if there is a net profit of $12 per instrument when an exchange has 800 subscribers or less and if, when there are over 800 subscribers, the profits per instrument decrease $1\frac{1}{2}$ cents for each subscriber above that number?

17. How far from the foot of an electric-light pole should the anchor for the guy wire be placed if the guy wire is to be 26 ft. long and fastened to the pole and anchor so that the moment about the foot of the pole is a maximum?

18. It is desired to cut out a small square from each corner of a 14- by 28-in. rectangular sheet of copper so that when the sides are turned up to form an open box it shall have a maximum volume. What should be the side of the small square to be cut out?

19. How many tons of structural steel should be produced per day for maximum profit if the price of structural steel is 0.65 that of vanadium steel and the plant is capable of producing x tons per day of structural steel and y tons per day of vanadium steel and $y = \dfrac{30 - 4x}{8 - x}$?

CHAPTER XI

DIFFERENTIALS

214. What is meant by a differential?

If two variables are so related that one is dependent and the other independent, then for corresponding values of the variables the differential of the independent variable is the same as its increment, change, or growth and the differential of the dependent variable or function is what would be its increment if its change or growth became and remained uniform with respect to the independent variable at the point in question.

215. How is a differential of a variable or function denoted?

By writing d before the variable or function.

dx means "differential of x."

$d(x)^2$ means "differential of x^2."

$d(x^2 + x + 3)$ means "differential of $(x^2 + x + 3)$."

$df(x)$ means "differential of $f(x)$."

216. When is the differential of a function (dependent variable) also equal to its increment?

Only when the function $y = f(x)$ represents a straight line (in which case there is a uniform rate of change or growth in the function) is the differential of the function equal to its increment.

Example

a. If $y = mx + b$, the equation of a straight line of slope $= m$.

Then $dx = \Delta x$ and $dy = \Delta y$ because y changes uniformly with respect to x.

b. If the area *A* of rectangle *CBDE* of constant height *c* is increased by increasing its base by length *EG*, then *A* changes uniformly with respect to *x*, and

$$EG = \Delta x = dx$$

while rectangle $DEGF = \Delta A = dA$.

c. If the rate of travel of a train remained uniform at 30 m.p.h. and if the increment or differential of the time *t* is taken as 1 hr., then the increment of the distance covered would be equal to its differential, which is 30 miles, or

$$dt = 1 \text{ hr.}, \qquad ds = 30 \text{ miles}$$

For smaller changes there would be smaller corresponding differentials.

If $dt = 1$ min.,
$$ds = \tfrac{1}{2} \text{ mile}$$

If $dt = 1$ sec.,
$$ds = 44 \text{ ft.}$$

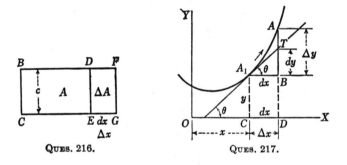

QUES. 216.   QUES. 217.

217. How can we show graphically the relation between the differential and the increment of a function (dependent variable) when it is changing nonuniformly?

At any point as at A_1 the direction of the curve is along the tangent line at the point A_1. Therefore, the instantaneous rate at which the function *y* is changing at point A_1 is the same as if the point were moving along the tangent line at A_1, in which case the change in y becomes and remains uniform with respect to x.

Then at A_1 the increment of x is $\Delta x = CD = dx$ while $dy = BT$ (as for the case of uniform rate of change). But if the point A_1 continues on the curve to A, the position corresponding to Δx, then the corresponding increment of y is Δy or BA.

Therefore, at point A_1 the differential of y is BT while, for the corresponding Δx, the increment of y is BA.

218. What is the relation between differentials and derivatives?

In the figure of Question 217, for the right triangle BA_1T,

$$\tan BA_1T = \tan \theta = \text{the slope of the tangent line}$$
$$\therefore dy = dx \tan \theta$$

Since the slope of the tangent line at any point is equal to the value of the derivative for that point,

then $dy = \dfrac{dy}{dx} \cdot dx =$ the differential of the function

Since ordinarily $\dfrac{dy}{dx}$ is the first derivative and not a fraction,

then $$dy = f'(x)\, dx$$

and $$\frac{dy}{dx} = f'(x)$$

Here the left-hand member is a fraction and represents the quotient of two differentials. This last relation shows that the first derivative and the ratio of the differentials may be used interchangeably. A derivative is sometimes called a *differential coefficient*.

219. How may the differentials of the independent and dependent variables now be defined?

The differential of the independent variable is its increment. The differential of the dependent variable is the product of a derivative and the increment of the independent variable.

220. What is the practical significance of the statement that the first derivative and the ratio of the differentials may be used interchangeably?

The parts may be separated and operated upon in the same manner as for any other fraction.

Example

a. If $\dfrac{dy}{dx} = 5x$, then $dy = 5x \cdot dx$.

b. If $\dfrac{dy}{dv} \cdot \dfrac{dv}{dx}$, then we may cancel the dv and the expression reduces to $\dfrac{dy}{dx}$.

c. If $y = u + v$, we may get $dy = du + dv$ and

$$\frac{dy}{dx} = \frac{du}{dx} + \frac{dv}{dx}$$

221. How may we readily find the differential of a function?

Merely find the derivative of the function, and multiply by dx, if $y = f(x)$.

222. If $y = 3x^4 - 2x^2 + 3x + 5$, what is dy?

$$\frac{dy}{dx} = 12x^3 - 4x + 3$$

and $dy = (12x^3 - 4x + 3)\, dx$ or
$$dy = f'(x)\, dx = (12x^3 - 4x + 3)\, dx$$

223. If A is the area of a square of side x, what are ΔA, dA, and $\Delta A - dA$ for any x and Δx and for $x = 5$ and $\Delta x = 0.0002$ in.?

Given $A = x^2$. Now assume x grows to $x + \Delta x$.

Then $A + \Delta A = (x + \Delta x)^2 = x^2 + 2x \cdot \Delta x + (\Delta x)^2$
Subtract A $=$ x^2 to get ΔA
$$\therefore \Delta A = \qquad\qquad\qquad 2x \cdot \Delta x + (\Delta x)^2 \quad (1)$$

Divide by Δx to get $\dfrac{\Delta A}{\Delta x} = 2x + \Delta x$ from which

$$\frac{dA}{dx} = \lim_{\Delta x \to 0} \frac{\Delta A}{\Delta x} = 2x$$

and
$$dA = 2x \cdot dx \tag{2}$$

Eq. (1) $-$ Eq. (2) $= \Delta A - dA = 2x \cdot \Delta x + (\Delta x)^2$
$$- 2x \cdot dx = (\Delta x)^2$$

NOTE.—$\Delta x = dx$

For $x = 5$ and $\Delta x = 0.0002$ in.,
$$\Delta A = 2 \cdot 5 \cdot 0.0002 + (0.0002)^2 = 0.00200004$$
$$dA = 2 \cdot 5 \cdot 0.0002 = 0.002$$
$$\Delta A - dA = 0.00200004 - 0.002 = 0.00000004$$

Because Δx is so small, dA closely approximates ΔA.

224. Of what practical use is the knowledge that the differential of a function closely approximates the increment of the function when the increment of the independent variable is comparatively small?

When only an approximate value of the increment of a function is desired, it is usually easier to calculate the value of the corresponding differential and use this value.

225. By using differentials, how would you find an approximation to the value of $y = x^3 - 3x^2 + 2x - 1$ when $x = 1.998$?

Consider the value 1.998 as the result of applying an increment of -0.002 to an original value of 2.

Then $x = 2$ and $\Delta x = -0.002 = dx$

Now $\dfrac{dy}{dx} = 3x^2 - 6x + 2$ and $dy = (3x^2 - 6x + 2)\,dx$

Substitute $x = 2$ and $dx = -0.002$

We get $dy = [3 \cdot (2)^2 - 6 \cdot 2 + 2] \cdot -0.002 = -0.004$

which is approximately the change in y caused by changing from $x = 2$ to $x = 1.998$.

Now, when $x = 2$ (in the original equation)

$$y = 2^3 - 3 \cdot 2^2 + 2 \cdot 2 - 1 = -1$$
$$\therefore y + dy = -1 - 0.004 = -1.004$$

which is approximately the value of the polynomial for $x = 1.998$.

To check the closeness of this method, substitute 1.998 in the original function to get the exact value for this point, and get $y + \Delta y = -1.003988008$. This shows that the method of approximations by differentials holds very closely.

226. What is the volume, approximately, of a spherical shell of outside diameter 12 inches and thickness $\frac{3}{32}$ inch?

$$V = \text{volume of sphere} = \tfrac{1}{6}\pi x^3 \quad (x = \text{diameter})$$

The exact volume of the shell $= \Delta V$, which is the difference between the volume of a sphere of diameter $= 12$ in. and the volume of a sphere of diameter $= 11\frac{13}{16}$ in.

For an approximate value of ΔV, find dV.

$$\frac{dV}{dx} = \frac{\pi x^2}{2} \quad \text{and} \quad dV = \frac{\pi x^2}{2} \cdot dx$$

Now, for $x = 12$ in. and $dx = -\frac{3}{16}$ in.,

$$dV = \frac{\pi}{2} \cdot 12^2 \cdot (-\tfrac{3}{16}) = -42.4116 \text{ cu. in.} = 42.41 \text{ cu. in.}$$

approximately

neglecting the sign, which merely means that V decreases as x decreases.

The exact value of ΔV is

$$\tfrac{1}{6}\pi[12^3 - (11\tfrac{13}{16})^3] = 41.75237 = 41.75 \text{ cu. in.}$$

The approximation is close, for dx is relatively small as compared with x $(= 12$ in.$)$.

227. By using differentials, what is approximately the greatest possible error in the calculated value of the volume of a right circular cone if its altitude, which is the same as the radius of its base, is measured as 10 inches, with a possible error in measurement of 0.01 inch?

Given $h = r$.

$$\therefore V = \frac{1}{3} \cdot h \cdot \pi h^2 = \frac{1}{3}\pi h^3 = \frac{1}{3}\pi \cdot 10^3 = \frac{1,000\pi}{3} \text{ cu. in.}$$

The error to which the altitude h is subject may be taken as $\Delta h = dh = \pm 0.01$.

Since Δh is small, the computed value of the volume is subject to an error approximately given by dV.

Now $\quad \dfrac{dV}{dh} = \pi h^2 \quad$ and $\quad dV = \pi h^2 \cdot dh$

For $h = 10$ and $dh = \pm 0.01$,

$$dV = \pi \cdot 10^2 \cdot (\pm 0.01) = \pm \pi$$

This means that the computed volume $\dfrac{1,000\pi}{3}$ cu. in. may be too great or too small by π cu. in. approximately.

228. What is the approximate maximum error in the area of a circle if the diameter is found by measurement to be 10 inches, with a maximum error of 0.01 inch?

$$A = \tfrac{1}{4}\pi x^2 \qquad\qquad (x = \text{diameter}) \qquad\qquad (1)$$
$$= \frac{\pi}{4} \cdot 10^2 = 25\pi = 78.54 \text{ sq. in.}$$

The approximate error is dA.

$$\frac{dA}{dx} = \frac{1}{2}\pi x \qquad \text{or} \qquad dA = \frac{\pi x}{2} \cdot dx \quad (dx = 0.01 \text{ as given})$$
$$\therefore dA = \frac{\pi}{2} \cdot 10 \cdot 0.01 = 0.15708 \text{ sq. in.}$$

The exact maximum error in A will be the change (ΔA) in its value as found by Eq. (1) when x changes from 10 to 10.01 in., which is determined to be 0.1572 sq. in.

229. What is sin 46° approximately, by using differentials, given sin 45° = 0.7071 = cos 45° and 1° = 0.01745 radian?

Let $$y = \sin \theta$$

then $$\frac{dy}{d\theta} = \cos \theta \quad \text{or} \quad dy = \cos \theta \cdot d\theta \tag{1}$$

When θ changes to $\theta + d\theta$, y will change to $y + dy$ approximately.

In Eq. (1) substitute 0.7071 for $\cos \theta$ and 0.0175 for $d\theta$.

$$\therefore \ dy = 0.7071 \cdot 0.0175 = 0.01237425$$

Since $y = \sin \theta = 0.7071$,

$$y + dy = 0.7071 + 0.01237 = 0.7195 = \sin \ 46° \text{ approximately}$$

From four-place tables,

$$\sin 46° = 0.7193$$

This shows a close approximation by using the differential.

230. What is meant by relative error and percentage error?

Relative error in y is defined by $\dfrac{\Delta y}{y}$ if Δy is the error in y.

Percentage error is defined as $100 \cdot \dfrac{\Delta y}{y}$.

These may be approximated by $\dfrac{dy}{y}$ and $100 \cdot \dfrac{dy}{y}$, respectively, if the increment of the variable on which y depends is small enough.

231. What are the relative error and percentage error in the volume of the cone in Question 227?

Assume the error is positive.

Then
$$\frac{dV}{V} = \frac{\pi}{\dfrac{1,000\pi}{3}} = 0.003$$

which means the volume is subject to a relative error of approximately 0.003

or
$$100 \cdot 0.003 = 0.3 \text{ per cent}$$

232. What are the relative error and percentage error in the area of the circle in Question 228?

$$\frac{dA}{A} = \frac{0.1571}{78.54} = 0.002$$

$$100 \cdot \frac{dA}{A} = 100 \cdot 0.002 = 0.2 = \text{percentage error}$$

PROBLEMS

1. If $y = x^5 - 5x^2 + 3x$, what is dy for any x?

2. If $\frac{x^2}{9} - \frac{y^2}{4} = 1$, what is dy for any x?

3. If the distance that a body will fall in t sec. is given by $s = \frac{1}{2}gt^2$, what is ds (a) for any value of t; (b) for $t = 3$ and $dt = 1$?

4. If V = the volume in cubic feet of gas issuing from a 3-in.-diameter pipe at a uniform rate of 5 ft./sec., what is dV corresponding to (a) $dt = 1$ sec.; (b) $dt = 30$ sec.?

5. If a point is moving on the circle $x^2 + y^2 = 25$, what are dy and ds corresponding to a change in x of $dx = 0.1$ at $x = 3$ in the first quadrant? Use the relation $(ds)^2 = (dx)^2 + (dy)^2$.

6. What is the differential of $y = \dfrac{x}{\sqrt{4 - x^2}}$?

7. What is the differential of $y = x\sqrt{4 - x^2}$?

8. What is the approximate error in the volume and surface of a cube whose edge is 8 in. if there is an error of 0.01 in. in measuring the edge?

9. What is the expression (a) for the approximate area of a circular ring of radius r and width dr; (b) for the exact area?

10. If the radius of a sphere is measured to be 7 in., with a maximum error in measurement of 0.02 in., (a) what is the approximate maximum error in the calculated values of the surface and the volume; (b) what are the maximum percentage errors in the surface and the volume? Given S = surface = $4\pi r^2$ and V = volume = $\frac{4}{3}\pi r^3$.

11. What is the approximate value of $2x^4 - 3x^3 + 2x^2 - 3x - 1$ for $x = 1.98$?

12. What is the approximate value of $\sqrt[3]{108}$?

13. What are Δy and dy for $x = 2$ and $\Delta x = 0.1$ if

$$y = 2x^3 - 2x^2 + 3x - 5?$$

14. What is the allowable percentage error in the diameter of a sphere if the volume is to be correct to within 2 per cent?

Chapter XII

SPLITTING FRACTIONS TO AID DIFFERENTIATION

233. What is a "proper" algebraic fraction?

A fraction whose highest exponent of the variable in the numerator *is less than* that in the denominator, as $\dfrac{4x + 1}{x^2 - 2}$ = a proper fraction. It cannot be simplified.

234. Is $\dfrac{x^2 + 4}{x^2 - 2}$ a proper algebraic fraction?

It is not. It can be simplified to $1 + \dfrac{6}{x^2 - 2}$, which is a proper algebraic fraction.

235. How would you obtain the denominator of the final resulting fraction if the denominators of two or more proper fractions to be added contain only terms in x and no higher powers?

Take the product of denominators of the proper fractions concerned.

Example

$$\frac{1}{x + 2} + \frac{3}{x - 1} = \frac{(x - 1) + (3x + 6)}{(x + 2)(x - 1)} = \frac{4x + 5}{(x + 2)(x - 1)}$$
$$= \frac{4x + 5}{(x^2 + x - 2)}$$

236. How would you perform the reverse operation of splitting up the above resulting fraction, $\dfrac{4x + 5}{(x^2 + x - 2)}$?

Obviously, to get the denominators of the partial fractions, factor the denominator first. Whereas the partial fractions are proper fractions, the numerators are mere numbers without x at all.

$$\frac{4x + 5}{x^2 + x - 2} = \frac{A}{x + 2} + \frac{B}{x - 1} = \frac{A(x - 1) + B(x + 2)}{(x + 2)(x - 1)} \qquad (1)$$

Since the denominators are equal, the numerators must be equal.

$$\therefore \; 4x + 5 = A(x - 1) + B(x + 2) \qquad (2)$$

The equation must be true for all values of x and therefore true for values that make $(x - 1) = 0$ and $(x + 2) = 0$.

For $x = 1$ in Eq. (2),

$$4 \cdot 1 + 5 = (A)(1 - 1) + B(1 + 2)$$
$$9 = 0 + 3B$$
$$\therefore \; B = 3$$

For $x = -2$ in Eq. (2),

$$4 \cdot (-2) + 5 = A(-2 - 1) + B(-2 + 2)$$
$$-8 + 5 = -3A + 0$$
$$-3 = -3A$$
$$\therefore \; A = 1$$

Now place these values of A and B in Eq. (1).

$$\therefore \; \frac{4x + 5}{x^2 + x - 2} = \frac{1}{x + 2} + \frac{3}{x - 1}$$

The original fraction is easier to differentiate when split up.

237. How would you split up $\dfrac{3x^2 + 4x - 15}{x^3 + 5x^2 - x - 5}$?

First factor the denominator.

$$(x^3 + 5x^2 - x - 5) = (x + 1)(x - 1)(x + 5)$$
$$\therefore \; \frac{3x^2 + 4x - 15}{x^3 + 5x^2 - x - 5} = \frac{A}{(x + 1)} + \frac{B}{(x - 1)} + \frac{C}{(x + 5)}$$

Now equate the numerators (since the denominators are equal).

$$3x^2 + 4x - 15 = A(x-1)(x+5) + B(x+1)(x+5) \\ + C(x+1)(x-1) \quad (1)$$

This equation must be true for values of x that make $(x-1) = 0$, $(x+5) = 0$, and $(x+1) = 0$ or for values of $x = 1$, $x = -5$, and $x = -1$.

For $x = -1$ in Eq. (1),

$$3 \cdot 1 + 4(-1) - 15 = A(-2) \cdot 4 + B \cdot (0) + C \cdot (0)$$
$$-16 = -8A$$
$$\therefore A = 2$$

For $x = 1$ in Eq. (1),

$$3 \cdot 1 + 4 \cdot 1 - 15 = A \cdot (0) + B \cdot 2 \cdot 6 + C \cdot (0)$$
or
$$-8 = 12B$$
$$\therefore B = -\tfrac{2}{3}$$

For $x = -5$ in Eq. (1),

$$3 \cdot 25 + 4(-5) - 15 = A(0) + B(0) \\ + C(-5+1)(-5-1)$$
or
$$40 = 24C$$
$$\therefore C = \tfrac{5}{3}$$

Finally, $\dfrac{3x^2 + 4x - 15}{x^3 + 5x^2 - x - 5} = \dfrac{2}{x+1} - \dfrac{2}{3(x-1)}$

$$+ \dfrac{5}{3(x+5)}$$

which is easier to differentiate.

238. How would you express the numerator when a factor of the denominator contains a term in x^2?

The numerator must contain a term in x as well as a simple number.

$$\therefore \text{Numerator} = Ax + B$$

239. How would you split up $\dfrac{-3x^2 - 8}{(x^2 + 1)(x + 2)}$?

$$\frac{-3x^2 - 8}{(x^2 + 1)(x + 2)} = \frac{Ax + B}{x^2 + 1} + \frac{C}{x + 2}$$

$$= \frac{(Ax + B)(x + 2) + C(x^2 + 1)}{(x^2 + 1)(x + 2)}$$

Equating the numerators on both sides,

$$-3x^2 - 8 = (Ax + B)(x + 2) + C(x^2 + 1) \qquad (1)$$

The equation must hold for all values of x, including $x = -2$ (from $x + 2 = 0$).

For $x = -2$ in Eq. (1),

$$-3 \cdot 4 - 8 = 0 + C(4 + 1)$$
$$-20 = 5C$$
$$\therefore C = -4$$

Substitute this in Eq. (1) to get

$$-3x^2 - 8 = (Ax + B)(x + 2) - 4(x^2 + 1) \qquad (2)$$

For $x = 0$, in Eq. (2)

$$-8 = B \cdot 2 - 4$$
$$-4 = 2B$$
$$\therefore B = -2$$

Substitute this in Eq. (2) above, and solve.

$$-3x^2 - 8 = (Ax - 2)(x + 2) - 4(x^2 + 1)$$
$$-3x^2 - 8 + 4x^2 + 4 = (Ax - 2)(x + 2)$$
$$x^2 - 4 = (Ax - 2)(x + 2)$$
$$(x - 2)(x + 2) = (Ax - 2)(x + 2)$$
$$x - 2 = Ax - 2$$
$$0 = Ax - x$$
$$= x(A - 1)$$
$$\therefore A = 1$$

Finally, $\dfrac{-3x^2 - 8}{(x^2 + 1)(x + 2)} = \dfrac{x - 2}{x^2 + 1} - \dfrac{4}{x + 2}$

which is easier to differentiate.

240. How would you split up $\dfrac{x^3 + x^2 + 5x + 7}{(x^2 + 3)(x^2 + 4)}$?

$$\frac{x^3 + x^2 + 5x + 7}{(x^2 + 3)(x^2 + 4)} = \frac{Ax + B}{(x^2 + 3)} + \frac{Cx + D}{(x^2 + 4)}$$

$$= \frac{(Ax + B)(x^2 + 4) + (Cx + D)(x^2 + 3)}{(x^2 + 3)(x^2 + 4)}$$

$$= \frac{Ax^3 + Bx^2 + 4Ax + 4B + Cx^3 + Dx^2 + 3Cx + 3D}{(x^2 + 3)(x^2 + 4)}$$

$$= \frac{(A + C)x^3 + (B + D)x^2 + (4A + 3C)x + (4B + 3D)}{(x^2 + 3)(x^2 + 4)}$$

The coefficients of like powers of x in the numerators must be equal and of the same sign.

$$
\begin{array}{llll}
x^3, & A + C = 1, & A = 1 - C, \\
 & & A = 1 - (-1), & \text{or} \quad A = 2 \\
x^2, & (B + D) = 1, & B = 1 - D, \\
 & & B = 1 - (-3), & \text{or} \quad B = 4 \\
x, & 4A + 3C = 5, & 4(1 - C) + 3C = 5, \\
 & & 4 - 4C + 3C = 5 & \therefore \; C = -1
\end{array}
$$

Constants, $4B + 3D = 7,\quad 4(1 - D) + 3D = 7,$

$$4 - 4D + 3D = 7 \qquad \therefore \; D = -3$$

$$\therefore \; \frac{x^3 + x^2 + 5x + 7}{(x^2 + 3)(x^2 + 4)} = \frac{2x + 4}{x^2 + 3} - \frac{x + 3}{x^2 + 4}$$

and this is easier to differentiate.

241. How does the method of equating the numerators compare with that of equating the coefficients of like powers of x?

The method of equating the coefficients of like powers of x in the numerators may always be used. But equating the numerators is quicker in cases where there are only factors in x alone.

242. What would you do when the denominator contains a factor raised to a power?

Allow for that many denominators containing that factor up to the highest power.

Example

In $\dfrac{4x^2 - 3x + 2}{(x + 2)^2(x - 1)}$, allow for denominators $(x + 2)$, $(x + 2)^2$, and $(x - 1)$.

243. Must there be an x term in the numerator when there is an $(x + 1)^2$ term in the denominator?

No numerator of the $Ax + B$ type is necessary. A mere letter A is sufficient; otherwise, there will be more unknowns than relations.

244. How would you split up $\dfrac{4x^2 - 3x + 3}{(x + 1)^2(x - 1)}$?

$$\frac{4x^2 - 3x + 3}{(x + 1)^2(x - 1)} = \frac{A}{(x + 1)^2} + \frac{B}{(x + 1)} + \frac{C}{(x - 1)}$$

Equating the numerators of both sides,

$$4x^2 - 3x + 3 = A(x - 1) + B(x + 1)(x - 1)$$
$$+ C(x + 1)^2 \quad (1)$$

For $x = 1$ in Eq. (1),

$$4 - 3 + 3 = 0 + 0 + 4C$$
$$4 = 4C$$
$$\therefore C = 1$$

Substitute this in Eq. (1).

Then $4x^2 - 3x + 3 = A(x - 1) + B(x + 1)(x - 1)$
$$+ 1(x + 1)^2$$
$4x^2 - 3x + 3 - x^2 - 2x - 1 = A(x - 1)$
$$+ B(x + 1)(x - 1)$$
$$3x^2 - 5x + 2 = A(x - 1) + B(x + 1)(x - 1)$$
$$(3x - 2)(x - 1) = A(x - 1) + B(x + 1)(x - 1)$$
$$(3x - 2) = A + B(x + 1) \quad (2)$$

For $x = -1$, in Eq. (2),

$$-3 - 2 = A + 0$$
$$\therefore A = -5$$

Substitute this in Eq. (2).

Then $$(3x - 2) = -5 + B(x + 1) \qquad (3)$$

For $x = 0$, in Eq. (3),

$$-2 = -5 + B$$
$$\therefore B = 3$$

Finally, $\dfrac{4x^2 - 3x + 3}{(x + 1)^2(x - 1)} = \dfrac{3}{(x + 1)} - \dfrac{5}{(x + 1)^2} + \dfrac{1}{(x - 1)}$

and this is easier to differentiate.

245. When must the numerator be of the form $(Ax + B)$?

When there is a power of a factor containing x^2 in the denominator.

246. How would you split $\dfrac{4x - 2}{(3x^2 - 2)^2(x - 1)}$?

$$\frac{4x - 2}{(3x^2 - 2)^2(x - 1)} = \frac{Ax + B}{(3x^2 - 2)^2} + \frac{Cx + D}{(3x^2 - 2)} + \frac{E}{(x - 1)}$$
$$4x - 2 = (Ax + B)(x - 1) + (Cx + D)(3x^2 - 2)(x - 1)$$
$$+ E(3x^2 - 2)^2 \quad (1)$$

For $x = 1$ in Eq. (1),

$$4 - 2 = 0 + 0 + E(3 - 2)^2$$
$$\therefore E = 2$$

Substitute this in Eq. (1).

Then $4x - 2 = (Ax + B)(x - 1)$
$$+ (Cx + D)(3x^2 - 2)(x - 1) + 2(3x^2 - 2)^2$$
$$4x - 2 - 18x^4 + 24x^2 - 8 = (Ax + B)(x - 1)$$
$$+ (Cx + D)(3x^2 - 2)(x - 1)$$
$$-18x^4 + 24x^2 + 4x - 10 = (Ax + B)(x - 1)$$
$$+ (Cx + D)(3x^2 - 2)(x - 1)$$

Divide both sides by $(x - 1)$.

$$
\begin{aligned}
-18x^3 - 18x^2 + 6x + 10 &= (Ax + B) \\
&\quad + (Cx + D)(3x^2 - 2) \\
&= Ax + B + 3Cx^3 + 3Dx^2 \\
&\quad - 2Cx - 2D \\
&= 3Cx^3 + 3Dx^2 + (A - 2C)x \\
&\quad + (B - 2D) \quad (2)
\end{aligned}
$$

Equate coefficients of like powers of x in Eq. (2).

$$
\begin{aligned}
x^3, &\quad -18 = 3C &&\therefore C = -6 \\
x^2, &\quad -18 = 3D &&\therefore D = -6 \\
x, &\quad 6 = A - 2C = A - 2(-6) = A + 12 \\
& &&\therefore A = -6 \\
\text{Constants} &\quad 10 = B - 2D = B - 2(-6) = B + 12 \\
& &&\therefore B = -2
\end{aligned}
$$

Finally,
$$
\begin{aligned}
\frac{4x - 2}{(3x^2 - 2)^2(x - 1)} &= \frac{-6x - 2}{(3x^2 - 2)^2} + \frac{-6x - 6}{(3x^2 - 2)} \\
&\quad + \frac{2}{(x - 1)} \\
&= \frac{2}{x - 1} - \frac{2(3x + 1)}{(3x^2 - 2)^2} \\
&\quad - \frac{6(x + 1)}{(3x^2 - 2)}
\end{aligned}
$$

247. How would you check the results of splitting?

Replace x by a single value, say (-1), in the given expression and in the partial fractions obtained, and see whether or not an identity results.

248. What is a quick method of splitting a proper fraction when the denominator contains only a single factor raised to a power?

Substitute a single letter for the factor in the denominator.

249. How would you split up $\dfrac{(3x + 2)}{(x + 2)^3}$?

Let $x + 2 = z$. Then $x = z - 2$.

Then $\dfrac{3(z-2)+2}{z^3} = \dfrac{3z-4}{z^3} = \dfrac{3z}{z^3} - \dfrac{4}{z^3} = \dfrac{3(x+2)}{(x+2)^3}$

$$- \dfrac{4}{(x+2)^3}$$

$$= \dfrac{3}{(x+2)^2} - \dfrac{4}{(x+2)^3}$$

250. Can you show that splitting after differentiation is more difficult than splitting before differentiation?

Let
$$y = \dfrac{6-3x}{5x^2+3x-2}$$

$$\dfrac{dy}{dx} = \dfrac{(5x^2+3x-2)(-3) - (6-3x)(10x+3)}{(5x^2+3x-2)^2}$$

$$= \dfrac{-15x^2 - 9x + 6 - 51x + 30x^2 - 18}{(5x^2+3x-2)^2}$$

$$= \dfrac{15x^2 - 60x - 12}{(5x^2+3x-2)^2}$$

This result is difficult to split up.
But

$$\dfrac{6-3x}{5x^2+3x-2} = \dfrac{6-3x}{(5x-2)(x+1)} = \dfrac{A}{(5x-2)} + \dfrac{B}{(x+1)}$$
$$6 - 3x = A(x+1) + B(5x-2) \qquad (1)$$

For $x = -1$ in Eq. (1),

$$6 + 3 = 0 + B(-5-2)$$
$$9 = -7B$$
$$\therefore B = -\tfrac{9}{7}$$

Substitute this in Eq. (1).

$$6 - 3x = A(x+1) - \tfrac{9}{7}(5x-2) = A(x+1) - \dfrac{45x}{7}$$
$$+ \tfrac{18}{7}$$

$$6 - \dfrac{18}{7} - 3x + \dfrac{45x}{7} = A(x+1)$$

$$\dfrac{24}{7} + \dfrac{24x}{7} = A(x+1) \qquad (2)$$

For $x = 0$ in Eq. (2),

$$\tfrac{24}{7} = A$$

Finally, $y = \dfrac{6 - 3x}{5x^2 + 3x - 2} = \dfrac{24}{7(5x - 2)} - \dfrac{9}{7(x + 1)}$

Now $\dfrac{dy}{dx} = \dfrac{24}{7} \cdot \dfrac{-5}{(5x - 2)^2} - \dfrac{9}{7} \cdot - \dfrac{1}{(x + 1)^2}$

or $\dfrac{dy}{dy} = -\dfrac{120}{7(5x - 2)^2} + \dfrac{9}{7(x + 1)^2}$

The derivative is already split up.

PROBLEMS

Split into partial fractions to aid differentiation:

1. $\dfrac{(2x + 3)}{(x - 2)(x + 3)}$

2. $\dfrac{(2x + 3)}{(x^2 + 3x - 9)}$

3. $\dfrac{(2x - 5)}{(x - 1)(x - 3)}$

4. $\dfrac{x + 2}{x^2 - 5x + 8}$

5. $\dfrac{x - 6}{(x + 3)(4x - 1)}$

6. $\dfrac{x^2 - 9x + 14}{(x - 3)(x - 4)(x - 5)}$

7. $\dfrac{6x^2 + 5x + 1}{(3x + 1)(4x - 3)(4x + 5)}$

8. $\dfrac{2x^2}{3x^3 - 1}$

9. $\dfrac{x^4 + 1}{x^3 + 1}$

10. $\dfrac{x}{(2x^2 - 3)(x + 1)}$

11. $\dfrac{x}{(x + 1)(x + 2)^2}$

12. $\dfrac{6x^2 + 4x - 7}{(x + 2)^3}$

13. $\dfrac{x^2}{(x^3 - 8)(x - 2)}$

14. $\dfrac{4x^2 + 3x - 1}{(2x - 1)^4}$

Chapter XIII

TYPES OF GROWTH

251. What is the simple-interest type of growth?

It is that growth where the original capital or principal remains fixed ($= C$) and the interest is fixed $\left(=\dfrac{C}{n} \right)$ and accumulates independently.

The interest fraction $= \dfrac{1}{n}$ of the original capital C.

252. What will be the property value at the end of, say, t periods for simple-interest growth?

The value of the property at the end of t periods is P_t.

P_t = original capital + accumulated interest for t periods

or
$$P_t = C + \frac{C}{n} \cdot t$$

253. What will be the property value at the end of n periods for simple-interest growth, when the interest fraction is $\dfrac{1}{n}$?

The original capital is doubled in n periods of time or each dollar is doubled in n periods of time

because $P_n = C + \dfrac{C}{n} \cdot n = 2C =$ twice the original capital

254. How would you show simple-interest growth graphically?

C = original capital = constant

$\dfrac{C}{n}$ = interest fraction = constant growth

109

$$P_n = C + \frac{C}{n} \cdot n = 2C = \text{value of property in } n \text{ periods of time}$$

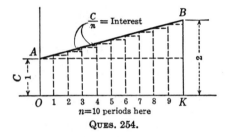

QUES. 254.

255. What is the compound-interest type of growth?

The capital is not constant but increases by successive additions of the interest so that the rate of increase of capital during any period is proportional to the value of the capital at the beginning of that period, or each dollar becomes $\left(1 + \frac{1}{n}\right)$ dollars in each period of time where $\frac{1}{n}$ = the interest fraction.

256. What is the capital at the end of the first, second, third, and xth periods in compound-interest growth?

At the end of the first period,

$$C_1 = \left(C + \frac{C}{n}\right) = C\left(1 + \frac{1}{n}\right)$$

Each dollar becomes $\left(1 + \frac{1}{n}\right)$.

At the end of the second period,

$$C_2 = \left(C_1 + C_1 \cdot \frac{1}{n}\right) = C_1\left(1 + \frac{1}{n}\right) = C\left(1 + \frac{1}{n}\right)^2$$

Each dollar again becomes $\left(1 + \frac{1}{n}\right)$.

At the end of the third period,

$$C_3 = \left(C_2 + \frac{C_2}{n}\right) = C_2\left(1 + \frac{1}{n}\right) = C\left(1 + \frac{1}{n}\right)^3$$

Each dollar again becomes $\left(1 + \frac{1}{n}\right)$.

At the end of the xth period, $C_x = C\left(1 + \frac{1}{n}\right)^x$.

257. What is the expression for the capital at the end of the nth period when the interest fraction is $\frac{1}{n}$?

$$C_n = C\left(1 + \frac{1}{n}\right)^n$$

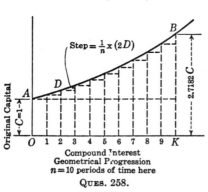

Compound Interest
Geometrical Progression
$n = 10$ periods of time here
QUES. 258.

258. How would you show compound interest graphically?

$$C_1 = \left(1 + \frac{1}{n}\right) = \text{capital at end of first period}$$

$$C_2 = \left(1 + \frac{1}{n}\right)^2 = \text{capital at end of second period}$$

$$C_{10} = \left(1 + \frac{1}{n}\right)^{10} = \text{capital at end of tenth period}$$

The steps of increase of capital are not equal. Each step is $\frac{1}{n}$ of the ordinate to that part of the curve.

Each successive ordinate is $\left(1 + \dfrac{1}{n}\right)$ times as high as its predecessor.

259. What is the numerical ultimate value of $\left(1 + \dfrac{1}{n}\right)^n$ when n is taken infinitely large?

$$2.71828 \cdots = \text{epsilon} = \epsilon = \left(1 + \frac{1}{n}\right)^n$$

or $1 becomes $2.71828 when the number of periods of growth is infinitely large or when there is continual growth.

260. What is logarithmic growth?

$$C_n = C\left(1 + \frac{1}{n}\right)^n = \text{the value of the function at the } n\text{th}$$
period in logarithmic growth

It is that growth which is proportional at every instant to the magnitude of the function at that instant. The successive ordinates are in geometrical progression.

261. What is a unit logarithmic rate of growth?

The rate at which a unit grows continually to 2.718281 in unit time.

262. What is an organic rate of growth?

That type where the growth of the organism in a given time is proportional to the magnitude of the organism itself at the time.

263. What is the value of unity, or 1, growing arithmetically at unit rate compared with 1 growing logarithmically at unit rate?

Assume 100 per cent = unit of rate.

Arithmetically, 1 grows at unit rate for unit time to 2.

Logarithmically, 1 grows at unit rate for same time to $2.71828 = \epsilon$.

264. How would you arrive at a formula for the speed v of a chemical reaction at any temperature T if the speed increases 10 per cent with every degree rise in temperature?

This is a compound-interest type of growth because the rate of increase of the speed at any time is proportional to the speed at the time considered. It is of the form

$$C_x = C\left(1 + \frac{1}{n}\right)^x, \text{ where } C_x = v, \frac{1}{n} = \frac{1}{10}, x = T, \text{ and}$$

$C = v_0 =$ the velocity of the reaction at 0°.

Then
$$v = v_0(1 + \tfrac{1}{10})^T$$

Now, from a table of natural logarithms,

$$1.10 = \epsilon^{0.0953}$$

$\therefore v = v_0(\epsilon^{0.0953})^T = v_0\epsilon^{0.0953T} =$ the formula for the speed of reaction at any temperature

265. What is the amount of \$500 at compound interest of 6 per cent per annum for 10 years if compounded (a) annually; (b) semiannually; (c) quarterly?

(a) $C_{10} = C\left(1 + \frac{1}{n}\right)^{10} = 500(1 + 0.06)^{10} = \$895.43$

(b) $C_{10} = C\left(1 + \frac{1}{2n}\right)^{2\times10} = 500\left(1 + \frac{0.06}{2}\right)^{20} = \$903.05$

(c) $C_{10} = C\left(1 + \frac{1}{4n}\right)^{4\times10} = 500\left(1 + \frac{0.06}{4}\right)^{40} = \$907.01$

PROBLEMS

1. What is the property value of \$1,000 at the end of (a) 1 year; (b) 2 years; (c) 3 years at simple interest of 5 per cent per annum?

2. How long will it take to double the original capital in the above problem?

3. What is the property value of \$1,000 at 5 per cent at the end of (a) 1 year; (b) 2 years; (c) 3 years at compound interest?

4. How long will it take to double the original capital in the above, and how long will it take to multiply the value by $\epsilon = 2.71828$?

5. What is the amount of $1,000 at compound interest at 10 per cent per annum for 20 years, compounded quarterly?

6. How fast is a chemical reaction taking place at 30°C. if its speed increases 6 per cent with every degree rise in temperature and it is one-eighth transformed at 0°C.?

EPSILON = ϵ—EXPONENTIAL FUNCTIONS

266. What is the expression for $\left(1 + \dfrac{1}{n}\right)^n$ expanded by the binomial theorem?

By rule,

$$(a + b)^n = a^n + n\,\frac{a^{n-1}b}{\underline{1}} + n(n - 1)\,\frac{a^{n-2}b^2}{\underline{2}}$$
$$+ n(n - 1)(n - 2)\,\frac{a^{n-3} \cdot b^3}{\underline{3}} + \cdots$$

and when $\qquad a = 1 \qquad$ and $\qquad b = \dfrac{1}{n}$

then $\left(1 + \dfrac{1}{n}\right)^n = 1 + 1 + \dfrac{1}{\underline{2}}\left(\dfrac{n - 1}{n}\right)$

$+ \dfrac{1}{\underline{3}}\dfrac{(n - 1)(n - 2)}{n^2} + \dfrac{1}{\underline{4}}\dfrac{(n - 1)(n - 2)(n - 3)}{n^3} + \cdots$

267. What is the value of the above series when n is indefinitely large?

When n is indefinitely large,

then $(n - 1), \qquad (n - 2), \qquad$ and $\qquad (n - 3)$ are sensibly $= n$

$$\therefore \epsilon = 1 + \frac{1}{\underline{1}} + \frac{1}{\underline{2}} + \frac{1}{\underline{3}} + \frac{1}{\underline{4}} + \cdots = \text{epsilon}$$

268. What is the value of ϵ resulting from working out 10 terms of the above series?

One divided by 1	= 1.00 00 00
One divided by $\lfloor 1$	= 1.00 00 00
One divided by $\lfloor 2$	= 0.50 00 00
One divided by $\lfloor 3$	= 0.16 66 67
One divided by $\lfloor 4$	= 0.04 16 67
One divided by $\lfloor 5$	= 0.00 83 33
One divided by $\lfloor 6$	= 0.00 13 89
One divided by $\lfloor 7$	= 0.00 01 98
One divided by $\lfloor 8$	= 0.00 00 25
One divided by $\lfloor 9$	= 0.00 00 02

$$\text{Total} = \epsilon = 2.71\ 82\ 81 = \left(1 + \frac{1}{n}\right)^n = \text{epsilon}$$

269. What are the chief characteristics of ϵ?

ϵ is incommensurable with 1. It is an interminable nonrecurrent decimal, resembling π.

270. What is the value of $\left(1 + \dfrac{1}{n}\right)^{nx}$ when n is indefinitely great?

It is $\epsilon^x = 2.718281^x = \text{(epsilon)}^x$.

271. What is $\left(1 + \dfrac{1}{n}\right)^{nx}$ called?

The exponential series.

272. What is the expression of the exponential series obtained from $\left(1 + \dfrac{1}{n}\right)^{nx} = \epsilon^x$ expanded by the binomial theorem?

By rule,

$$(a + b)^{nx} = a^{nx} + nx\frac{a^{nx-1} \cdot b}{\lfloor 1} + nx(nx - 1)\frac{a^{nx-2} \cdot b^2}{\lfloor 2}$$
$$+ nx(nx - 1)(nx - 2)\frac{a^{nx-3} \cdot b^3}{\lfloor 3} + \cdots$$

and when $\quad a = 1 \quad$ and $\quad b = \dfrac{1}{n}$

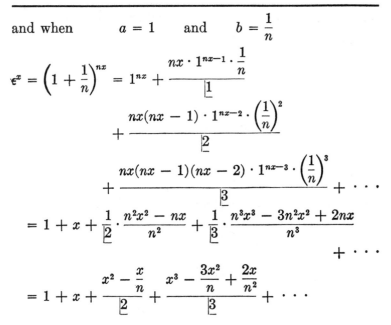

$$\epsilon^x = \left(1 + \frac{1}{n}\right)^{nx} = 1^{nx} + \frac{nx \cdot 1^{nx-1} \cdot \dfrac{1}{n}}{\underline{1}}$$

$$+ \frac{nx(nx-1) \cdot 1^{nx-2} \cdot \left(\dfrac{1}{n}\right)^2}{\underline{2}}$$

$$+ \frac{nx(nx-1)(nx-2) \cdot 1^{nx-3} \cdot \left(\dfrac{1}{n}\right)^3}{\underline{3}} + \cdots$$

$$= 1 + x + \frac{1}{\underline{2}} \cdot \frac{n^2x^2 - nx}{n^2} + \frac{1}{\underline{3}} \cdot \frac{n^3x^3 - 3n^2x^2 + 2nx}{n^3} + \cdots$$

$$= 1 + x + \frac{x^2 - \dfrac{x}{n}}{\underline{2}} + \frac{x^3 - \dfrac{3x^2}{n} + \dfrac{2x}{n^2}}{\underline{3}} + \cdots$$

And when n is indefinitely large, the above is simplified to

$$\epsilon^x = 1 + x + \frac{x^2}{\underline{2}} + \frac{x^3}{\underline{3}} + \cdots$$

This is called the *exponential series*.

273. What is the unique property of ε^x?

Its derivative is unchanged.

Example

If $\qquad \epsilon^x = 1 + x + \dfrac{x^2}{\underline{2}} + \dfrac{x^3}{\underline{3}} + \dfrac{x^4}{\underline{4}} + \cdots$

then $\dfrac{d}{dx}(\epsilon^x) = 0 + 1 + \dfrac{2x}{1 \cdot 2} + \dfrac{3x^2}{1 \cdot 2 \cdot 3} + \dfrac{4x^3}{1 \cdot 2 \cdot 3 \cdot 4} + \cdots$

or $\qquad = 1 + x + \dfrac{x^2}{\underline{2}} + \dfrac{x^3}{\underline{3}} + \cdots$ unchanged $= \epsilon^x$

274. What is the derivative of ϵ^y with respect to y?

Let $x = \epsilon^y$.

Then $$\frac{dx}{dy} = \epsilon^y \qquad \text{(unchanged)}$$

275. What is the derivative $\frac{dy}{dx}$ of $y = \epsilon^{ax}$?

If $$y = \epsilon^{ax} = \left(1 + \frac{1}{n}\right)^{nxa}$$

then $\epsilon^{ax} = 1^{nxa} + \dfrac{nxa}{\underline{1}} \cdot 1^{nxa-1} \cdot \dfrac{1}{n} + \dfrac{nxa(nxa-1)}{\underline{2}} \cdot \left(\dfrac{1}{n}\right)^2$

$$+ \frac{nxa(nxa-1)(nxa-2)}{\underline{3}} \cdot \left(\frac{1}{n}\right)^3 + \cdots$$

$$= 1 + \frac{nxa}{n} + \frac{n^2x^2a^2 - nxa}{n^2\underline{2}}$$

$$+ \frac{n^3x^3a^3 - 3n^2x^2a^2 + 2nxa}{n^3\underline{3}} + \cdots$$

$$= 1 + xa + \frac{x^2a^2 - \dfrac{xa}{n}}{\underline{2}} + \frac{x^3a^3 - \dfrac{3x^2a^2}{n} + \dfrac{2xa}{n^2}}{\underline{3}}$$

$$+ \cdots$$

And if n is infinitely great, then

$$\epsilon^{ax} = 1 + xa + \frac{x^2a^2}{\underline{2}} + \frac{x^3a^3}{\underline{3}} + \frac{x^4a^4}{\underline{4}} + \cdots$$

Now $\dfrac{d}{dx}(\epsilon^{ax}) = a + \dfrac{2xa^2}{1\cdot2} + \dfrac{3x^2a^3}{1\cdot2\cdot3} + \dfrac{4x^3a^4}{1\cdot2\cdot3\cdot4} + \cdots$

$$= a + \frac{xa^2}{\underline{1}} + \frac{x^2a^3}{\underline{2}} + \frac{x^3a^4}{\underline{3}} + \cdots$$

$$= a\left(1 + xa + \frac{x^2a^2}{\underline{2}} + \frac{x^3a^3}{\underline{3}} + \cdots\right)$$

or $\quad \dfrac{d}{dx}(\epsilon^{ax}) = a\epsilon^{ax}$

276. What is the derivative of $y = \varepsilon^{-ax}$?

In $y = \epsilon^{-ax}$

let $\qquad z = -ax,$ then $\qquad \dfrac{dz}{dx} = -a$

and $\quad y = \epsilon^z,$ while $\qquad \dfrac{dy}{dz} = \epsilon^z$ by substitution

Finally, $\dfrac{dy}{dx} = \dfrac{dy}{dz} \cdot \dfrac{dz}{dx} = \epsilon^z \cdot (-a) = -a\epsilon^z = -a\epsilon^{-ax}$

277. If $y = 5\varepsilon^{-\frac{x}{3x-1}}$, what is $\dfrac{dy}{dx}$?

Let $-\dfrac{x}{3x-1} = z,$ $\qquad \dfrac{dz}{dx} = -\left[\dfrac{(3x-1)\cdot 1 - x \cdot 3}{(3x-1)^2}\right]$

or $\qquad \dfrac{dz}{dx} = -\left[\dfrac{3x-1-3x}{(3x-1)^2}\right] = +\dfrac{1}{(3x-1)^2}$

Now $\qquad\qquad\qquad y = 5\epsilon^z, \qquad \dfrac{dy}{dz} = 5\epsilon^z$

$\therefore \dfrac{dy}{dx} = \dfrac{dy}{dz} \cdot \dfrac{dz}{dx} = 5\epsilon^z \cdot \dfrac{1}{(3x-1)^2} = \dfrac{5\epsilon^{-\frac{x}{3x-1}}}{(3x-1)^2}$

278. What is the derivative of $y = \varepsilon^{\frac{2x^2}{5}}$?

Let $\dfrac{2x^2}{5} = u$

Then $\qquad\qquad\qquad \dfrac{du}{dx} = \dfrac{2}{5} \cdot 2x = \dfrac{4}{5}x$

Now $\qquad\qquad y = \epsilon^u \qquad$ and $\qquad \dfrac{dy}{du} = \epsilon^u$

$\therefore \dfrac{dy}{dx} = \dfrac{dy}{du} \cdot \dfrac{du}{dx} = \epsilon^u \cdot \dfrac{4}{5}x = \epsilon^{\frac{2x^2}{5}} \cdot \dfrac{4}{5}x = \dfrac{4x}{5}\epsilon^{\frac{2x^2}{5}}$

279. What is the derivative of $y = \varepsilon^{\frac{3x}{x+2}}$?

Let $\dfrac{3x}{x+2} = u$

Then $\dfrac{du}{dx} = \dfrac{(x+2)\cdot 3 - 3x\cdot 1}{(x+2)^2} = \dfrac{3x+6-3x}{(x+2)^2} = \dfrac{6}{(x+2)^2}$

Now $\qquad\qquad y = \epsilon^u \qquad$ and $\qquad \dfrac{dy}{du} = \epsilon^u$

$\therefore \dfrac{dy}{dx} = \dfrac{dy}{du}\cdot\dfrac{du}{dx} = \epsilon^u \cdot \dfrac{6}{(x+2)^2} = \dfrac{6}{(x+2)^2}\cdot \epsilon^{\frac{3x}{x+2}}$

280. What is the derivative of $y = \epsilon^{\sqrt{x^2+1}}$?

Let $(x^2+1) = u \qquad$ and $\qquad (x^2+1)^{\frac12} = v, \qquad \dfrac{du}{dx} = 2x$

Then $\qquad\qquad v = u^{\frac12} \qquad$ and $\qquad \dfrac{dv}{du} = \dfrac12 u^{-\frac12}$

$\therefore \dfrac{dv}{dx} = \dfrac{dv}{du}\cdot\dfrac{du}{dx} = \dfrac12 u^{-\frac12}\cdot 2x = x\cdot u^{-\frac12} = \dfrac{x}{(x^2+1)^{\frac12}}$

Now $\qquad\qquad y = \epsilon^v \qquad$ or $\qquad \dfrac{dy}{dv} = \epsilon^v = \epsilon^{\sqrt{x^2+1}}$

$\therefore \dfrac{dy}{dx} = \dfrac{dy}{dv}\cdot\dfrac{dv}{dx} = \epsilon^{\sqrt{x^2+1}}\cdot\dfrac{x}{(x^2+1)^{\frac12}} = \dfrac{x}{(x^2+1)^{\frac12}}\cdot\epsilon^{\sqrt{x^2+1}}$

PROBLEMS

1. What is the derivative of $y = a(\epsilon^{2x} - \epsilon^{-2x})$?

2. Differentiate $y = 4\epsilon^{-\frac{2x}{x-1}}$.

3. If $y = \epsilon^{-4ax}$, what is $\dfrac{dy}{dx}$?

4. If $y = \epsilon^{\frac{5x^2}{9}}$, what is $\dfrac{dy}{dx}$?

5. What is the derivative of $y = \epsilon^{\frac{6x}{x+a}}$?
6. Differentiate $x = \epsilon^{\sqrt{y^2+4}}$.

7. Differentiate $x = 4\epsilon^{-\frac{3y}{x-5}}$.

CHAPTER XV

LOGARITHMS

281. What is a log of a number?

A logarithm is an exponent, or power, (x) that raises a base (ϵ) to equal a number (y)

or $$x = \log_\epsilon y$$

Example

a. If $y = \epsilon^x$, then $x = $ log of y to the base ϵ.
b. If $y = 10^x$, then $x = $ log of y to the base 10.

282. What is the Napierian, or natural, system of logarithms?

Here the base $= \epsilon = $ epsilon $= 2.7182; x = \log_\epsilon y$.

283. What is the common system of logarithms?

Here the base $= 10; x = \log_{10} y$, or, as commonly written, $x = \log y$.

284. What is the log of a product of two numbers in the natural system?

$$\log_\epsilon cd = \log_\epsilon c + \log_\epsilon d$$

which follows the same rule as for the common system.

285. What is the log of a quotient of two numbers in the natural system?

$$\log_\epsilon \frac{c}{d} = \log_\epsilon c - \log_\epsilon d$$

again similar to the common system.

286. What is the log of a number raised to a power?

$$\log_\epsilon b^x = x \log_\epsilon b$$

287. What is $\log_\epsilon \epsilon$?

$$\log_\epsilon \epsilon = 1$$

288. What is the form of the curve $y = \epsilon^x$?

$x =$	0	.5	1.0	1.5	2.0
$y =$	1	1.65	2.72	4.48	7.39

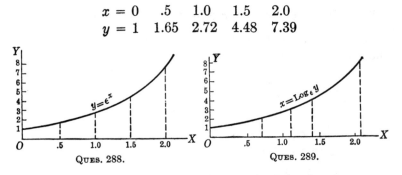

QUES. 288.   QUES. 289.

289. What is the form of the curve $x = \log_\epsilon y$?

$y =$	1	2.00	3.00	4.00	8.00
$x =$	0	0.69	1.10	1.39	2.08

We see that the two equations mean the same thing and have the same form.

290. How do you obtain a common log from a natural log?

Multiply the natural log by 0.4343

as $\qquad \log x = 0.4343 \log_\epsilon x$

291. How would you obtain a natural log from a common log?

Multiply the common log by 2.3026

as $\qquad \log_\epsilon x = 2.3026 \log x$

Chapter XVI

DIFFERENTIATION OF LOGARITHMIC FUNCTIONS

292. If $y = \log_\epsilon x$, what is $\dfrac{dy}{dx}$?

$$x = \epsilon^y \text{ by definition of a logarithm}$$

Then
$$\frac{dx}{dy} = \epsilon^y$$

and $\dfrac{dy}{dx} = \dfrac{1}{\dfrac{dx}{dy}} = \dfrac{1}{\epsilon^y} = \dfrac{1}{x}$ by rule for an inverse function

$$\therefore \frac{dy}{dx} = x^{-1} \qquad \text{or} \qquad \frac{d}{dx} (\log_\epsilon x) = x^{-1}$$

293. If $y = \log_\epsilon (x + a)$, what is $\dfrac{dy}{dx}$?

$$x + a = \epsilon^y \text{ by definition of a logarithm}$$

Now
$$\frac{d}{dy} (\epsilon^y) = \epsilon^y = x + a = \frac{dx}{dy}$$

and $\dfrac{dy}{dx} = \dfrac{1}{\dfrac{dx}{dy}} = \dfrac{1}{x + a}$ by rule for an inverse function

294. If $y = \log x$, what is $\dfrac{dy}{dx}$?

$$y = 2.3026 \log_\epsilon x$$

But
$$\frac{d}{dx} (\log_\epsilon x) = \frac{1}{x} \quad \text{(from Question 292)}$$

$$\therefore \frac{dy}{dx} = \frac{.4343}{x}$$

295. If $y = a^x$, what is $\dfrac{dy}{dx}$?

$\log_\epsilon y = x \log_\epsilon a$ by taking logs of both sides of the given
equation

$\log_\epsilon y = $ a function of x (Note this form of the equation.)

Then $\qquad x = \dfrac{\log_\epsilon y}{\log_\epsilon a}$ where $\dfrac{1}{\log_\epsilon a} = $ a constant

and $\qquad\qquad\qquad \dfrac{dx}{dy} = \dfrac{1}{y} \cdot \dfrac{1}{\log_\epsilon a}$

$\therefore \dfrac{dy}{dx} = \dfrac{1}{\dfrac{dx}{dy}} = y \log_\epsilon a \qquad$ or $\qquad \dfrac{1}{y} \cdot \dfrac{dy}{dx} = \log_\epsilon a$ (Note this
result)

296. What is the rule for differentiating expressions that can be put in the form $\log_\epsilon y = $ a function of x?

If $\log_\epsilon y = $ a function of x,

then $\dfrac{1}{y} \cdot \dfrac{dy}{dx} = $ the differential coefficient of the function of x

297. What form of expression is called the logarithmic derivative of y with respect to x?

The form $\dfrac{1}{y} \cdot \dfrac{dy}{dx}$ is called the logarithmic derivative of y
with respect to x.

298. If $y = f(x)$, what is the expression for the total rate of increase of y with respect to x?

$$\dfrac{dy}{dx} = \text{the total rate of increase}$$

299. What is the relative rate of increase of y with respect to x?

$\dfrac{1}{y} \cdot \dfrac{dy}{dx} = $ the relative rate of increase $ = $ the rate of increase
of the function per unit value of the function $ = $ the
logarithmic derivative

300. What is the percentage rate of increase?

The rate of increase per 100 units of the function or 100 times the relative rate of increase.

$$\therefore\ 100 \cdot \frac{1}{y} \cdot \frac{dy}{dx} = \text{the percentage rate of increase}$$

301. How is the expression $\frac{1}{y} \cdot \frac{dy}{dx}$ usually found?

It is found either by taking the log of both sides of $y = f(x)$ and then finding the derivative or by finding the ordinary derivative $\frac{dy}{dx}$ and then dividing by the value of the function.

302. What is the relative rate of increase and the percentage rate of increase of $y = 5\epsilon^{\frac{x}{4}}$?

Taking logs,

$$\log_\epsilon y = \log_\epsilon 5 + \frac{x}{4} \log_\epsilon \epsilon$$

or $\qquad\qquad \log_\epsilon y = \log_\epsilon 5 + \frac{x}{4} \quad (\text{as } \log_\epsilon \epsilon = 1)$

This is of the form $\log_\epsilon y = $ a function of x.

$$\therefore\ \frac{1}{y} \cdot \frac{dy}{dx} = \frac{1}{4} = \text{the differential coefficient of the function}$$
$$\text{of } x$$

This is then the relative rate of increase

and $\quad 100 \cdot \frac{1}{y} \cdot \frac{dy}{dx} = 25 = $ percentage rate of increase

303. What is the relation of the relative rate of increase in the area of a heated square sheet of metal to the coefficient of expansion of the material?

The coefficient of expansion is here the relative rate of increase because it is the increase in lengtn per unit length for 1 degree increase in temperature.

Let x = the length of a side of the square

and y = the area = x^2

and $\dfrac{1}{x} \cdot \dfrac{dx}{dT}$ = coefficient of expansion by definition

where T = temperature.

Now $y = x^2$, and, taking logs, we get

$$\log_\epsilon y = 2 \log_\epsilon x$$

and differentiating we get

$$\frac{1}{y} \cdot \frac{dy}{dT} = 2 \cdot \frac{1}{x} \cdot \frac{dx}{dT}$$

Therefore, the relative rate of increase of the area is twice the coefficient of expansion of the material.

304. If $y = \epsilon^{-bx}$, what is $\dfrac{dy}{dx}$?

This is of the form $\log_\epsilon y = -bx \log_\epsilon \epsilon = -bx =$ a function of x.

By rule, $\therefore \dfrac{1}{y} \cdot \dfrac{dy}{dx} = -b =$ the differential coefficient of

the function of x

or $$\frac{dy}{dx} = y(-b) = -b\epsilon^{-bx}$$

305. If $y = \epsilon^{\frac{x^2}{4}}$, what is $\dfrac{dy}{dx}$?

$$\log_\epsilon y = \frac{x^2}{4} \log_\epsilon \epsilon = \frac{x^2}{4} = \text{a function of } x$$

$$\therefore \frac{1}{y} \cdot \frac{dy}{dx} = \frac{2x}{4} = \text{the differential coefficient of the function}$$

of x

or $$\frac{dy}{dx} = y \cdot \frac{2x}{4} = \epsilon^{\frac{x^2}{4}} \cdot \frac{2x}{4} = \frac{x}{2} \epsilon^{\frac{x^2}{4}}$$

306. If $y = \epsilon^{\frac{3x}{x+2}}$, what is $\dfrac{dy}{dx}$?

$$\log_\epsilon y = \frac{3x}{x+2} \log_\epsilon \epsilon = \frac{3x}{x+2} = \text{a function of } x$$

By rule, $\dfrac{1}{y} \cdot \dfrac{dy}{dx} = \dfrac{(x+2) \cdot 3 - 3x \cdot 1}{(x+2)^2} = \dfrac{6}{(x+2)^2} = $ the

differential coefficient of the function of x

Finally, $\quad \dfrac{dy}{dx} = y \cdot \dfrac{6}{(x+2)^2} = \dfrac{6}{(x+2)^2} \cdot \epsilon^{\frac{3x}{x+2}}$

307. If $y = \epsilon^{\sqrt{x^3+b}}$, what is $\dfrac{dy}{dx}$?

$\log_\epsilon y = (x^3 + b)^{\frac{1}{2}} \log_\epsilon \epsilon = (x^3 + b)^{\frac{1}{2}} = $ a function of x

By rule, $\dfrac{1}{y} \cdot \dfrac{dy}{dx} = \dfrac{1}{2}(x^3 + b)^{-\frac{1}{2}} \cdot 3x^2 = \dfrac{3x^2}{2(x^3 + b)^{\frac{1}{2}}} = $ the

differential coefficient of the function of x

Finally, $\quad \dfrac{dy}{dx} = \dfrac{3x^2}{2\sqrt{x^3 + b}} \cdot \epsilon^{\sqrt{x^3+b}}$

308. If $y = \log_\epsilon (b + x^4)$, what is $\dfrac{dy}{dx}$?

Let $\qquad\qquad (b + x^4) = z$

Then $\qquad\qquad \dfrac{dz}{dx} = 4x^3$

and, by substitution,

$$y = \log_\epsilon z$$

Then $\qquad\qquad \dfrac{dy}{dz} = \dfrac{1}{z} = \dfrac{1}{b + x^4}$

$\therefore \dfrac{dy}{dx} = \dfrac{dy}{dz} \cdot \dfrac{dz}{dx} = \dfrac{1}{b + x^4} \cdot 4x^3 = \dfrac{4x^3}{b + x^4}$

309. If $y = \log_\epsilon (4x^3 + \sqrt{b + x^3})$, what is $\dfrac{dy}{dx}$?

Let $z = 4x^3 + \sqrt{b + x^3}$ $\qquad$ and $\qquad \dfrac{dz}{dx} = 12x^2$

$$+ \frac{3x^2}{2\sqrt{b + x^3}}$$

and, by substitution,

$$y = \log_e z$$

Then
$$\frac{dy}{dz} = \frac{1}{z} = \frac{1}{4x^3 + \sqrt{b + x^3}}$$

and
$$\frac{dy}{dx} = \frac{dy}{dz} \cdot \frac{dz}{dx} = \frac{1}{4x^3 + \sqrt{b + x^3}} \cdot \left(12x^2 + \frac{3x^2}{2\sqrt{b + x^3}} \right)$$

$$= \frac{3x^2(8\sqrt{b + x^3} + 1)}{(4x^3 + \sqrt{b + x^3})(2\sqrt{b + x^3})}$$

310. If $y = (x + 4)^2 \sqrt{x - 3}$, what is $\dfrac{dy}{dx}$?

$\log_e y = 2 \log_e (x + 4) + \frac{1}{2} \log_e (x - 3) =$ a function of x

By rule, $\dfrac{1}{y} \cdot \dfrac{dy}{dx} = \dfrac{2}{(x + 4)} + \dfrac{1}{2(x - 3)} =$ the differential

coefficient of the function of x

Finally, $\dfrac{dy}{dx} = (x + 4)^2 \sqrt{x - 3} \left[\dfrac{2}{x + 4} + \dfrac{1}{2(x - 3)} \right]$

$$= (x + 4)^2 \sqrt{x - 3} \left[\frac{4x - 12 + x + 4}{2(x + 4)(x - 3)} \right]$$

$$= (x + 4)^2 \sqrt{x - 3} \left[\frac{5x - 8}{2(x + 4)(x - 3)} \right]$$

$$= \frac{(x + 4)(5x - 8)}{2(x - 3)^{\frac{1}{2}}}$$

311. If $y = (x^2 + 4)^4 (x^3 - 3)^{\frac{3}{4}}$, what is $\dfrac{dy}{dx}$?

$\log_e y = 4 \log_e (x^2 + 4) + \frac{3}{4} \log_e (x^3 - 3) =$ a function of x

By rule, $\dfrac{1}{y} \cdot \dfrac{dy}{dx} = 4 \cdot \dfrac{2x}{x^2 + 4} + \dfrac{3}{4} \cdot \dfrac{3x^2}{x^3 - 3} = \dfrac{8x}{x^2 + 4}$

$+ \dfrac{9x^2}{4(x^3 - 3)} =$ the differential coefficient of the function

of x

Finally, $\dfrac{dy}{dx} = (x^2 + 4)^4 (x^3 - 3)^{\frac{3}{4}} \left[\dfrac{8x}{x^2 + 4} + \dfrac{9x^2}{4(x^3 - 3)} \right]$

The differential coefficient of $\log_\epsilon (x^2 + 4)$ is found by letting, say,

$$w = \log_\epsilon (x^2 + 4) \qquad \text{and} \qquad (x^2 + 4) = z$$

Then $$\frac{dz}{dx} = 2x$$

Now $$w = \log_\epsilon z \qquad \text{and} \qquad \frac{dw}{dz} = \frac{1}{z}$$

$$\therefore \frac{dw}{dx} = \frac{dw}{dz} \cdot \frac{dz}{dx} = \frac{1}{z} \cdot 2x = \frac{2x}{(x^2 + 4)}$$

The differential coefficient of $\log_\epsilon (x^3 - 3)$ is found by letting, say,

$$u = \log_\epsilon (x^3 - 3) \qquad \text{and} \qquad (x^3 - 3) = v$$

Then $$\frac{dv}{dx} = 3x^2$$

Now $$u = \log_\epsilon v \qquad \text{and} \qquad \frac{du}{dv} = \frac{1}{v}$$

$$\therefore \frac{du}{dx} = \frac{du}{dv} \cdot \frac{dv}{dx} = \frac{1}{v} \cdot 3x^2 = \frac{3x^2}{(x^3 - 3)}$$

312. If $y = \dfrac{\sqrt{x^2 + b}}{\sqrt[4]{x^3 - b}}$, what is $\dfrac{dy}{dx}$?

$\log_\epsilon y = \frac{1}{2}\log_\epsilon (x^2 + b) - \frac{1}{4}\log_\epsilon (x^3 - b) = $ a function of x

By rule, $\dfrac{1}{y} \cdot \dfrac{dy}{dx} = \dfrac{1}{2} \cdot \dfrac{2x}{(x^2 + b)} - \dfrac{1}{4} \cdot \dfrac{3x^2}{(x^3 - b)} = $ the differential coefficient of the function of x

Finally, $$\frac{dy}{dx} = \frac{\sqrt{x^2 + b}}{\sqrt[4]{x^3 - b}}\left[\frac{x}{x^2 + b} - \frac{3x^2}{4(x^3 - b)}\right]$$

313. If $y = \dfrac{1}{\log_\epsilon x}$, what is $\dfrac{dy}{dx}$?

$$\frac{dy}{dx} = \frac{\log_\epsilon x \cdot (0) - 1 \cdot \dfrac{1}{x}}{\log_\epsilon^2 x} = -\frac{1}{x \log_\epsilon^2 x}$$

by rule for the derivative of a quotient.

314. If $y = \sqrt[4]{\log_\epsilon x}$, what is $\dfrac{dy}{dx}$?

Let
$$u = \log_\epsilon x$$

Then
$$\frac{du}{dx} = \frac{1}{x} \qquad \text{by substitution}$$

$$y = u^{\frac{1}{4}} \quad \text{and} \quad \frac{dy}{du} = \frac{1}{4} u^{-\frac{3}{4}}$$

$$\frac{dy}{dx} = \frac{dy}{du} \cdot \frac{du}{dx} = \frac{1}{4} u^{-\frac{3}{4}} \cdot \frac{1}{x}$$

$$\therefore \frac{dy}{dx} = \frac{1}{4(\log_\epsilon x)^{\frac{3}{4}}} \cdot \frac{1}{x} = \frac{1}{4x \sqrt[4]{\log_\epsilon^3 x}}$$

315. If $y = \left(\dfrac{1}{b^{2x}}\right)^{2bx}$, what is $\dfrac{dy}{dx}$?

$$\log_\epsilon y = 2bx \cdot \log_\epsilon \frac{1}{b^{2x}}$$

But
$$\log_\epsilon \frac{1}{b^{2x}} = \log_\epsilon 1 - \log_\epsilon b^{2x}$$

$$= -\log_\epsilon b^{2x} = -2x \log_\epsilon b$$

$$\therefore \log_\epsilon y = 2bx \cdot (-2x \log_\epsilon b)$$

or
$$\log_\epsilon y = -4bx^2 \cdot \log_\epsilon b = \text{a function of } x$$

By rule, $\dfrac{1}{y} \cdot \dfrac{dy}{dx} = -8bx \log_\epsilon b = $ the differential coefficient

of the function of x

Finally,
$$\frac{dy}{dx} = -\left(\frac{1}{b^{2x}}\right)^{2bx} \cdot 8bx \log_\epsilon b$$

$$= -8bx \left(\frac{1}{b^{2x}}\right)^{2bx} \cdot \log_\epsilon b$$

or
$$\frac{dy}{dx} = -8xb^{1-4bx^2} \cdot \log_\epsilon b$$

PROBLEMS

1. If $y = \log_\epsilon (a + 2x^2)$, what is $\dfrac{dy}{dx}$?

2. If $y = \log_\epsilon (2x^2 + \sqrt{x^2 + 1})$, what is $\dfrac{dy}{dx}$?

3. If $y = (x + 1)^2 \sqrt{x - 3}$, what is $\dfrac{dy}{dx}$?

4. If $y = (x^2 + 4)^2(x^3 + 1)^{\frac{1}{4}}$, what is $\dfrac{dy}{dx}$?

5. If $y = \dfrac{\sqrt{x^2 + 1}}{\sqrt[3]{x^3 - 1}}$, what is $\dfrac{dy}{dx}$?

6. If $x = \dfrac{2}{\log_\epsilon y}$, what is $\dfrac{dx}{dy}$?

7. If $x = \sqrt[4]{\log_\epsilon y}$, what is $\dfrac{dx}{dy}$?

8. If $x = \left(\dfrac{1}{a^v}\right)^{av}$, what is $\dfrac{dx}{dy}$?

Differentiate the following:

9. $y = \dfrac{1}{a} \cdot \dfrac{b^{ax}}{\log_\epsilon b}$

10. $y = \log_\epsilon x^4$

11. $x = (4y^2 + 2)(\sqrt{y} + 3)$

12. $x = b^v \cdot y^b$

13. $y = \log_\epsilon (2x\epsilon^x)$

14. $y = (\log_\epsilon 3x)^4$

15. $y = \dfrac{\log_\epsilon (x + 2)}{x + 2}$

16. $y = (2x^2 + 1)\epsilon^{-3x}$

17. What is the logarithmic derivative of $y = 5\epsilon^{3x}$?

18. What is the logarithmic derivative of $y = x^{\sin x}$?

19. What is $\dfrac{dy}{dx}$ in the implicit function $\epsilon^{x-v} = xy$?

20. What is $\dfrac{dy}{dx}$ in the implicit function $x^v = y^x$?

21. What is the ratio of the relative rate of $\sin x$ to that of $\cos x$?

22. What is the ratio of the relative rate of ϵ^{ax} to that of 10^{ax}?

23. What are the relative rate of increase and the percentage rate of increase of the number of bacteria in a culture as expressed by $N = 2,000\epsilon^{0.35t}$ where N = number and t = time in hours?

Chapter XVII

CONDITIONS OF LOGARITHMIC OR ORGANIC GROWTH

316. What does the equation $y = ak^x$ represent?

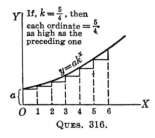

If, $k = \frac{5}{4}$, then each ordinate $= \frac{5}{4}$ as high as the preceding one

Ques. 316.

It is the general expression of a curve which has its successive ordinates in geometrical progression and which means that each ordinate bears a constant ratio k to the preceding one. This is called *logarithmic*, or *organic*, *growth*.

Example

If $y = ak^x$, then

When $x = 0$, $\qquad y = a = $ initial value
When $x = 1$, $\qquad y_1 = ak$
When $x = 2$, $\qquad y_2 = ak^2$
When $x = 3$, $\qquad y_3 = ak^3$

$\therefore \dfrac{y_2}{y_1} = \dfrac{ak^2}{ak} = k = $ constant ratio of any two ordinates.

Therefore, each ordinate is k times the height of the preceding ordinate. This is a criterion of geometric progression or logarithmic growth.

317. Under what condition is there a constant difference of the logs of any two successive ordinates in a curve?

When the ordinates of a curve are related in a constant ratio k, which means in logarithmic or organic growth.

Example

If $y = ak^x$,

then $\qquad\log_\epsilon y = \log_\epsilon a + x \log_\epsilon k$

or $\qquad\log_\epsilon y - \log_\epsilon a = x \log_\epsilon k = xb$

if we let $\qquad\log_\epsilon k = \text{constant} = b$

Now at $x = 1$, $\log_\epsilon y_1 - \log_\epsilon a = \log_\epsilon k = b$

and at $x = 2$, $\log_\epsilon y_2 - \log_\epsilon a = 2 \log_\epsilon k = 2b$

Then $\log_\epsilon y_2 - \log_\epsilon y_1 = 2b - b = b = $ a constant difference

So the logs of any two successive ordinates have a constant difference.

318. What is the form of the curve of $\log_\epsilon y = \log_\epsilon a + x \cdot \log_\epsilon k$?

A straight line, sloping up by equal steps.

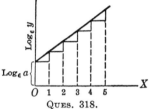

QUES. 318.

319. What other form does the equation for organic growth assume?

If $y = ak^x = $ original equation for organic growth

then $\qquad\log_\epsilon y = \log_\epsilon a + x \log_\epsilon k$

or $\qquad\log_\epsilon y - \log_\epsilon a = x \log_\epsilon k = xb$

when $k > 1$, $\log_\epsilon k$ is a positive constant equal to, say, b

or $\qquad\log_\epsilon \dfrac{y}{a} = bx$

and, from the definition of a logarithm, therefore,

$y = a\epsilon^{bx} = $ the new form of the equation for organic growth

320. What is the value of k (= constant ratio) in compound-interest growth?

$$k = \left(1 + \frac{1}{n}\right)$$

because $C_x = C\left(1 + \dfrac{1}{n}\right)^x$ is of the form $y = ak^x$.

In this case, each successive ordinate is $\left(1 + \dfrac{1}{n}\right) = \dfrac{n+1}{n}$ times (or k times) as high as its predecessor and is another example of logarithmic or organic growth.

321. What results when the ratio k of the successive ordinates is less than 1 in $y = ak^x$?

If $\qquad\qquad k = \tfrac{5}{8}$

then each successive ordinate $= \tfrac{5}{8}$ of the height of the preceding ordinate. Here the curve tends to sink downward.

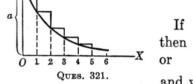

QUES. 321.

Example

If $\qquad y = ak^x$

then $\quad \log_\epsilon y = \log_\epsilon a + x \log_\epsilon k$

or $\quad \log_\epsilon y - \log_\epsilon a = x \log_\epsilon k$

and when $k = {<}1$, $\log_\epsilon k =$ negative and can be taken equal to $-b$.

$$\therefore \log_\epsilon \frac{y}{a} = -bx$$

$$\therefore y = a\epsilon^{-bx}$$

322. What does the equation $y = a\epsilon^{-bx}$ represent when the independent variable is time?

Here $\qquad\qquad y = a\epsilon^{-bt}$

It represents physical processes in which something is gradually dying away.

323. What is the die-away factor?

ϵ^{-bt} is the die-away factor.

324. How would you express the relation of the cooling of a hot body—Newton's law of cooling?

$$\theta_t = \theta_o \epsilon^{-bt} = \text{Newton's law of cooling}$$

where θ_o = original excess of temperature of the hot body
over that of its surroundings,

θ_t = excess of temperature at the end of time t,

b = constant of decrement, depending on the
amount of the surface exposed by the body and
on its coefficients of conductivity and emis-
sivity, etc.

**325. How would you express the relation of the charge of
an electrified body?**

$$Q_t = Q_o \epsilon^{-bt} = \text{charge of an electrified body}$$

where Q_o = original charge,

Q_t = charge at end of time t,

b = constant decrement at which charge is leaking
away depending upon capacity of the body
and resistance of leakage path.

**326. How would you express the relation of the oscillations
of a flexible spring?**

$$O_t = O_o \epsilon^{-bt}$$

where O_o = original oscillations imparted to the spring,

O_t = oscillations at end of time t,

b = constant decrement depending upon material,
modulus of elasticity, and form.

**327. In general, to what does the die-away factor ε^{-bt}
apply?**

To all phenomena in which the rate of decrease $\dfrac{dy}{dt}$
(= slope) is proportional to the magnitude of that which
is decreasing ($= y$).

**328. What is the relation of the derivative $\dfrac{dy}{dt}$ of $y = a\varepsilon^{-bt}$
to y and the constant b?**

If $$y = a\epsilon^{-bt}$$

then $\log_\epsilon y = \log_\epsilon a - bt \log_\epsilon \epsilon = \log_\epsilon a - bt = $ a function of t

By rule, $\dfrac{1}{y} \cdot \dfrac{dy}{dt} = -b = $ the differential coefficient of the

function of t

$$\therefore \frac{dy}{dt} = -yb$$

and the derivative is directly proportional to y and the constant b.

329. What is the time constant T?

$T = \dfrac{1}{b} = $ the reciprocal of the constant b in the die-away

factor ϵ^{-bt}

330. What is the expression for the die-away factor with the time constant T?

$\epsilon^{-\frac{t}{T}} = $ the die-away factor with the time constant T

331. What is the meaning of T when $t = T$?

The time required for the original quantity to die away to $\dfrac{1}{\epsilon}$ part $(= 0.3678)$ of its original value.   For when $t = T$,

then $\qquad \epsilon^{-\frac{t}{T}} = \epsilon^{-1} = \dfrac{1}{\epsilon} = \dfrac{1}{2.71828} = 0.3678$

332. What is the temperature of a cooling body at the end of 80 minutes $(= t)$, if the original temperature θ_o is 80° hotter than the surrounding objects and it cools to $\dfrac{1}{\epsilon}$ of 80° in 40 minutes $(= T)$?

$$\theta_t = \theta_o \epsilon^{-bt} = \text{Newton's law of cooling}$$

$$\frac{1}{b} = T = 40 \text{ min.} = \text{time constant}$$

$$t = 80 \text{ min.}$$

$$\therefore \frac{t}{T} = \frac{80}{40} = 2$$

Then $\qquad \theta_t = 80° \cdot \epsilon^{-2} = 80° \cdot 0.1353$

∴ Final temperature $\theta_t = 10.824°$ at end of 80 min.

333. What is the strength of an electric current I at the end of (a) 0.002 second, (b) 0.05 second after applying an electromotive force E?

$$I = \frac{E}{R}\left(1 - \epsilon^{-\frac{Rt}{L}}\right)$$

where I = the current at any time t sec. after the e.m.f. is applied,

R = the resistance,

L = the coefficient of self-induction.

Assume $E = 20$ volts, $R = 2$ ohms, $L = 0.02$ henry.

$$\frac{L}{R} = T = \text{the time constant} = \frac{0.02}{2} = 0.01$$

Then $\quad I = \frac{20}{2}(1 - \epsilon^{-\frac{t}{0.01}}) = 10 - 10\epsilon^{-\frac{t}{0.01}}$

The time constant 0.01 means it takes 0.01 second for the *variable* term to *fall* to $\frac{1}{\epsilon}$ (= 0.3679) of its initial value, which is $10\epsilon^{-\frac{0}{0.01}} = 10$. This is the initial value of the *variable self-inductance* against which the e.m.f. has to act, to decrease it, so that the *electric current* can go on *increasing* with time until it reaches its own ultimate value of 10.

After $t = 0.002$ sec.,

$$\frac{t}{T} = \frac{0.002}{0.01} = 0.2$$

and $I = 10 - 10\epsilon^{-0.2} = 10 - 10 \cdot 0.8187 = 1.813 =$

actual current

Now after $t = 0.05$ sec.,

$$\frac{t}{T} = \frac{0.05}{0.01} = 5$$

and $I = 10 - 10\epsilon^{-5} = 10 - 10 \cdot 0.0067 = 9.933 =$ actual
current

334. If the intensity of a beam of light is diminished by 40 per cent in passing through 20 inches of a certain transparent medium, what thickness is required to reduce the intensity to one-half its original value?

Let I_o = original intensity of beam,
  I = intensity, any thickness,
  B = constant of absorption, found by experiment,
  l = thickness of medium, inches.

Then $\qquad\qquad I = I_o\epsilon^{-Bl}$
Now $\qquad$ 60 per cent = 100 per cent $\epsilon^{-B\cdot 20}$
or $\qquad\qquad 0.6 = \epsilon^{-20B}$
and $\qquad\qquad B = 0.0255$

Now, to reduce the intensity to one-half its value,

$$50 \text{ per cent} = 100 \text{ per cent} \cdot \epsilon^{-0.0255l}$$
or, $\qquad\qquad 0.5 = \epsilon^{-0.0255l}$
and $\qquad\qquad \log 0.5 = -0.0255l \cdot \log \epsilon$
$$\therefore l = -\frac{\log 0.5}{0.0255 \log \epsilon} = \frac{\overline{1}.69897}{-0.0255 \cdot 0.43429} = 27.18 \text{ in. of}$$
transparent medium required

335. How long will it take to transform one-half of a radioactive substance, say, radium C?

$Q = Q_o\epsilon^{-bt}$ = a physical process in which something is
dying away
where Q = quantity not yet transformed,
  Q_o = original quantity of the substance,
  b = a constant = $4.45 \cdot 10^{-3}$ (found by experiment),
  t = time elapsed, seconds, since transformation began.

Then $$0.5 = \epsilon^{-0.00445t}$$
and $$\log 0.5 = -0.00445t \cdot \log \epsilon$$
$$\therefore t = -\frac{\log 0.5}{\log \epsilon \cdot 0.00445} = -\frac{\overline{1}.69897}{0.43429 \cdot 0.00445} = 155.77$$
sec. to transform one-half the substance, called the *mean life* of the substance

PROBLEMS

1. What is the derivative of $x = y^{2v}$?

2. What is the differential coefficient of $y = (\epsilon^x)^{2x}$?

3. Find the $\dfrac{dy}{dx}$ of $y = \epsilon^{x^{2x}}$.

4. Draw the curve $y = a\epsilon^{-\frac{t}{T}}$ where $a = 10$, $T = 6$, and t is taken at values from 0 to 10.

5. What is the minimum or maximum of $x = y^{2v}$?

6. What is the minimum or maximum of $y = xb^{\frac{3}{x}}$?

7. If the dying away of a current on a sudden removal of the e.m.f. from a circuit containing resistance and self-induction is expressed by $I = I_o\epsilon^{-\frac{Rt}{L}}$ where I = the current at any time t sec. after the e.m.f. is removed, R = the resistance, and L = the coefficient of self-induction, then plot a curve to show the current at any time from $t = 0$ to $t = 0.4$ sec., if $I_o = 20$ amp., $R = 0.3$ ohm, and $L = 0.02$ henry.

8. What is the time constant and how long will it take for a body to cool down to 10 per cent of the original excess of temperature if it takes 20 min. for the original excess of temperature to fall to one-half its original excess?

9. What is the potential E after 0.2 sec. of discharge if a condenser of capacity $C = 4 \cdot 10^{-5}$ charged to a potential $E_o = 30$ is discharging through a resistance R of 8,000 ohms, assuming that the fall of potential is according to $E = E_o\epsilon^{-\frac{t}{CR}}$?

10. If the pressure p of the atmosphere at an altitude of H miles is expressed by $p = p_o\epsilon^{-bH}$, where p_o = the pressure at sea level (14.7 lb./sq. in.), what is the value of b when the pressure at 6 miles has dropped to $\frac{1}{4}$ atm. and what is the percentage error?

CHAPTER XVIII

PARTIAL DIFFERENTIATION

336. How would you denote a quantity that is a function of more than one independent variable?

If u and v are independent variables and y is a variable dependent upon them, then

$$y = f(u, v)$$

337. How would you express a *partial differential* or growth when the function contains two independent variables?

If
$$y = u \cdot v$$
then $d_v y = u \cdot dv = $ a *partial differential* or growth

This means that $d_v y$ is the growth of y, due to a growth of v, while u is considered constant.

$$d_u y = v \cdot du = \text{a *partial differential* or growth}$$

This means that $d_u y$ is the growth of y, due to a growth of u, while v is considered constant. (The subscript indicates which variable is considered as varying.)

338. How would you express a *partial derivative* or *ratio of growths*, using deltas?

If
$$y = u \cdot v$$
then $\dfrac{\partial y}{\partial v} = u = $ a *partial derivative* of y with respect to v

   (u remaining constant)

   = rate of change of y with respect to v if v is the only independent variable considered varying

$\dfrac{\partial y}{\partial u} = v =$ a *partial derivative* of y with respect to u (v remaining constant)

$=$ rate of change of y with respect to u if u is the only independent variable considered varying

339. How would you express partial differentials with relation to partial derivatives?

Substitute the above values of the partial derivatives in the equations of the partial differentials in Question 337, obtaining

$$\left.\begin{array}{l} d_v y = \dfrac{\partial y}{\partial v} \cdot dv \\[2ex] d_u y = \dfrac{\partial y}{\partial u} \cdot du \end{array}\right\} \text{These equations are generally known as partial differentials.}$$

340. What is a total differential?

The sum of the two partial differentials is the **total** differential, because the total growth of y depends upon the growths of the two independent variables at the same time.

Then $dy = \dfrac{\partial y}{\partial v} \cdot dv + \dfrac{\partial y}{\partial u} \cdot du =$ the total differential

This is sometimes written as $dy = \left(\dfrac{dy}{du}\right) du + \left(\dfrac{dy}{dv}\right) dv$.

341. What are the partial derivatives and the total differential of $y = 4bu^3 + 5au^2v + 6cv^2$?

$\dfrac{\partial y}{\partial u} = 12bu^2 + 10auv =$ the partial derivative of y with respect to u, v considered as constant

$\dfrac{\partial y}{\partial v} = 5au^2 + 12cv =$ the partial derivative of y with respect to v, u considered as constant

Then $dy = (12bu^2 + 10auv)\, du + (5au^2 + 12cv)\, dv =$ the total differential or growth of y, due to a growth of both u and v

342. What are the partial derivatives and the total differential of $z = y^x$?

$\dfrac{\partial z}{\partial x} = y^x \log_e y$ = the partial derivative of z with respect

to x, y considered as constant

$\dfrac{\partial z}{\partial y} = xy^{x-1}$ = the partial derivative of z with respect to y,

x considered as constant

Then $dz = (y^x \log_e y)\, dx + (xy^{x-1})\, dy$ = the total differential or the growth of z, due to a growth in both x and y

343. How would you express partial derivatives of the second order?

If $\qquad\qquad y = f(u, v)$

$\dfrac{\partial^2 y}{\partial u^2} = \dfrac{\partial}{\partial u}\left(\dfrac{\partial y}{\partial u}\right)$ = the second partial derivative of y

with respect to u

$\dfrac{\partial^2 y}{\partial v^2} = \dfrac{\partial}{\partial v}\left(\dfrac{\partial y}{\partial v}\right)$ = the second partial derivative of y with

respect to v

$\dfrac{\partial^2 y}{\partial u \cdot \partial v} = \dfrac{\partial}{\partial u}\left(\dfrac{\partial y}{\partial v}\right)$ = the second partial derivative of y

first with respect to v (because the part in the parenthesis was first obtained) and second with respect to u

$\dfrac{\partial^2 y}{\partial v \cdot \partial u} = \dfrac{\partial}{\partial v}\left(\dfrac{\partial y}{\partial u}\right)$ = the second partial derivative of y

first with respect to u and second with respect to v

In the last two expressions it is practically immaterial which partial differentiation is taken first and which second.

344. What are the $\dfrac{\partial y}{\partial u}$, $\dfrac{\partial y}{\partial v}$, $d_u y$, $d_v y$, and dy of

$$y = u^2 + uv + v^2?$$

$\dfrac{\partial y}{\partial u} = 2u + v$ = partial derivative of y with respect to u,

considering v as constant

$$\frac{\partial y}{\partial v} = u + 2v = \text{partial derivative of } y \text{ with respect to } v,$$

considering u as constant

$$d_u y = \frac{\partial y}{\partial u} \cdot du = (2u + v) \cdot du = \text{partial differential or}$$

growth of y corresponding to du

$$d_v y = \frac{\partial y}{\partial v} \cdot dv = (u + 2v) \cdot dv = \text{partial differential or}$$

growth of y corresponding to dv

$$dy = d_u y + d_v y = (2u + v) \cdot du + (u + 2v) \cdot dv = \text{the}$$
total differential of y due to a growth in both u and v
together

345. What are the partial derivatives of $y = \cos(2u + 3v)$?

$$\frac{\partial y}{\partial u} = -2 \sin(2u + 3v) \quad \text{(considering } v \text{ as constant)}$$

$$\frac{\partial y}{\partial v} = -3 \sin(2u + 3v) \quad \text{(considering } u \text{ as constant)}$$

346. What are the $\dfrac{\partial^2 y}{\partial x^2}$, $\dfrac{\partial^2 y}{\partial t^2}$, $\dfrac{\partial^2 y}{\partial t \cdot \partial x}$, and $\dfrac{\partial^2 y}{\partial x \cdot \partial t}$ of $y = \varepsilon^{xt}$?

$$\frac{\partial y}{\partial x} = t\epsilon^{xt} \quad \text{and} \quad \frac{\partial^2 y}{\partial x^2} = \frac{\partial}{\partial x}\left(\frac{\partial y}{\partial x}\right) = t^2 \epsilon^{xt} \quad (t \text{ is constant;}$$

x varies)

$$\frac{\partial y}{\partial t} = x\epsilon^{xt} \quad \text{and} \quad \frac{\partial^2 y}{\partial t^2} = \frac{\partial}{\partial t}\left(\frac{\partial y}{\partial t}\right) = x^2 \epsilon^{xt} \quad (x \text{ is constant;}$$

t varies)

$$\frac{\partial^2 y}{\partial t \cdot \partial x} = \frac{\partial}{\partial t}\left(\frac{\partial y}{\partial x}\right) = \frac{\partial}{\partial t}(t\epsilon^{xt}) = t\frac{\partial}{\partial t}(\epsilon^{xt}) + \epsilon^{xt} \cdot \frac{\partial t}{\partial t}$$
$$= xt\epsilon^{xt} + \epsilon^{xt} = \epsilon^{xt}(xt + 1) \quad (x \text{ is constant})$$

$$\frac{\partial^2 y}{\partial x \cdot \partial t} = \frac{\partial}{\partial x}\left(\frac{\partial y}{\partial t}\right) = \frac{\partial}{\partial x}(x\epsilon^{xt}) = x \cdot \frac{\partial}{\partial x}(\epsilon^{xt}) + \epsilon^{xt}\frac{\partial x}{\partial x}$$
$$= xt\epsilon^{xt} + \epsilon^{xt} = \epsilon^{xt}(xt + 1) \quad (t \text{ is constant})$$

It is to be noted that in $\dfrac{\partial^2 y}{\partial t \cdot \partial x}$ it is immaterial which
partial differentiation is carried through first; the final
result is the same.

347. What are $\frac{\partial y}{\partial x}$ and $\frac{\partial y}{\partial z}$ of $xy = 5z$?

If $$xy = 5z$$

then $$y = \frac{5z}{x}$$

and $$\frac{\partial y}{\partial x} = -\frac{5z}{x^2} \text{ (considering } z \text{ as constant)}$$

and $$\frac{\partial y}{\partial z} = \frac{5}{x} \text{ (considering } x \text{ as constant)}$$

The partial derivatives can also be found from the implicit function $xy = 5z$.

$$x\frac{\partial y}{\partial x} + y = 0 \text{ (considering } z \text{ as constant)}$$

$$\therefore \frac{\partial y}{\partial x} = -\frac{y}{x} = -\frac{5z}{x^2}$$

and $$x \cdot \frac{\partial y}{\partial z} = 5 \text{ (considering } x \text{ as constant)}$$

$$\therefore \frac{\partial y}{\partial z} = \frac{5}{x}$$

348. How would you illustrate partial differentiation graphically?

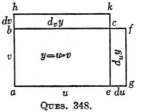

QUES. 348.

Let figure $abce$ be a variable rectangle of variable base u and variable height v. Then the area $y = u \cdot v$. And $d_u y = egfc =$ the growth in area y if v is considered constant while u grows to $eg = du$. And $d_v y = bhkc$ is the growth in area y if u is considered constant while v grows to $bh = dv$. Then the total differential of area y is

$$egfc + bhkc = d_u y + d_v y = dy$$

Also, if $y = u \cdot v$, then

$$\frac{\partial y}{\partial u} = v \quad \text{and} \quad \frac{\partial y}{\partial v} = u$$

$$\therefore d_u y = \frac{\partial y}{\partial u} \cdot du = v \cdot du = \text{area } egfc$$

and
$$d_v y = \frac{\partial y}{\partial v} \cdot dv = u \cdot dv = \text{area } bhkc$$

349. **What is the rate of change of the area of a right circular cone with respect to its radius r when the slant height s remains constant and with respect to its slant height when the radius r remains constant, and what is the total change of area when both the radius and slant height are varying at the same time?**

Given $A = \text{area} = \pi rs + \pi r^2$.

Then $\dfrac{\partial A}{\partial r} = \pi s + 2\pi r = $ rate of change of area with respect

to r (considering s constant)

$\dfrac{\partial A}{\partial s} = \pi r = $ rate of change of area with respect to s

(r is constant)

$$dA = \frac{\partial A}{\partial s} \cdot ds + \frac{\partial A}{\partial r} \cdot dr = (\pi r) \cdot ds + (\pi s + 2\pi r)\, dr$$

= the total change of area when both r and s are varying at the same time

350. **How would you express the variation of the volume of a cone if the radius of its base R and its height H vary at the same time?**

$$\text{Volume} = V = \frac{\pi R^2 H}{3}$$

$$\text{Partial derivative } \frac{\partial V}{\partial R} = \frac{2\pi RH}{3} \quad (H = \text{constant})$$

$$\text{Partial derivative } \frac{\partial V}{\partial H} = \frac{\pi R^2}{3} \quad (R = \text{constant})$$

and $dV = \dfrac{2\pi RH}{3} \cdot dR + \dfrac{\pi R^2}{3} \cdot dH = $ total differential

351. **How would you arrive at the dimensions of a rectangular box having no top for the condition that the**

area A of its sides and base together is a **minimum** for a given volume ($= V$)?

Let $x =$ the length and $y =$ the width. Then the depth $= \dfrac{V}{xy}$, and the surface area

$$A = xy + 2 \cdot x \cdot \frac{V}{xy} + 2 \cdot y \cdot \frac{V}{xy} = xy + \frac{2V}{y} + \frac{2V}{x} \quad (1)$$

Now $dA = \dfrac{\partial A}{\partial x} \cdot dx + \dfrac{\partial A}{\partial y} \cdot dy =$ the total differential, as the area depends upon x and y at the same time

or $\quad dA = \left(y - \dfrac{2V}{x^2} \right) dx + \left(x - \dfrac{2V}{y^2} \right) dy$

For a minimum

$$\left(y - \frac{2V}{x^2} \right) = 0 \quad \text{and} \quad \left(x - \frac{2V}{y^2} \right) = 0$$

$$2V = yx^2 = xy^2$$

$$\therefore x = y$$

Substitute this in Eq. (1) and get

$$A = x^2 + \frac{4V}{x}$$

Now $\quad \dfrac{dA}{dx} = 2x - \dfrac{4V}{x^2} = 0 \quad$ for a minimum

$\therefore x = \sqrt[3]{2V} = y \quad$ for a minimum

and $\quad \dfrac{V}{xy} = \dfrac{V}{(2V)^{\frac{1}{3}} \cdot (2V)^{\frac{1}{3}}} = \dfrac{V}{(2V)^{\frac{2}{3}}} = \dfrac{V}{\sqrt[3]{4V^2}}$

352. What would be the lengths of the sides of a triangle of maximum area formed from a piece of string 60 inches long?

Let $\qquad x =$ length of one side,

$y =$ length of another side,

$60 - (x + y) =$ length of third side.

Now the area of the triangle

$$A = \sqrt{s(s - x)(s - y)(s - 60 + x + y)}$$

where s = one-half the perimeter of the triangle = $\frac{60}{2}$
$$= 30 \text{ in.}$$

$\therefore A$ = area = $\sqrt{30M}$

where $M = (30 - x)(30 - y)(x + y - 30)$.

$$M = xy^2 + x^2y - 30x^2 - 30y^2 - 90xy + 1,800x + 1,800y$$
$$- 27,000 \quad (1)$$

The area = A is a maximum when M is a maximum.

$$\text{Total differential } dM = \frac{\partial M}{\partial x} \cdot dx + \frac{\partial M}{\partial y} \cdot dy$$

as M depends upon x and y at the same time.

For a maximum,

$$\frac{\partial M}{\partial x} = 0 \quad \text{and} \quad \frac{\partial M}{\partial y} = 0$$

Now $\dfrac{\partial M}{\partial x} = y^2 + 2xy - 60x - 90y + 1,800 = 0$

and $\dfrac{\partial M}{\partial y} = 0 = 2xy + x^2 - 60y - 90x + 1,800$

An immediate solution is $x = y$. Substitute this in Eq. (1).

$\therefore M = x^3 + x^3 - 30x^2 - 30x^2 - 90x^2 + 1,800x$
$$+ 1,800x - 27,000$$

or $M = 2x^3 - 150x^2 + 3,600x - 27,000$

Now $\dfrac{dM}{dx} = 6x^2 - 300x + 3,600 = 0$ for a maximum or
$$\text{minimum}$$

or $\dfrac{dM}{dx} = 6(x^2 - 50x + 600) = 0; \quad (x - 30)(x - 20)$
$$= 0; \quad x = 30 \quad \text{and} \quad x = 20$$
$$\frac{d^2M}{dx^2} = 12x - 300.$$

For $x = 30$, $\dfrac{d^2M}{dx^2} = +$ a minimum is indicated; for $x = 20$,

$\dfrac{d^2M}{dx^2} = -$ a maximum is indicated.

Therefore, lengths of sides = 20 in. for each side for a maximum.

353. How would you obtain the total derivative from the total differential?

If $\qquad\qquad u = f(x, y)$

then $du = \dfrac{\partial u}{\partial x} \cdot dx + \dfrac{\partial u}{\partial y} \cdot dy =$ the total differential

Now, if x and y and therefore u are functions of another variable t, then, dividing by dt,

$\dfrac{du}{dt} = \dfrac{\partial u}{\partial x} \cdot \dfrac{dx}{dt} + \dfrac{\partial u}{\partial y} \cdot \dfrac{dy}{dt} =$ the total derivative of u with

respect to t

where $\dfrac{dx}{dt}$ and $\dfrac{dy}{dt}$ are derivatives that represent the rates

of change of x and y, respectively, with respect to t.

Ques. 354.

354. What is the rate at which the hypotenuse u of a variable right triangle is changing when $x = 10$ and $y = 7$ if x is increasing at the rate of 4 inches per minute and y at the rate of 6 inches per minute?

Given $\dfrac{dx}{dt} = 4$ and $\dfrac{dy}{dt} = 6$ where $t =$ time in minutes.

Now $\qquad u^2 = x^2 + y^2 \qquad$ or $\qquad u = \sqrt{x^2 + y^2}$

Then $\dfrac{\partial u}{\partial x} = \dfrac{x}{\sqrt{x^2 + y^2}} \qquad$ and $\qquad \dfrac{\partial u}{\partial y} = \dfrac{y}{\sqrt{x^2 + y^2}}$

and $\dfrac{du}{dt} = \dfrac{\partial u}{\partial x} \cdot \dfrac{dx}{dt} + \dfrac{\partial u}{\partial y} \cdot \dfrac{dy}{dt} =$ the total derivative or rate

of change of u with respect to t

or $\dfrac{du}{dt} = \dfrac{10}{\sqrt{(10)^2 + (7)^2}} \cdot 4 + \dfrac{7}{\sqrt{(10)^2 + (7)^2}} \cdot 6$

$$= \frac{40}{12.2066} + \frac{42}{12.2066}$$

or $\dfrac{du}{dt} = 6.72$ in./min.

Therefore, at the instant when $x = 10$ and $y = 7$ the hypotenuse is increasing at the rate of 6.72 in./min.

355. What is the rate at which the pressure of a certain gas is changing, given the relation $p \cdot v = kT$, if $k = 50$ at the instant that the volume $v = 16$ cubic feet and the temperature T is 320° absolute and also if at that instant the volume is increasing at 0.8 cubic foot per minute and the temperature is increasing at 0.8 degree per minute?

Given $k = 50$, $v = 16$, $T = 320$.

Then $pv = kT$ or $p = \dfrac{kT}{v} = \dfrac{50 \cdot 320}{16} = 1{,}000$

lb./sq. ft.

Now $\dfrac{\partial p}{\partial T} = \dfrac{k}{v} = \dfrac{50}{16}$

and $\dfrac{\partial p}{\partial v} = -\dfrac{kT}{v^2} = -\dfrac{50 \cdot 320}{(16)^2} = -\dfrac{16{,}000}{(16)^2}$

Given $\dfrac{dv}{dt} = 0.8$ and $\dfrac{dT}{dt} = 0.8$ (where t = time). Therefore the total derivative

$$\frac{dp}{dt} = \frac{\partial p}{\partial T} \cdot \frac{dT}{dt} + \frac{\partial p}{\partial v} \cdot \frac{dv}{dt}$$

or $\dfrac{dp}{dt} = \dfrac{50}{16} \cdot 0.8 + \left(-\dfrac{16{,}000}{(16)^2} \cdot 0.8 \right) = 2.5 - 50$

$$= -47.5 \text{ lb./sq. ft.}$$

Therefore, the pressure is decreasing at the rate of **47.5** lb./sq. ft. each minute.

PROBLEMS

1. What is the total differential of $z = y^{\frac{x}{3}}$?

2. Find the total differential of $y = x^3 \sin \theta$.

3. Find the total differential of $y = (\sin \theta)^x$.

4. Find the total differential of $z = \dfrac{\log_\epsilon x}{y}$.

5. Differentiate $\dfrac{2x^3}{5} + 3x^3y - 5xy^2 + \dfrac{y}{6}$ with respect to x alone and with respect to y alone.

6. What is the maximum or minimum of $z = \dfrac{\epsilon^{x-y}}{xy}$?

7. What is the maximum or minimum of

$$z = -2y - 3x + 5 \log_\epsilon x + 3 \log_\epsilon y?$$

8. What are the partial derivatives, the partial differentials, and the total differential of $z = 5xy$?

9. What are the partial derivatives, the partial differentials, and the total differential of $y = \sin (3\theta + 4\alpha)$?

What are the partial derivatives of the following?

10. $y = \epsilon^x \sin \theta$

11. $y = u^{\log_\epsilon v}$

12. $z = \log (\epsilon^{2x} + \epsilon^{2y})$

13. $z = 3x^3y^3 + 4xy^2 - 5xy$

14. If $z = x^2 + y^2$, what are the second partial differentials $\dfrac{\partial^2 z}{\partial x^2}$ and $\dfrac{\partial^2 z}{\partial y^2}$?

15. If $z = \sqrt{x^2 + y^2}$, show that $\dfrac{\partial^2 z}{\partial y \cdot \partial x} = \dfrac{\partial^2 z}{\partial x \cdot \partial y}$.

16. The altitude of a variable right circular cone at a certain instant is 60 in. and is decreasing at the rate of 3 in. a second, while the radius of the base is 6 in. and is increasing at the rate of 3 in. a second. What is the rate at which the volume is changing?

17. If the variable base of a right triangle is x, the variable hypotenuse is z, and the variable angle is θ between x and z and if $z = x \sec \theta$, what are $\dfrac{\partial z}{\partial x}$ and $\dfrac{\partial z}{\partial \theta}$?

18. What is the greatest volume that can be sent by parcel post in the case of a package of circular cross section if the regulations state that the length plus the girth shall not exceed 6 ft.?

CHAPTER **XIX**

CURVATURE OF CURVES

356. What is meant by the curvature of a curve?

As a point moves along a curve, there is a continual change of direction. This change of direction or deviation from a straight line is called the curvature of a curve.

357. Upon what does the shape of a curve at a point depend?

The shape of a curve at a point depends upon the rate of change of direction, which is called the *curvature at the point*.

358. What is meant by the total curvature of an arc?

The total curvature of an arc is its total change in direction.

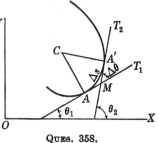

Example

The total curvature of arc AA' is the difference in its directions at A and at A', or

$\Delta\theta = \theta_2 - \theta_1 =$ the change in inclination of the tangent to the curve

QUES. 358.

359. What is meant by the average curvature of an arc?

The average curvature is the ratio of the total curvature to the length of arc.

$\dfrac{\Delta\theta}{\Delta s} =$ average curvature of arc AA' (see figure, Question 358)

$\Delta\theta$ is in radians and Δs is in linear units.

360. What is meant by the curvature of a curve at a point?

The curvature of a curve at a point is represented by K and is defined as $K = \lim\limits_{\Delta s \to 0} \left[\dfrac{\Delta \theta}{\Delta s} \right] = \dfrac{d\theta}{ds} =$ curvature at A. This means that the curvature at A is the limiting value of the average curvature when A' approaches A as a limiting position (see figure, Question 358).

361. What is the curvature at any point of a circle?

$\Delta \theta$ between the tangent lines at A and A' equals the central angle ACA'. Now, in any circle the central angle in radians times the radius in linear units equals the included arc in linear units, or

$$\text{Angle } ACA' \cdot R = \Delta s$$

Then
$$\frac{\Delta \theta}{\Delta s} = \frac{\text{angle } ACA'}{\Delta s} = \frac{\dfrac{\Delta s}{R}}{\Delta s} = \frac{1}{R}$$

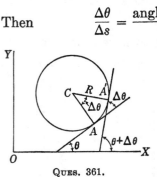

QUES. 361.

Passing to the limit,

$$K = \frac{d\theta}{ds} = \frac{1}{R}$$

From this it is seen that

a. The average curvature of an arc of a circle is the same as the curvature at any point of that circle.

b. The circle is a curve of constant or uniform curvature which is equal to the reciprocal of the radius.

c. The greater the radius of the circle the less the curvature, and the greater the curvature the less the radius.

362. What is zero curvature?

A straight line is of zero curvature because it is an arc of a circle of infinite radius.

$$K = \frac{1}{R} = \frac{1}{\infty} = \text{zero}$$

363. What is a unit of curvature?

In a circle, if R = radius = 10, the curvature K is $\frac{1}{10}$ radian for each unit length of arc.

If $R = 1$, the curvature K is 1 radian for each unit length of arc.

Therefore, unit curvature is the curvature of a circle of unit radius.

364. How is curvature determined in rectangular coordinates?

Since $\tan \theta$ is the slope of the tangent line,

$$\frac{dy}{dx} = \tan \theta$$

and

$$\theta = \text{arc} \tan \frac{dy}{dx}$$

Differentiating this expression we get

$$\frac{d\theta}{dx} = \frac{y''}{1 + (y')^2}$$

where $\qquad y' = \dfrac{dy}{dx} \qquad$ and $\qquad y'' = \dfrac{d^2y}{dx^2}$

or

$$d\theta = \frac{y''}{1 + (y')^2} \cdot dx \qquad (1)$$

But $\qquad (ds)^2 = (dx)^2 + (dy)^2$

Dividing this by $(dx)^2$, we get

$$\left(\frac{ds}{dx}\right)^2 = 1 + \left(\frac{dy}{dx}\right)^2 = 1 + (y')^2$$

or

$$ds = \sqrt{1 + (y')^2} \cdot dx \qquad (2)$$

Now divide Eq. (1) by Eq. (2).

Then

$$\frac{d\theta}{ds} = \frac{y'' \cdot dx}{1 + (y')^2} \cdot \frac{1}{[1 + (y')^2]^{\frac{1}{2}} \cdot dx}$$

$$\therefore K = \text{curvature} = \frac{d\theta}{ds} = \frac{y''}{[1 + (y')^2]^{\frac{3}{2}}} \qquad (3)$$

where $\qquad y' = \dfrac{dy}{dx},$ and $\qquad y'' = \dfrac{d^2y}{dx^2}$

Where the differentiation with respect to y is simpler,

$$K = \frac{-x''}{[1 + (x')^2]^{\frac{3}{2}}}$$

where $\qquad x' = \dfrac{dx}{dy}$ and $\qquad x'' = \dfrac{d^2x}{dy^2}$

365. What is the sign of K?

If we choose the positive sign in the denominator of Eq. (3) in Question 364, K and y'' will have like signs. This means that K is positive or negative according as the curve is concave upward or concave downward.

366. What is meant by the circle of curvature or the osculating circle?

A circle having a radius equal to the reciprocal of the curvature will have the same curvature as the curve at the

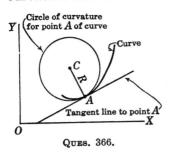

QUES. 366.

point considered. This circle is called the circle of curvature or osculating circle if it is placed tangent to the curve at the point considered and if it is so placed that the center of the circle is on the concave side of the curve. Each point of the curve has a different circle of curvature.

The circle of curvature can also be defined as the limiting position of a secant circle.

367. What is meant by the center of curvature and radius of curvature?

The center of the circle of curvature is called the center of curvature of the curve for the point.

The radius of the circle of curvature is called the radius of curvature of the curve for the point.

368. What is the expression for the radius of curvature?

$$R = \frac{1}{K} = \frac{[1 + (y')^2]^{\frac{3}{2}}}{y''}$$

The radius of curvature at a point on a curve is equal to the reciprocal of the curvature at that point.

369. How do we determine the coordinates x_1 and y_1 of the center of curvature?

x_1 and y_1 are constants. $(x - x_1)^2 + (y_1 - y)^2 = R^2$ is the equation of the circle of curvature.

Expanding this equation, we get

$$(x^2 - 2x_1x + x_1{}^2) + (y_1{}^2 - 2y_1y + y^2) = R^2$$

Differentiating to get rid of the constant R, the total differential is

$$(2x - 2x_1)\, dx + (2y - 2y_1)\, dy = 0$$
Factoring, $2(x - x_1)\, dx + 2(y - y_1)\, dy = 0$

Dividing by 2 dx,

$$(x - x_1) + (y - y_1)\frac{dy}{dx} = 0 \tag{1}$$

To eliminate one of the two remaining constants x_1 and y_1, differentiate again.

$$\frac{d}{dx}(x) + \frac{d}{dx}\left[(y - y_1)\frac{dy}{dx}\right] = 0$$

Now differentiate the product of the two terms.

$$1 + (y - y_1)\frac{d}{dx}\left(\frac{dy}{dx}\right) + \frac{dy}{dx}\cdot\frac{d}{dx}(y - y_1) = 0$$

$$1 + (y - y_1)\frac{d^2y}{dx^2} + \left(\frac{dy}{dx}\right)^2 = 0$$

or $$-(y - y_1)\frac{d^2y}{dx^2} = 1 + \left(\frac{dy}{dx}\right)^2$$

And now divide each side by $\dfrac{d^2y}{dx^2}$.

$$\therefore y_1 = y + \frac{1 + \left(\dfrac{dy}{dx}\right)^2}{\dfrac{d^2y}{dx^2}} = \text{the } y \text{ coordinate}$$

Now substitute in Eq. (1).

$$x - x_1 + \left[y - y - \frac{1 + \left(\dfrac{dy}{dx}\right)^2}{\dfrac{d^2y}{dx^2}} \right] \cdot \frac{dy}{dx} = 0$$

$$\therefore x_1 = x - \frac{\left[1 + \left(\dfrac{dy}{dx}\right)^2\right] \cdot \dfrac{dy}{dx}}{\dfrac{d^2y}{dx^2}} = \text{the } x \text{ coordinate}$$

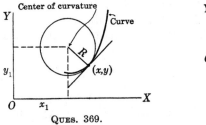

QUES. 369.

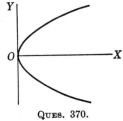

QUES. 370.

370. What is the curvature of the parabola $y^2 = 2x$, when $x = 2$?

$$\text{Curvature} = K = \frac{1}{R} = \frac{\dfrac{d^2y}{dx^2}}{\left[1 + \left(\dfrac{dy}{dx}\right)^2\right]^{\frac{3}{2}}}$$

Given $y^2 = 2x$, or $y = 2^{\frac{1}{2}} \cdot x^{\frac{1}{2}}$.   Then

$$\frac{dy}{dx} = 2^{\frac{1}{2}} \cdot \frac{1}{2} x^{-\frac{1}{2}}$$

or
$$= 2^{-\frac{1}{2}} \cdot x^{-\frac{1}{2}}$$

and $\qquad \dfrac{d^2y}{dx^2} = 2^{-\frac{1}{2}} \cdot -\dfrac{1}{2} \cdot x^{-\frac{3}{2}} = -(2x)^{-\frac{3}{2}}$

$$\therefore K = \dfrac{-(2x)^{-\frac{3}{2}}}{\left[1 + \dfrac{1}{2x}\right]^{\frac{3}{2}}} = \dfrac{(-2x)^{-\frac{3}{2}}}{\dfrac{(2x+1)^{\frac{3}{2}}}{(2x)^{\frac{3}{2}}}} = \dfrac{-(2x)^{-\frac{3}{2}} \cdot (2x)^{\frac{3}{2}}}{(2x+1)^{\frac{3}{2}}}$$

or $\qquad K = \dfrac{-1}{(2x+1)^{\frac{3}{2}}}$

and at $x = 2$

$$K = -\dfrac{1}{(2\cdot2+1)^{\frac{3}{2}}} = -\dfrac{1}{\sqrt{5^3}} = -\dfrac{1}{\sqrt{125}} = -\dfrac{1}{11.18}$$

or $K = -0.0894$ radian per unit $\qquad$ at $\qquad x = 2$

371. What is the radius of curvature of the above?

$$\text{Curvature} = K = \dfrac{1}{R} = -\dfrac{1}{11.18}$$
$$\therefore R = 11.18 \text{ units} = \text{radius of curvature}$$

372. What are the coordinates of the center of curvature of the above?

$$x_1 = x - \dfrac{\left[1 + \left(\dfrac{dy}{dx}\right)^2\right]\dfrac{dy}{dx}}{\dfrac{d^2y}{dx^2}} = 2 - \dfrac{\left(1 + \dfrac{1}{2x}\right)\dfrac{1}{(2x)^{\frac{1}{2}}}}{-(2x)^{-\frac{3}{2}}}$$

$$= 2 - \dfrac{\dfrac{(2x+1)}{2x} \cdot \dfrac{1}{(2x)^{\frac{1}{2}}}}{-(2x)^{-\frac{3}{2}}}$$

$$\therefore x_1 = 2 - \dfrac{(2x+1)}{(2x)^{\frac{3}{2}}} \cdot -\dfrac{1}{(2x)^{-\frac{3}{2}}} = 2 + (2x+1) = 2$$
$$+ 2\cdot2 + 1 = 7$$

$$y_1 = y + \dfrac{1 + \left(\dfrac{dy}{dx}\right)^2}{\dfrac{d^2y}{dx^2}} = \pm2 + \dfrac{1 + \dfrac{1}{2x}}{-(2x)^{-\frac{3}{2}}} \qquad \text{(If } x = 2,$$

$$y = \pm2 \text{ in } y^2 = 2x)$$

$$= \pm 2 + \frac{(2x + 1)}{(2x)} \cdot - \frac{1}{(2x)^{-\frac{1}{2}}} = \pm 2 - \frac{(2x + 1)}{(2x)^{-\frac{1}{2}}}$$
$$= \pm 2 - (2x + 1)(2x)^{\frac{1}{2}}$$
$$= \pm 2 - (2 \cdot 2 + 1)(2 \cdot 2)^{\frac{1}{2}} = \pm 2 - 5 \cdot (\pm 2)$$
$$= \pm 2 \mp 10$$

$$\therefore \ y_1 = \pm 8$$

373. What is a point of inflection?

A point where the curve changes from convex to concave, or vice versa.

374. What is the criterion for a point of inflection?

A point of inflection is at $\frac{d^2y}{dx^2} = 0$ in the figure where the curve is a bit of a straight line and the radius of curva-

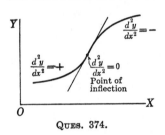

Ques. 374.

ture is infinitely great and when, to the left of the point, the curve is *convex downward* and $\frac{d^2y}{dx^2}$ (= the rate of change of slope) is *positive*, indicating that the slope is getting *greater upward* to the right; while, to the right of the point, the curve is *concave downward* and $\frac{d^2y}{dx^2}$ is *negative*, indicating that the slope is getting *smaller upward* to the right.

These conditions of $\frac{d^2y}{dx^2}$ on either side of the point of inflection may be interchanged.

375. What is the radius of curvature and what are the coordinates of the center of curvature of $y = x^2 - 2x + 1$ when $x = 1$?

$$\frac{dy}{dx} = 2x - 2 \quad \text{and} \quad \frac{d^2y}{dx^2} = 2$$

Then Radius $= R = \dfrac{\left[1 + \left(\dfrac{dy}{dx}\right)^2\right]^{\frac{3}{2}}}{\dfrac{d^2y}{dx^2}} = \dfrac{[1 + (2x - 2)^2]^{\frac{3}{2}}}{2}$

$$= \dfrac{[1 + (2 \cdot 1 - 2)^2]^{\frac{3}{2}}}{2} = \dfrac{1}{2}$$

and Coordinate $x_1 = x - \dfrac{\dfrac{dy}{dx}\left[1 + \left(\dfrac{dy}{dx}\right)^2\right]}{\dfrac{d^2y}{dx^2}}$

$$= x - \dfrac{(2x - 2)[1 + (2x - 2)^2]}{2}$$

$$= 1 - \dfrac{(2 \cdot 1 - 2)[1 + (2 \cdot 1 - 2)^2]}{2}$$

$$= 1 \qquad \text{(when } x = 1,\ y = 0)$$

Coordinate $y_1 = y + \dfrac{1 + \left(\dfrac{dy}{dx}\right)^2}{\dfrac{d^2y}{dx^2}} = 0 + \dfrac{1 + (2x - 2)^2}{2}$

$$= 0 + \dfrac{1 + (2 \cdot 1 - 2)^2}{2} = \dfrac{1}{2}$$

Check by substituting these values in the equation of the circle of curvature

or $(x - x_1)^2 + (y - y_1)^2 = (1 - 1)^2 + (0 - \frac{1}{2})^2 = 0$
$$+ \tfrac{1}{4} = R^2$$

$$\therefore R = \tfrac{1}{2} \text{ (as above)}$$

376. Where is the point of inflection of

$$y = 2x^3 - x^2 + 3x - 5?$$

$$\frac{dy}{dx} = 6x^2 - 2x + 3$$

$$\frac{d^2y}{dx^2} = 12x - 2, \qquad 12x - 2 = 0, \qquad \text{and} \qquad x = \frac{1}{6}$$

$$= 0.1667$$

Substitute this in the original equation, obtaining

$$y = 2(\tfrac{1}{6})^3 - (\tfrac{1}{6})^2 + 3(\tfrac{1}{6}) - 5 = 2(\tfrac{1}{216}) - \tfrac{1}{36} + \tfrac{1}{2} - 5$$

$$\text{or } y = \frac{1}{108} - \frac{1}{36} + \frac{1}{2} - 5 = \frac{1 - 3 + 54 - 540}{108}$$

$$= -4.5185+$$

Therefore, the point of inflection is at $x = \tfrac{1}{6}$, $y = -4.52$.

Now, for $x = \tfrac{5}{32} = 0.156$, which is a little less than $x = 0.166$ (the point of inflection),

$$\frac{d^2y}{dx^2} = 12x - 2 = 12 \cdot \frac{5}{32} - 2 = -$$

For $x = \tfrac{3}{16} = 0.1875$, which is a little greater than $x = 0.1667$,

$$\text{then} \qquad \frac{d^2y}{dx^2} = 12x - 2 = 12 \cdot \frac{3}{16} - 2 = +$$

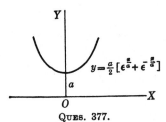

QUES. 377.

Therefore the criterion for a point of inflection is satisfied.

377. What is the radius and where is the center of curvature of $y = \frac{a}{2}\left(\varepsilon^{\frac{x}{a}} + \varepsilon^{-\frac{x}{a}}\right)$ when $x = 0$?

$$y = \frac{a}{2}\,\epsilon^{\frac{x}{a}} + \frac{a}{2}\,\epsilon^{-\frac{x}{a}}$$

$$\text{Then} \qquad \frac{dy}{dx} = \frac{a}{2} \cdot \frac{1}{a}\,\epsilon^{\frac{x}{a}} + \frac{a}{2} \cdot -\frac{1}{a}\,\epsilon^{-\frac{x}{a}} = \frac{1}{2}\left(\epsilon^{\frac{x}{a}} - \epsilon^{-\frac{x}{a}}\right)$$

$$\text{and} \qquad \frac{d^2y}{dx^2} = \frac{1}{2} \cdot \frac{1}{a}\,\epsilon^{\frac{x}{a}} - \frac{1}{2} \cdot -\frac{1}{a}\,\epsilon^{-\frac{x}{a}} = \frac{1}{2a}\left(\epsilon^{\frac{x}{a}} + \epsilon^{-\frac{x}{a}}\right)$$

$$\text{or} \qquad \frac{d^2y}{dx^2} = \frac{1}{2a} \cdot \frac{a}{a}\left(\epsilon^{\frac{x}{a}} + \epsilon^{-\frac{x}{a}}\right) = \frac{1}{2a} \cdot \frac{2y}{a} = \frac{y}{a^2}$$

$$\text{since} \qquad \left[2y = 2 \cdot \frac{a}{2}\left(\epsilon^{\frac{x}{a}} + \epsilon^{-\frac{x}{a}}\right)\right]$$

$$\text{Radius} = R = \frac{\left[1 + \left(\frac{dy}{dx}\right)^2\right]^{\frac{3}{2}}}{\frac{d^2y}{dx^2}} = \frac{\left\{1 + \left[\frac{1}{2}\left(\epsilon^{\frac{x}{a}} - \epsilon^{-\frac{x}{a}}\right)\right]^2\right\}^{\frac{3}{2}}}{\frac{y}{a^2}}$$

$$= \frac{a^2\left\{1 + \frac{1}{4}\left(\epsilon^{\frac{2x}{a}} - 2 + \epsilon^{-\frac{2x}{a}}\right)\right\}^{\frac{3}{2}}}{y}$$

$$= \frac{a^2\left(\dfrac{4 + \epsilon^{\frac{2x}{a}} - 2 + \epsilon^{-\frac{2x}{a}}}{4}\right)^{\frac{3}{2}}}{y}$$

$$= \frac{a^2}{8y}\sqrt{\left(2 + \epsilon^{\frac{2x}{a}} + \epsilon^{-\frac{2x}{a}}\right)^3}$$

$$= \frac{a^2}{8y}\sqrt{\left(2\epsilon^{\frac{x}{a}-\frac{x}{a}} + \epsilon^{\frac{2x}{a}} + \epsilon^{-\frac{2x}{a}}\right)^3} \text{ as } 2 = 2\epsilon^{\frac{x}{a}-\frac{x}{a}} = 2\epsilon^0 = 2$$

$$= \frac{a^2}{8y}\sqrt{\left(\epsilon^{\frac{x}{a}} + \epsilon^{-\frac{x}{a}}\right)^6} = \sqrt{\frac{a^4}{64y^2}\left(\epsilon^{\frac{x}{a}} + \epsilon^{-\frac{x}{a}}\right)^6}$$

$$= \sqrt{\frac{a^4 \cdot \left(\epsilon^{\frac{x}{a}} + \epsilon^{-\frac{x}{a}}\right)^6}{64 \cdot \dfrac{a^2}{4}\left(\epsilon^{\frac{x}{a}} + \epsilon^{-\frac{x}{a}}\right)^2}}$$

$$= \sqrt{\frac{a^2}{16}\left(\epsilon^{\frac{x}{a}} + \epsilon^{-\frac{x}{a}}\right)^4} = \sqrt{\frac{a^2}{16} \cdot \frac{a^2}{a^2}\left(\epsilon^{\frac{x}{a}} + \epsilon^{-\frac{x}{a}}\right)^4}$$

$$= \sqrt{\frac{a^4}{16}\left(\epsilon^{\frac{x}{a}} + \epsilon^{-\frac{x}{a}}\right)^4 \cdot \frac{1}{a^2}}$$

Now

$$y^4 = \frac{a^4}{16}\left(\epsilon^{\frac{x}{a}} + \epsilon^{-\frac{x}{a}}\right)^4$$

$$\therefore R = \sqrt{\frac{y^4}{a^2}} = \frac{y^2}{a}$$

When $x = 0$,

$$y = \frac{a}{2}\left(\epsilon^{\frac{x}{a}} + \epsilon^{-\frac{x}{a}}\right) = \frac{a}{2}\left(\epsilon^{\frac{0}{a}} + \epsilon^{-\frac{0}{a}}\right) = \frac{a}{2}\left(\epsilon^0 + \epsilon^0\right)$$

$$= \frac{a}{2}(1 + 1) = a$$

and $\dfrac{dy}{dx} = \dfrac{1}{2}\left(\epsilon^{\frac{x}{a}} - \epsilon^{-\frac{x}{a}}\right) = \dfrac{1}{2}\left(\epsilon^0 - \epsilon^0\right) = 0$

Also, $\dfrac{d^2y}{dx^2} = \dfrac{1}{2a}\left(\epsilon^{\frac{x}{a}} + \epsilon^{-\frac{x}{a}}\right) = \dfrac{1}{2a}\left(\epsilon^0 + \epsilon^0\right) = \dfrac{1}{2a} \cdot 2 = \dfrac{1}{a}$

Finally, $R = \dfrac{y^2}{a} = \dfrac{a^2}{a} = a = $ radius of curvature at $x = 0$

and $x_1 = x - \dfrac{\dfrac{dy}{dx}\left[1 + \left(\dfrac{dy}{dx}\right)^2\right]}{\dfrac{d^2y}{dx^2}} = 0 - \dfrac{0(1 + 0)}{\dfrac{1}{a}} = 0$

$$= x \text{ coordinate}$$

$y_1 = y + \dfrac{1 + \left(\dfrac{dy}{dx}\right)^2}{\dfrac{d^2y}{dx^2}} = a + \dfrac{1 + 0}{\dfrac{1}{a}} = a + a = 2a$

$$= y \text{ coordinate}$$

PROBLEMS

1. What is the radius of curvature and what are the coordinates of the center of curvature of $y = 3x\left(\dfrac{4x}{7} - 2\right)$ at $x = 1$?

2. Find the radius and center of curvature of $y = x^3$ for $x = -1$ and $x = 2$.

3. Find the radius of curvature and the coordinates of the center of curvature of $y = 2x^3 - 3x + 1$ at $x = 1$.

4. Find the coordinates of the point or points of inflection of $y = 2x^3 + 3x^2 + 1$.

5. What are the radius and center of curvature of

$$y = x^3 + 2x^2 + x + 1$$

at $x = 2$, and where is the point of inflection?

6. What is the radius of curvature and where is the center of curvature for $y = \epsilon^x$ for $x = 1$?

7. Find the radius and center of curvature of $y = \sin \theta$ at $\theta = \dfrac{\pi}{2}$, and find the position of the point of inflection.

8. Find the radius and center of curvature for $y = \cos \theta$ at $\theta = \dfrac{\pi}{2}$.

9. Find the curvature of the circle $a^2 = x^2 + y^2$.

10. If the radius of a circle $= 10$, what is the curvature for each unit length of arc? A circle of unit radius has what curvature?

11. What is the radius of curvature for the parabola $y^2 = 2x$ at $(x = 3)$?

12. Examine the curve $y = 2x^4 + 3x^3 + 4$ for points of inflection.

SECTION II

SIMPLE INTEGRAL CALCULUS

CHAPTER XX

INTEGRATION—FUNDAMENTAL IDEAS

378. What is the primary consideration of integral calculus?

Given the rate of change or growth of the dependent variable or function with respect to the independent variable and a value of the function at a particular instant, to find the function at any other value of the independent variable.

In symbols, this means: Given $\frac{dy}{dx} = f(x)$ or $dy = f(x)\ dx$ and a value of y corresponding to a particular value of x, to find the relation $y = F(x)$.

379. What is integration?

Integration is the process of finding the function when its derivative or its differential is given. It is the inverse process of differentiation.

Example

Given $\frac{dy}{dx} = 3x^2$, what is the function whose derivative is $3x^2$ or whose differential $dy = 3x^2 \cdot dx$?

Evidently, $y = x^3$ is such a function.

380. What is an integral?

The result of integration is the integral. In the above question, x^3 is the integral.

381. Would calculus be helpful in finding the value of a function when the rate of change of the function is constant?

Arithmetic is all that is required in such a case.

Example

a. If an automobile travels at a constant rate of 30 m.p.h., then in 3 hr. it will travel

$3 \times 30 = 90$ miles   or $s = rt =$ rate $\times$ time

b. If a storage battery is being charged at a constant rate of r amp./hr. and it had an initial charge of C amp., then in t hr. the total charge is

$$I = rt + C$$

382. What is the usual symbol for integration?

$\int$ is the symbol for integration. It is called the *integral sign.* It is an elongated S and means that the differential following it is to be integrated.

$\int dy$ means that a function is to be found of which dy is the differential.

$\int x^2 \cdot dx$ means that a function is to be found of which $x^2 \cdot dx$ is the differential.

$\int f(x) \cdot dx$ is read "the integral of $f(x) \cdot dx$."

The integral sign is placed before the differential and not the derivative.

383. What is meant by the integrand?

The function after the integral sign is known as the integrand.

Example

In $y = \int 3x^2 \cdot dx$, $3x^2$ is the integrand.

384. What does integral imply?

It implies the total amount.

Integral of the rate means the total made from the rate.

385. What is meant by an indefinite integral?

The integral of $\int 3x^2 \cdot dx$ is x^3. But x^3 is not the only function of which the differential is $3x^2 \cdot dx$; $x^3 + 5$, $x^3 + \sin 45°$, $x^3 + $ (any constant) C all have $3x^2 \cdot dx$ for a differential.

$$\therefore \int 3x^2 \cdot dx = x^3 + C$$

and, in general,

$$\int f'(x)\ dx = f(x) + C$$

This is called an indefinite integral and the constant C is known as the *constant of integration*.

386. How is the constant of integration determined?

It is necessary to have some other information besides the rate of change or the differential of the function to find the constant of integration.

Example

What is the equation of a straight line having a slope of 3 and passing through the point (2, 4)?

The slope of any line is $\dfrac{dy}{dx}$.

Then $\qquad \dfrac{dy}{dx} = 3 \qquad$ or $\qquad dy = 3 \cdot dx$

and $y = \int 3 \cdot dx = 3x + C = $ the general equation of all
lines having a slope of 3

But only one line can pass through (2, 4). Therefore, substitute $x = 2$, $y = 4$ in $y = 3x + C$ to get C

or $\qquad 4 = 3 \cdot 2 + C \qquad$ and $\qquad C = -2$

Therefore $y = 3x - 2$ is the equation of the particular line that is required.

387. What is the equation of a curve that passes through (3, 4) and whose tangent line at any point has a slope equal to twice the abscissa of the point?

Given $\dfrac{dy}{dx} = 2x$, or $dy = 2x \cdot dx$.

Then $y = \int 2x \cdot dx = x^2 + C = $ the equation of all curves meeting the general condition

But only one curve can pass through $(3, 4)$ in the plane. Therefore, substitute $x = 3$, $y = 4$ in $y = x^2 + C$.

$$4 = 3^2 + C \qquad \text{and} \qquad C = -5$$

Therefore $y = x^2 - 5$ is the equation satisfying both conditions.

388. If the velocity of a falling body is $32t$ feet per second for any time t seconds after it starts to fall, what is the distance from the surface of the earth 5 seconds after it starts to fall from a height of 1,000 feet above the earth's surface?

Let $s = $ the distance in feet the body is above the earth's surface at any time. And since s is decreasing as t is increasing,

$$\frac{ds}{dt} = -32t \qquad \text{or} \qquad ds = -32t \cdot dt$$

and $\qquad\qquad s = \int -32t \cdot dt = -16t^2 + C \qquad\qquad$ (1)

To determine C we know that $s = 1,000$ when $t = 0$. Substitute in Eq. (1).

$$1,000 = -0 + C \qquad \text{or} \qquad C = 1,000$$

Then, at any time after the body starts to fall,

$$s = -16t^2 + 1,000$$

Now, when $t = 5$,

$$s = -16 \cdot 5^2 + 1,000 = 600 \text{ ft.}$$

Therefore at the end of 5 sec. the body is 600 ft. above the earth's surface.

389. In general, how is an indefinite integral found?

a. By inspection. Recognizing the function at once as belonging to the given differential.

b. By reversing the rules of differentiation.

c. By formulas.

d. By reference to a table of integrals.

PROBLEMS

What is the indefinite integral of the following? Check by differentiating.

1. $\int dx$

2. $\int dy$

3. $\int y \cdot dy$

4. $\int 5x \cdot dx$

5. $\int 8x \cdot dx$

6. $\int 5x^2 \cdot dx$

7. $\int 6x^2 \cdot dx$

8. $\int 6x^3 \cdot dx$

9. $\int 6x^5 \cdot dx$

10. $\int 2(x + 1)\, dx$

11. $\int 3t \cdot dt$

12. $\int 3(x + 1)^2 \cdot dx$

13. $\int 2(x^2 + 1)2x \cdot dx$

14. $\int 3(x^2 - 1)^2 \cdot 2x \cdot dx$

In the following what is y (the dependent variable)?

15. $dy = 6x \cdot dx$

16. $dy = -4x^3 \cdot dx$

17. $dy = -5x \cdot dx$

18. $dy = 3(x^2 - 3)^2 \cdot 2x \cdot dx$.   Check by differentiating.

19. What is the equation of a straight line that has a slope of $-\frac{1}{2}$ and passes through point (1, 3)?

20. What is the equation of a curve that passes through point (2, 3) and whose tangent line at any point has a slope equal to three times the abscissa of the point?

21. If the velocity of a falling body is $32t$ ft./sec. for any time t sec. after it starts to fall, what is the distance from the surface of the earth 10 sec. after it starts to fall from a height of 2,000 ft. above the earth's surface?

Chapter XXl

INTEGRATION—THE INVERSE OF DIFFERENTIATION

ELEMENTARY GENERAL RULES AND FORMS

390. What is the rule for the integral when the independent variable is raised to a power?

Increase the power by 1, and divide by the increased power.

In general, if $\dfrac{dy}{dx} = x^n$ or $dy = x^n \cdot dx$,

then $\quad y = \displaystyle\int x^n \cdot dx = \dfrac{x^{n+1}}{n+1} + C =$ the integral

This can be proved by working backward and differentiating.

$$y = \frac{x^{n+1}}{n+1} + C \qquad \text{or} \qquad \frac{dy}{dx} = \frac{n+1}{n+1} \cdot x^{n+1-1} = x^n$$

391. If $\dfrac{dy}{dx} = x^{-\frac{1}{2}}$, what is the integral?

$$dy = x^{-\frac{1}{2}} \cdot dx \text{ and } y = \frac{x^{-\frac{1}{2}+1}}{-\frac{1}{2}+1} = \frac{x^{\frac{1}{2}}}{\frac{1}{2}} = 2x^{\frac{1}{2}} + C$$
$$= \text{ the integral}$$

392. What is the rule for the integral when there is a multiplying constant in the derivative?

The constant remains as a multiplier of the integral.

In general, if $\dfrac{dy}{dx} = ax^n$ or $dy = ax^n \cdot dx$,

then $\quad y = a \displaystyle\int x^n \cdot dx = \dfrac{ax^{n+1}}{n+1} + C =$ the integral

393. If $\dfrac{dy}{dx} = 3x^2$, what is the integral?

$$dy = 3x^2 \cdot dx \quad\text{or}\quad y = \int 3x^2 \cdot dx = 3 \int x^2 \cdot dx$$

$$= 3 \cdot \frac{x^{2+1}}{2+1} = x^3 + C$$

Working backward and differentiating $y = x^3 + C$, we get

$$\frac{dy}{dx} = 3x^2$$

394. What is the form of the expression usually set up before integrating?

Before integrating, we usually set up the expression to be integrated in the differential-equation form.

Example

If $\dfrac{dy}{dx} = x^3$ is the expression to be integrated, then we set $dy = x^3 \cdot dx =$ the differential-equation form before integrating. This reads "Differential y equals x^3 times differential x." Now in integrating we add up all the dy's to get the whole, or y for the left-hand member, and we indicate by the integral sign that we desire to add up all the $x^3 \cdot dx$'s to get their whole, or, in symbols, $y = \int x^3 \cdot dx$.

395. If $\dfrac{dy}{dx} = x^{-3}$, what is the integral?

$$dy = x^{-3} \cdot dx = \text{differential-equation form}$$
$$\text{Then } y = \int x^{-3} \cdot dx = \frac{x^{-3+1}}{-3+1} = \frac{x^{-2}}{-2} = -\frac{1}{2x^2} + C$$
$$= \text{the integral}$$

396. What is the integral of $\dfrac{dy}{dx} = ax^9$?

$$dy = ax^9 \cdot dx \quad\text{or}\quad y = a\int x^9 \cdot dx$$
$$\text{Then } y = a\frac{x^{9+1}}{9+1} = \frac{ax^{10}}{10} + C = \text{the integral}$$

397. What is the rule for integrating a sum or difference of two or more differentials?

Take the sum or difference of the integrals of these differentials.

Example

If $\dfrac{dy}{dx} = x^3 \pm x^4$,

then $dy = x^3 \cdot dx \pm x^4 \cdot dx =$ the differential-equation form

and $y = \displaystyle\int x^3 \cdot dx \pm \int x^4 \cdot dx = \dfrac{x^4}{4} \pm \dfrac{x^5}{5} + C =$ the integral

398. What is the rule for the integral when there is an added constant term?

The constant reappears multiplied by the independent variable, in the integral.

In general, if $\dfrac{dy}{dx} = x^n \pm a,$

then $dy = x^n \cdot dx \pm a \cdot dx =$ the differential-equation form
and $y = \dfrac{x^{n+1}}{n+1} \pm ax + C =$ the integral

399. What is the integral if $\dfrac{dy}{dx} = 12x^9 + 3$?

$dy = 12x^9 \cdot dx + 3 \cdot dx$

$y = 12 \displaystyle\int x^9 \cdot dx + 3 \int dx = 12\dfrac{x^{10}}{10} + 3x = \dfrac{6x^{10}}{5}$
$+ 3x + C$

400. What is the integral of $\dfrac{dy}{dx} = (a+b)(x+2)$?

$dy = (a+b)(x+2)\,dx$
$y = (a+b)\!\int(x+2)\,dx = (a+b)(\!\int x \cdot dx + 2\!\int dx)$
$= (a+b)\left(\dfrac{x^2}{2} + 2x\right) = (a+b)\left(\dfrac{x^2+4x}{2}\right) + C =$ the integral

401. If $\dfrac{dy}{dx} = 8.60x^{1.15}$, what is the integral?

$$dy = 8.60x^{1.15} \cdot dx$$
$$y = 8.60 \int x^{1.15} \cdot dx = 8.60 \cdot \frac{x^{1.15+1}}{1.15 + 1} = \frac{8.60x^{2.15}}{2.15}$$
$$= 4x^{2.15} + C = \text{the integral}$$

402. If $\dfrac{dy}{dx} = x^{-1}$, what is the integral?

$$dy = x^{-1} \cdot dx$$
$$y = \int x^{-1} \cdot dx = \log_\epsilon x + C = \text{the integral}$$

This is *the only exception to the rule for a power, i.e.,* where the exponent of the independent variable is -1. This is readily seen, because when we differentiate $y = \log_\epsilon x$ we get x^{-1}.

403. What is the integral of the expression $\dfrac{dy}{dx} = a^{-x^2}$?

This expression cannot be integrated because a^{-x^2} has not yet been found to result from differentiating any function. Nothing can be integrated before the inverse process of differentiation has produced that which must be integrated.

404. If $\dfrac{dy}{dx} = \dfrac{1}{x + a}$, what is the integral?

$$dy = \frac{1}{x + a} \cdot dx$$
$$y = \int \frac{1}{x + a} \cdot dx = \log_\epsilon (x + a) + C = \text{the integral}$$

because, working backward, if $y = \log_\epsilon (x + a)$,

then $$x + a = \epsilon^y$$
and $$\frac{d(x + a)}{dy} = \epsilon^y = \frac{dx}{dy} = (x + a)$$

Then
$$\frac{dy}{dx} = \frac{1}{\dfrac{dx}{dy}} = \frac{1}{x + a}$$

which is the function to be integrated.

405. If $\dfrac{dy}{dx} = \epsilon^x$, what is the integral?

$$dy = \epsilon^x \cdot dx$$
$$y = \int \epsilon^x \cdot dx = \epsilon^x + C = \text{the integral}$$

which is the same as the derivative in this unique function.

406. If $\dfrac{dy}{dx} = \epsilon^{-x}$, what is the integral?

$$dy = \epsilon^{-x} \cdot dx$$
$$y = \int \epsilon^{-x} \cdot dx = -\epsilon^{-x} + C = \text{the integral}$$

Working backward, if $y = -\epsilon^{-x} = -\dfrac{1}{\epsilon^x}$

then $\dfrac{dy}{dx} = -\dfrac{\epsilon^x \cdot 0 - 1 \cdot \epsilon^x}{\epsilon^{2x}} = \epsilon^{-x}$, the function to be
integrated

407. If $\dfrac{dy}{dx} = \sin \dfrac{x}{2}$, what is the integral?

$$dy = \sin \frac{x}{2} \cdot dx = \text{the differential-equation form}$$

$$y = \int \sin \frac{x}{2} \cdot dx = -2 \cos \frac{x}{2} + C = \text{the integral}$$

408. If $\dfrac{dy}{dx} = \sin x$, what is the integral?

$dy = \sin x \cdot dx = \text{the differential-equation form}$
$y = \int \sin x \cdot dx = -\cos x + C = \text{the function that when}$
differentiated will give the function under the integral sign

409. If $\dfrac{dy}{dx} = \cos x$, what is the integral?

$dy = \cos x \cdot dx =$ the differential-equation form
$y = \int \cos x \cdot dx = \sin x + C =$ the integral

410. If $\dfrac{dy}{dx} = \cos \dfrac{x}{2}$, what is the integral?

$dy = \cos \dfrac{x}{2} \cdot dx$

$y = \int \cos \dfrac{x}{2} \cdot dx = 2 \sin \dfrac{x}{2} + C =$ the integral

411. If $\dfrac{dy}{dx} = \sec^2 x$, what is the integral?

$dy = \sec^2 x \cdot dx$
$y = \int \sec^2 x \cdot dx = \tan x + C =$ the integral

412. If $\dfrac{dy}{dx} = \operatorname{cosec}^2 x$, what is the integral?

$dy = \operatorname{cosec}^2 x \cdot dx$
$y = \int \operatorname{cosec}^2 x \cdot dx = - \cot x + C =$ the integral

413. If $\dfrac{dy}{dx} = \dfrac{1}{\sqrt{1 - x^2}}$, what is the integral?

$dy = \dfrac{1}{\sqrt{1 - x^2}} \cdot dx$

$y = \int \dfrac{1}{\sqrt{1 - x^2}} \cdot dx = \arcsin x + C =$ the integral

See the reverse operation for this function under Differential Calculus, Question 125.

414. If $\dfrac{dy}{dx} = 3 \cos^2 x \cdot \sin x$, what is the integral?

$y = \int 3 \cos^2 x \cdot \sin x \cdot dx = -(\cos x)^3 + C =$ the integral

This can also be proved by differentiating $-(\cos x)^3$ to get the expression after the integral sign.

415. If $\dfrac{dy}{dx} = 3 \sec^2 (3x)$, what is the integral?

$$y = \int 3 \sec^2 (3x) \cdot dx = \tan 3x + C = \text{the integral}$$

416. If $\dfrac{dy}{dx} = \cos (x + b)$, what is the integral?

$$y = \int \cos (x + b) \cdot dx = \sin (x + b) + C = \text{the integral}$$

417. If $\dfrac{dy}{dx} = \cos^2 x - \sin^2 x$, what is the integral?

$$y = \int (\cos^2 x - \sin^2 x) \cdot dx = \sin x \cdot \cos x + C = \text{the integral}$$

418. If $\dfrac{dy}{dx} = \log_e x$, what is the integral?

$$y = \int \log_e x \cdot dx = x(\log_e x - 1) + C = \text{the integral}$$

Working backward, if $y = x \log_e x - x$,

$$\frac{dy}{dx} = x \cdot \frac{1}{x} + 1 \cdot \log_e x - 1 = \log_e x$$

419. If $\dfrac{dy}{dx} = b^x$, what is the integral?

$$y = \int b^x \cdot dx = \frac{b^x}{\log_e b} + C = \text{the integral}$$

Working backward, if $y = \dfrac{b^x}{\log_e b}$,

$$\frac{dy}{dx} = \frac{1}{\log_e b} \cdot b^x \cdot \log_e b = b^x$$

420. If $\dfrac{dy}{dx} = \cos bx$, what is the integral?

$$y = \int \cos bx \cdot dx = \frac{1}{b} \cdot \sin bx + C = \text{the integral}$$

Working backward, if $y = \sin bx$,

then $$\frac{dy}{dx} = b \cos bx$$

and, to get $\cos bx$, differentiate $y = \dfrac{1}{b} \sin bx$

421. If $\dfrac{dy}{dx} = \cos^2 x$, what is the integral?

$$y = \int \cos^2 x \cdot dx$$

Now $\cos 2x = \cos^2 x - \sin^2 x = 2 \cos^2 x - 1$

or $\cos^2 x = \tfrac{1}{2}(\cos 2x + 1)$

and
$$y = \int \cos^2 x \cdot dx = \tfrac{1}{2}\int (\cos 2x + 1)\, dx$$
$$= \tfrac{1}{2}\int \cos 2x \cdot dx + \tfrac{1}{2}\int dx$$
$$= \frac{\sin 2x}{4} + \frac{x}{2} + C = \text{the integral}$$

PROBLEMS

Find the integrals of the following, and work backward to check:

1. $\int x^{-\frac{1}{2}} \cdot dx$

2. $\int 6x^3 \cdot dx$

3. $\int 5x^{-4} \cdot dx$

4. $\int (3x^2 - 4x^3)\, dx$

5. $\int (8x^3 - 7)\, dx$

6. $\int (a - c)^{\frac{1}{2}}(2x - 3)\, dx$

7. $\int \dfrac{4.3}{x^2} \cdot dx$

8. $\int \left(\dfrac{3}{x} + 4\right) dx$

9. $\int \left(\dfrac{5}{x} - 3x\right) dx$

10. $\int \dfrac{3}{x - b} \cdot dx$

11. $\int \left(\dfrac{2}{x + 3} + 4c\right) dx$

12. $\int 3\epsilon^x \cdot dx$

13. $\int 2\epsilon^{-x} \cdot dx$

14. $\int 5\epsilon^{-3x} \cdot dx$

15. $\int 4\epsilon^{-\frac{x}{2}} \cdot dx$

16. $\int (3 \sin x + 2)\, dx$

17. $\int \left(\dfrac{\sin x}{2} - 3x\right) dx$

18. $\int \left(4 \cos \dfrac{x}{2} - 3\right) dx$

19. $\int 2\pi n \sin 2\pi nt \cdot dt$

20. $\int 2\pi n \cos 2\pi nt \cdot dt$

21. $\int \sqrt{2} \sec^2 x \cdot dx$

22. $\int \dfrac{5}{\sin^2 x} \cdot dx$

23. $\int \dfrac{3}{\sqrt{1 - x^2}} \cdot dx$

24. $\int \sqrt[3]{27} \cos^2 x \cdot \sin x \cdot dx$

25. $\int 8 \sec^2 (4x) \cdot dx$

26. $\int 4 \cos^3 x \cdot \sin x \cdot dx$

27. $\int 4 \sec^2 (4x) \cdot dx$

28. $\int \sqrt{3} \sec^2 (\sqrt{3}x) \cdot dx$

29. $\int 3 \cos (x + 4) \cdot dx$

30. $\int 8 \sin (x - 2) \cdot dx$

31. $\int (3 \cos^2 x - 3 \sin^2 x) \, dx$

32. $\int \sqrt{2} \log_e x \cdot dx$

33. $\int 3a^x \cdot dx$

34. $\int 6 \cos 3x \cdot dx$

35. $\int 4 \cos^2 x \cdot dx$

36. A body is moving at the rate $\dfrac{dx}{dt} = \dfrac{5}{3} t^2$, where x is the distance in feet and t is time in seconds.   What is the distance it will move in 3 sec. if $x = 0$ when $t = 0$?

37. If a body is thrown vertically downward with a velocity of 20 ft./sec., how far will it have fallen at the end of 5 sec. if $v = gt + v_o$, where v_o = initial velocity?

NOTE.—$v = \dfrac{ds}{dt} = gt + v_o$, and $s = 0$ when $t = 0$.

38. The velocity $v = \dfrac{ds}{dt}$ with which a car starts and continues is given by $v = \frac{1}{6}t^2$.   What would be the formula for the distance s when it has moved t sec., and how far will it go in 5 seconds?

39. A wheel revolves with an angular velocity ω of $0.0012t^2$ radian/sec. How long will it take to make 2 revolutions?

NOTE.—$\dfrac{d\theta}{dt} = \omega = 0.0012t^2$, where θ = radians turned through. Also, $\theta = 0$ when $t = 0$.

CHAPTER XXII

INTEGRATION BY FUNDAMENTAL
FORMULAS

422. What are the fundamental formulas for integration of algebraic functions?

(I) $\quad \int u^n \cdot du = \dfrac{u^{n+1}}{n+1} + C$ when n is not -1

(II) $\quad \int u^{-1} \cdot du = \int \dfrac{du}{u} = \log_e u + C$

(III) $\int \dfrac{du}{u^2 + a^2} = \dfrac{1}{a} \tan^{-1} \dfrac{u}{a} \quad$ or $\quad -\dfrac{1}{a} \cot^{-1} \dfrac{u}{a} + C$

(IV) $\int \dfrac{du}{u^2 - a^2} = \dfrac{1}{2a} \cdot \log_e \dfrac{u-a}{u+a} + C \quad$ when $\quad u^2 > a^2$

(V) $\int \dfrac{du}{\sqrt{a^2 - u^2}} = \sin^{-1} \dfrac{u}{a} \quad$ or $\quad -\cos^{-1} \dfrac{u}{a} + C$

(VI) $\int \dfrac{du}{\sqrt{u^2 + a^2}} = \log_e (u + \sqrt{u^2 + a^2}) + C$

(VII) $\int \sqrt{a^2 - u^2} \cdot du = \dfrac{u}{2} \sqrt{a^2 - u^2} + \dfrac{a^2}{2} \sin^{-1} \dfrac{u}{a} + C$

(VIII) $\int \sqrt{u^2 \pm a^2} \cdot du = \dfrac{u}{2} \sqrt{u^2 \pm a^2}$

$\pm \dfrac{a^2}{2} \log_e (u + \sqrt{u^2 \pm a^2}) + C$

(IX) $\int \dfrac{du}{u \sqrt{u^2 - a^2}} = \dfrac{1}{a} \sec^{-1} \dfrac{u}{a} \quad$ or

$-\dfrac{1}{a} \operatorname{cosec}^{-1} \dfrac{u}{a} + C$

(X) $\int \dfrac{du}{\sqrt{2au - u^2}} = \sin^{-1} \dfrac{u-a}{a} \quad$ or $\quad \operatorname{vers}^{-1} \dfrac{u}{a}$

NOTE.—The constant of integration C of an indefinite integral may be either shown or omitted and understood.

423. What is the integration of $\int \left(\dfrac{3b}{\sqrt{x}} - \dfrac{2a}{x^2} + 4c \sqrt[3]{x^2} \right) dx$?

$$y = \int 3bx^{-\frac{1}{2}} \cdot dx - \int 2ax^{-2} \cdot dx + \int 4cx^{\frac{2}{3}} \cdot dx$$

By formula (I) and the rule for multiplying constants

$$y = 3b\int x^{-\frac{1}{2}} \cdot dx - 2a\int x^{-2} \cdot dx + 4c\int x^{\frac{2}{3}} \cdot dx$$

$$= 3b \cdot \frac{x^{-\frac{1}{2}+1}}{-\frac{1}{2}+1} - 2a \cdot \frac{x^{-2+1}}{-2+1} + 4c \cdot \frac{x^{\frac{2}{3}+1}}{\frac{2}{3}+1}$$

$$= 3b \cdot \frac{x^{\frac{1}{2}}}{\frac{1}{2}} - 2a \cdot \frac{x^{-1}}{-1} + 4c \cdot \frac{x^{\frac{5}{3}}}{\frac{5}{3}}$$

$$= 6b \sqrt{x} + \frac{2a}{x} + \frac{12c}{5} \sqrt[3]{x^5} + C$$

424. If $\dfrac{dy}{dx} = (2x^2 + 1)(x^2 - 3x + 2)$, what is the integral?

By formula (I). First multiply out the two factors.

$$\frac{dy}{dx} = 2x^4 - 6x^3 + 5x^2 - 3x + 2$$

$$\text{or } dy = \int (2x^4 - 6x^3 + 5x^2 - 3x + 2) \, dx$$

$$= 2\int x^4 \cdot dx - 6\int x^3 \cdot dx + 5\int x^2 \cdot dx - 3\int x \cdot dx$$
$$+ 2\int dx$$

$$\therefore y = \frac{2x^5}{5} - \frac{3x^4}{2} + \frac{5x^3}{3} - \frac{3x^2}{2} + 2x = \text{the integral}$$

425. What is the integral of $\displaystyle\int \dfrac{2x \cdot dx}{\sqrt[3]{5 - 4x^2}}$?

By formula (I), $\int u^n \cdot du$.

$$\int \frac{2x \cdot dx}{\sqrt[3]{5 - 4x^2}} = \frac{1}{4} \int \frac{4 \cdot 2x \cdot dx}{\sqrt[3]{5 - 4x^2}}$$

$$= -\frac{1}{4} \int (5 - 4x^2)^{-\frac{1}{3}} \cdot (-8x) \, dx$$

Here $u = (5 - 4x^2)$ and $du = -8x \cdot dx$.

$$\therefore \int \frac{2x \cdot dx}{\sqrt[3]{5 - 4x^2}} = -\frac{1}{4} \left[\frac{(5 - 4x^2)^{-\frac{1}{3}+1}}{-\frac{1}{3} + 1} \right]$$

$$= -\frac{3}{8} (5 - 4x^2)^{\frac{2}{3}} + C$$

To prove the integral, differentiate it

or $d[-\frac{3}{8}(5 - 4x^2)^{\frac{1}{3}} + C] = -\frac{3}{8} \cdot \frac{2}{3}(5 - 4x^2)^{\frac{2}{3}-1} \cdot (-8x)\,dx$

$$= \frac{2x \cdot dx}{\sqrt[3]{5 - 4x^2}}$$

which was to be integrated.

426. If $\dfrac{dy}{dx} = \dfrac{3x^4 + 12x^2 - 2x + 9}{x^2 + 3}$, what is the integral?

By formulas (I) and (II). First divide the numerator by the denominator

getting $dy = \displaystyle\int \left(3x^2 + 3 - \frac{2x}{x^2 + 3}\right) dx$

$$= 3 \int x^2 \cdot dx + 3 \int dx - \int \frac{2x}{x^2 + 3} \cdot dx$$

Here $u = x^2 + 3$ and $du = 2x \cdot dx$.

$$y = 3 \cdot \frac{x^3}{3} + 3x - \log_e (x^2 + 3)$$
$$\therefore y = x^3 + 3x - \log_e (x^2 + 3) + C = \text{the integral}$$

427. What is the integral of $\displaystyle\int \frac{2ax \cdot dx}{b^2 + 3c^2x^2}?$

By formula (II).

$$\int \frac{2ax \cdot dx}{b^2 + 3c^2x^2} = 2a \int \frac{x \cdot dx}{b^2 + 3c^2x^2}$$

Now multiply by $\dfrac{6c^2}{6c^2}$.

$$2a \int \frac{x \cdot dx}{b^2 + 3c^2x^2} = \frac{2a}{6c^2} \int \frac{6c^2x \cdot dx}{b^2 + 3c^2x^2}$$

$$= \frac{a}{3c^2} \log_e (b^2 + 3c^2x^2) + C$$

Here $u = (b^2 + 3c^2x^2)$ and $du = 6c^2x \cdot dx$.

428. What is the integral of $\int \dfrac{x^3 \cdot dx}{x+1}$?

By formulas (I) and (II).   First divide the numerator by the denominator.

$$\frac{x^3}{x+1} = x^2 - x + 1 - \frac{1}{x+1}$$

$$\therefore \int \frac{x^3 \cdot dx}{x+1} = \int x^2 \cdot dx - \int x \cdot dx + \int dx - \int \frac{dx}{x+1}$$

$$= \frac{x^3}{3} - \frac{x^2}{2} + x - \log_e (x+1) + C$$

429. What is the integral of $\int \dfrac{3x-1}{3x+4} \cdot dx$?

By formula (II).   Dividing,

$$\frac{3x-1}{3x+4} = 1 - \frac{5}{3x+4}$$

$$\int \frac{3x-1}{3x+4} \cdot dx = \int dx - 5 \int \frac{dx}{3x+4} = x - \frac{5}{3} \int \frac{3\,dx}{3x+4}$$

$$= x - \tfrac{5}{3} \log_e (3x+4) + C = \text{the integral}$$

430. What is the integral of $\int \dfrac{dx}{9x^2 + 16}$?

By formula (III), $\int \dfrac{du}{u^2 + a^2}$.   Here $u^2 = 9x^2$ and $a^2 = 16$; then $u = 3x$, and $du = 3 \cdot dx$, and $a = 4$.
Now multiply by $\tfrac{3}{3}$.

$$\therefore \int \frac{dx}{9x^2 + 16} = \frac{1}{3} \int \frac{3 \cdot dx}{(3x)^2 + (4)^2} = \frac{1}{12} \tan^{-1} \frac{3x}{4} + C$$

$$= \text{the integral}$$

431. What is the integral of $\dfrac{dy}{dx} = \dfrac{1}{x^2 + 6x + 25}$?

By formula (III).   Complete the square of the denominator to get the form $\int \dfrac{du}{u^2 + a^2}$.

Then $dy = \int \dfrac{dx}{x^2 + 6x + 9 + 16} = \int \dfrac{dx}{(x+3)^2 + (4)^2}$

Here $u = x + 3$, $du = dx$, and $a = 4$.

$y = \dfrac{1}{a} \tan^{-1} \dfrac{u}{a} = \dfrac{1}{4} \tan^{-1} \dfrac{x+3}{4} + C =$ the integral

432. What is the integral of $\int \dfrac{dx}{9x^2 - 4}$?

By formula (IV), $\int \dfrac{du}{u^2 - a^2}$. Here $u = 3x$, $du = 3 \cdot dx$, $a = 2$.

$\therefore \int \dfrac{dx}{9x^2 - 4} = \dfrac{1}{3} \int \dfrac{3 \cdot dx}{(3x)^2 - (2)^2} = \dfrac{1}{3} \cdot \dfrac{1}{2 \cdot 2} \cdot \log_\epsilon \dfrac{3x - 2}{3x + 2}$

$\qquad = \dfrac{1}{12} \log_\epsilon \dfrac{3x - 2}{3x + 2} + C =$ the integral

433. What is the integral of $\int \dfrac{dx}{4x^2 + 4x - \frac{21}{4}}$?

By formula (IV), $\int \dfrac{du}{u^2 - a^2}$.

Now $4x^2 + 4x - \frac{21}{4} = 4(x^2 + x - \frac{21}{16})$
$\qquad\qquad\qquad\qquad = 4(x^2 + x + \frac{1}{4} - \frac{25}{16})$
$\qquad\qquad\qquad\qquad = 4[(x + \frac{1}{2})^2 - (\frac{5}{4})^2]$

Here $u = (x + \frac{1}{2})$, $du = dx$, and $a = \frac{5}{4}$.

Then $\int \dfrac{dx}{4x^2 + 4x - \frac{21}{4}} = \int \dfrac{dx}{4[(x + \frac{1}{2})^2 - (\frac{5}{4})^2]}$

$\qquad\qquad\qquad = \dfrac{1}{4} \int \dfrac{du}{u^2 - a^2} = \dfrac{1}{4} \cdot \dfrac{1}{2a} \log_\epsilon \dfrac{u - a}{u + a}$

$\therefore \int \dfrac{dx}{4[(x + \frac{1}{2})^2 - (\frac{5}{4})^2]} = \dfrac{1}{4} \cdot \dfrac{1}{2 \cdot \frac{5}{4}} \cdot \log_\epsilon \dfrac{(x + \frac{1}{2}) - \frac{5}{4}}{(x + \frac{1}{2}) + \frac{5}{4}}$

$\qquad\qquad\qquad\qquad\qquad = \dfrac{1}{10} \log_\epsilon \dfrac{x - \frac{3}{4}}{x + \frac{7}{4}}$

$\qquad\qquad\qquad\qquad = \dfrac{1}{10} \log_\epsilon \dfrac{4x - 3}{4x + 7} + C$

434. What is the integral of $\dfrac{dy}{dx} = \dfrac{1}{\sqrt{\dfrac{14}{25} + \dfrac{12x}{5} - 2x^2}}$?

By formula (V), $\displaystyle\int \frac{du}{\sqrt{a^2 - u^2}}$. First factor out the coefficient of x^2, and then complete the square.

$$y = \int \frac{dx}{\sqrt{\dfrac{14}{25} + \dfrac{12x}{5} - 2x^2}}$$

$$= \frac{1}{\sqrt{2}} \int \frac{dx}{\sqrt{\dfrac{7}{25} + \dfrac{9}{25} - \left(x^2 - \dfrac{6x}{5} + \dfrac{9}{25}\right)}}$$

$$= \frac{1}{\sqrt{2}} \int \frac{dx}{\sqrt{(\tfrac{4}{5})^2 - (x - \tfrac{3}{5})^2}}$$

Here $a = \tfrac{4}{5}$ and $u = (x - \tfrac{3}{5})$.

$$\therefore\ y = \frac{1}{\sqrt{2}} \sin^{-1} \frac{(x - \tfrac{3}{5})}{\tfrac{4}{5}} = \frac{1}{\sqrt{2}} \sin^{-1} \frac{(5x - 3)}{4} + C$$

435. What is the integral of $\displaystyle\int \frac{2x - 1}{\sqrt{9x^2 + 16}} \cdot dx$?

By formula (VI), $\displaystyle\int \frac{du}{\sqrt{u^2 + a^2}}$.

$$\int \frac{2x - 1}{\sqrt{9x^2 + 16}} \cdot dx = \int \frac{2x \cdot dx}{\sqrt{9x^2 + 16}} - \int \frac{dx}{\sqrt{9x^2 + 16}}$$

Now $\displaystyle\int \frac{2x \cdot dx}{\sqrt{9x^2 + 16}} = \int 2x(9x^2 + 16)^{-\frac{1}{2}} \cdot dx$

$$= \frac{1}{9} \int (9x^2 + 16)^{-\frac{1}{2}} \cdot 9 \cdot 2x \cdot dx$$

$$= \frac{1}{9} \cdot \frac{(9x^2 + 16)^{-\frac{1}{2}+1}}{-\frac{1}{2} + 1} = \frac{2}{9} \sqrt{9x^2 + 16}$$

and $\displaystyle\int \frac{dx}{\sqrt{9x^2 + 16}} = \frac{1}{3} \int \frac{3 \cdot dx}{\sqrt{9x^2 + 16}}$

$$= \frac{1}{3} \log_\epsilon (3x + \sqrt{9x^2 + 16})$$

Here $u = 3x$, and $a = 4$.

$$\therefore \int \frac{2x - 1}{\sqrt{9x^2 + 16}} \cdot dx = \frac{2}{9} \sqrt{9x^2 + 16}$$

$$- \frac{1}{3} \log_e (3x + \sqrt{9x^2 + 16}) + C$$

436. What is the integral of $\int \sqrt{16 - 9x^2} \cdot dx$?

By formula (VII), $\int \sqrt{a^2 - u^2} \cdot du$. Here $a^2 = 16$, $u = 3x$, and $du = 3\ dx$.

Then $\int \sqrt{16 - 9x^2} \cdot dx = \frac{1}{3}\int \sqrt{16 - 9x^2} \cdot 3 \cdot dx$

$$= \frac{1}{3}\int \sqrt{a^2 - u^2} \cdot du$$

$$= \frac{1}{3} \cdot \frac{3x}{2} \sqrt{16 - 9x^2}$$

$$+ \frac{1}{3} \cdot \frac{16}{2} \cdot \sin^{-1} \cdot \frac{3x}{4}$$

$$= \frac{x}{2} \sqrt{16 - 9x^2} + \frac{8}{3} \sin^{-1} \frac{3x}{4}$$

$$+ C = \text{the integral}$$

437. What is the integral of $\int \sqrt{9x^2 - 16} \cdot dx$?

By formula (VIII), $\int \sqrt{u^2 - a^2} \cdot du$. Here $u^2 = 9x^2$, $u = 3x$, $du = 3 \cdot dx$, and $a^2 = 16$.

$\int \sqrt{9x^2 - 16} \cdot dx = \frac{1}{3}\int \sqrt{9x^2 - 16} \cdot 3\ dx$

$$= \frac{1}{3}\int \sqrt{u^2 - a^2} \cdot du$$

$$= \frac{1}{3} \cdot \frac{u}{2} \sqrt{u^2 - a^2}$$

$$- \frac{1}{3} \cdot \frac{a^2}{2} \log_e (u + \sqrt{u^2 - a^2})$$

$$= \frac{1}{3} \cdot \frac{3x}{2} \sqrt{9x^2 - 16}$$

$$- \frac{1}{3} \cdot \frac{16}{2} \log_e (3x + \sqrt{9x^2 - 16})$$

$$= \frac{x}{2} \sqrt{9x^2 - 16}$$

$$- \frac{8}{3} \log_e (3x + \sqrt{9x^2 - 16}) + C$$

438. What is the integral of $\displaystyle\int \frac{dx}{x\sqrt{9x^2 - 16}}$?

By formula (IX), $\displaystyle\int \frac{du}{u\sqrt{u^2 - a^2}}$. Here $u^2 = 9x^2$, $u = 3x$, $du = 3 \cdot dx$, and $a^2 = 16$.

$$\therefore \int \frac{3 \cdot dx}{3x\sqrt{9x^2 - 16}} = \int \frac{du}{u\sqrt{u^2 - a^2}} = \frac{1}{a}\sec^{-1}\frac{u}{a}$$

$$= \frac{1}{4}\sec^{-1}\frac{3x}{4} + C = \text{the integral}$$

439. What is the integral of $\displaystyle\frac{dy}{dx} = \frac{x}{\sqrt{4x - x^2}}$?

By formula (X), $\displaystyle\int \frac{du}{\sqrt{2au - u^2}} = \sin^{-1}\frac{u - a}{a}$.

$$y = \int \frac{x \cdot dx}{\sqrt{4x - x^2}} = \int \frac{[2 - (2 - x)]\,dx}{\sqrt{4x - x^2}}$$

$$= \int \frac{2\,dx}{\sqrt{4x - x^2}} - \int \frac{(2 - x)\,dx}{\sqrt{4x - x^2}}$$

$$= 2\int \frac{dx}{\sqrt{4x - x^2}} - \frac{1}{2}\int (4x - x^2)^{-\frac{1}{2}}(4 - 2x)\,dx$$

Here $a = 2$, $u = x$, and $du = dx$.

$$\therefore y = 2\sin^{-1}\frac{x - 2}{2} - \frac{1}{2}\cdot\frac{(4x - x^2)^{-\frac{1}{2}+1}}{-\frac{1}{2} + 1}$$

$$= 2\sin^{-1}\frac{x - 2}{2} - (4x - x^2)^{\frac{1}{2}} + C$$

PROBLEMS

Find the integrals of the following, using formulas (I) to (X):

1. $\displaystyle\int \left(\frac{4}{\sqrt[3]{x}} - \frac{3}{5x^2} - 8a\,\sqrt[4]{x^2}\right)dx$

1. $\int (4x^2 + 3)(2x^2 - 5x + 3)\,dx$

3. $\int \dfrac{2x \cdot dx}{\sqrt[4]{3x^2 - 7}}$

4. $\int \dfrac{5x^4 + 22x^2 - 2x + 8}{x^2 + 4} \cdot dx$

5. $\int \dfrac{3a \cdot dx}{4 + 5b^2x^2}$

6. $\int \dfrac{2x^3}{x + 1} \cdot dx$

7. $\int \dfrac{12x - 15}{4x - 3} \cdot dx$

8. $\int \dfrac{dx}{4x^2 + 25} \cdot dx$

9. $\int \dfrac{dx}{x^2 + 2x + 20}$

10. $\int \dfrac{dx}{4x^2 - \frac{25}{4}}$

11. $\int \dfrac{dx}{3x^2 + 2x - 8}$

12. $\int \dfrac{dx}{\sqrt{\frac{17}{48} + 4x - 3x^2}}$

13. $\int \dfrac{4x}{\sqrt{75x^4 + 12}} \cdot dx$

14. $\int \sqrt{12 - 48x^2} \cdot dx$

15. $\int \sqrt{36x^2 - 100} \cdot dx$

16. $\int \dfrac{dx}{x\sqrt{4x^2 - 25}}$

17. $\int \dfrac{x \cdot dx}{\sqrt{8x^2 - 9x^4}}$

440. What are the fundamental formulas for integration of exponential functions?

(XI) $\qquad \int a^u \cdot du = \dfrac{a^u}{\log_\epsilon a} + C$

(XII) $\qquad \int \epsilon^u \cdot du = \epsilon^u + C$

441. What is the integral of $\int ba^{3x} \cdot dx$?

By formula (XI), $\int a^u \cdot du$. Here $u = 3x$ and $du = 3 \cdot dx$. Multiply the function by $\frac{2}{3}$ to bring it into form.

$$\therefore \int ba^{3x} \cdot dx = \frac{b}{3} \int a^{3x} \cdot 3 \cdot dx = \frac{b}{3} \int a^{3x} \cdot d(3x)$$

$$= \frac{b}{3} \int a^u \cdot du = \frac{b}{3} \cdot \frac{a^u}{\log_\epsilon a} = \frac{b}{3} \cdot \frac{a^{3x}}{\log_\epsilon a} + C$$

442. What is the integral of $\int \dfrac{\epsilon^{3x} \cdot dx}{\epsilon^x + 1}$?

By formulas (XII) and (II). First divide ϵ^{3x} by $\epsilon^x + 1$, getting

$$\epsilon^{2x} - \epsilon^x + \frac{\epsilon^x}{\epsilon^x + 1}$$

$$\therefore \int \frac{\epsilon^{3x} \cdot dx}{\epsilon^x + 1} = \int \epsilon^{2x} \cdot dx - \int \epsilon^x \cdot dx + \int \frac{\epsilon^x}{\epsilon^x + 1} \cdot dx$$

$$= \frac{1}{2} \int \epsilon^{2x} \cdot 2dx - \int \epsilon^x \cdot dx + \int \frac{\epsilon^x}{\epsilon^x + 1} \cdot dx$$

$$= \tfrac{1}{2}\epsilon^{2x} - \epsilon^x + \log_\epsilon (\epsilon^x + 1) + C$$

443. What is the result of $\int a^x b^x \cdot dx$?

By formula (XI).

$$\int a^x \cdot b^x \cdot dx = \int (ab)^x \cdot dx = \frac{(ab)^x}{\log_\epsilon (ab)}$$

$$= \frac{a^x \cdot b^x}{\log_\epsilon a + \log_\epsilon b} + C$$

444. What is $\int \epsilon^{\cos x} \cdot \sin x \cdot dx$?

By formula (XII), $\int \epsilon^u \cdot du = \epsilon^u$. Here $u = \cos x$, and $du = -\sin x \cdot dx$.

$$\therefore -\int \epsilon^{\cos x} (-\sin x \cdot dx) = -\int \epsilon^u \cdot du = -\epsilon^u = -\epsilon^{\cos x} + C$$

445. What is $\int a^{6x} \cdot dx$?

By formula (XI), $\displaystyle \int a^u \cdot du = \frac{a^u}{\log_\epsilon a}.$ Here $u = 6x$, and $du = 6 \cdot dx$.

$$\therefore \int a^{6x} \cdot dx = \frac{1}{6} \int a^{6x} \cdot 6dx = \frac{1}{6} \cdot \frac{a^{6x}}{\log_\epsilon a} + C$$

PROBLEMS

Integrate the following by formulas (XI) and (XII):

1. $\int 6c^{5x} \cdot dx$

2. $\int c^{3x} \cdot d^{3x} \cdot dx$

3. $\int \dfrac{\epsilon^{4x} \cdot dx}{\epsilon^x + 1}$

4. $\int \epsilon^{\sin x} \cdot \cos x \cdot dx$

5. $\int a^{4x} \cdot dx$

6. $\int \epsilon^{\tan x} \cdot \sec^2 x \cdot dx$

7. $\int a^{bx+c} \cdot dx$

8. $\int \epsilon^{x^4} \cdot x^3 \cdot dx$

9. $\int \epsilon^{-x^2} \cdot x \cdot dx$

10. $\int \epsilon^{\sin 3x} \cdot \cos 3x \cdot dx$

446. What are the fundamental formulas for integration of trigonometric functions?

(XIII) $\int \sin u \cdot du = -\cos u + C$

(XIV) $\int \cos u \cdot du = \sin u + C$

(XV) $\quad \int \tan u \cdot du = - \log_\epsilon \cos u = \log_\epsilon \sec u + C$

(XVI) $\quad \int \cot u \cdot du = \log_\epsilon \sin u = - \log_\epsilon \operatorname{cosec} u + C$

(XVII) $\quad \int \sec u \cdot du = \log_\epsilon (\sec u + \tan u) \qquad$ or

$$\log_\epsilon \tan \left(\frac{u}{2} + \frac{\pi}{4}\right) + C$$

(XVIII) $\quad \int \operatorname{cosec} u \cdot du = \log_\epsilon (\operatorname{cosec} u - \cot u) \qquad$ or

$$\log_\epsilon \tan \frac{u}{2} + C$$

(XIX) $\qquad \int \sec^2 u \cdot du = \tan u + C$

(XX) $\qquad \int \operatorname{cosec}^2 u \cdot du = - \cot u + C$

(XXI) $\qquad \int \sec u \cdot \tan u \cdot du = \sec u + C$

(XXII) $\qquad \int \operatorname{cosec} u \cdot \cot u \cdot du = - \operatorname{cosec} u + C$

447. What is $\int \sin 3ax \cdot dx$?

By formula (XIII), $\int \sin u \cdot du = - \cos u$. Here $u = 3ax$, and $du = 3a \cdot dx$.

$$\therefore \int \sin 3ax \cdot dx = \frac{1}{3a} \int \sin 3ax \cdot 3a \cdot dx$$

$$= - \frac{1}{3a} \cdot \cos 3ax + C$$

448. What is $\int \tan (4x + 1) \, dx$?

By formula (XV), $\int \tan u \cdot du = - \log_\epsilon \cos u$. Here $u = (4x + 1)$, and $du = 4 \cdot dx$.

$$\therefore \int \tan (4x + 1) \cdot dx = \tfrac{1}{4} \int \tan (4x + 1) \cdot 4 \, dx$$
$$= - \tfrac{1}{4} \log_\epsilon \cos (4x + 1) + C$$

449. What is $\int \sin^2 x \cdot dx$?

By formula (XIV), $\int \cos u \cdot du = \sin u$.

Now $\qquad \sin^2 x = \tfrac{1}{2}(1 - \cos 2x)$

Then $\int \sin^2 x \cdot dx = \tfrac{1}{2}\int (1 - \cos 2x) \cdot dx$

$$= \tfrac{1}{2} \int dx - \tfrac{1}{2} \int \cos 2x \cdot dx$$

or $\quad \displaystyle\int \sin^2 x \cdot dx = \frac{x}{2} - \frac{1}{4} \sin 2x = \frac{x}{2} - \frac{1}{2} \sin x \cdot \cos x + C$

Since $\sin 2x = 2 \sin x \cdot \cos x$.

450. What is $\int (\tan 2\theta - 1)^2 \cdot d\theta$?

By formulas (XIX) and (XV).

Now $(\tan 2\theta - 1)^2 = \tan^2 2\theta - 2 \tan 2\theta + 1$ (by squaring) and substituting

$$\tan^2 2\theta = \sec^2 2\theta - 1$$

or $\quad \sec^2 2\theta - 1 - 2 \tan 2\theta + 1 = \sec^2 2\theta - 2 \tan 2\theta$

then $\int (\tan 2\theta - 1)^2 \cdot d\theta = \int (\sec^2 2\theta - 2 \tan 2\theta) \cdot d\theta$

$$= \int \sec^2 2\theta \cdot d\theta - 2\int \tan 2\theta \cdot d\theta$$

Now $\qquad \int \sec^2 u \cdot du = \tan u$

and $\qquad \int \tan u \cdot du = -\log_e \cos u$

Here $\qquad u = 2\theta \quad$ and $\quad du = 2 \cdot d\theta$.

Then $\quad \int \sec^2 2\theta \cdot d\theta = \frac{1}{2}\int \sec^2 2\theta \cdot d(2\theta) = \frac{1}{2} \tan 2\theta$

and $\quad -2\int \tan 2\theta \cdot d\theta = -2 \cdot \frac{1}{2}\int \tan 2\theta \cdot d(2\theta)$

$$= \log_e \cos 2\theta$$

$\therefore \int (\tan 2\theta - 1)^2 \cdot d\theta = \frac{1}{2} \tan 2\theta + \log_e \cos 2\theta$

PROBLEMS

Integrate the following by formulas (XIII) to (XXII):

1. $\int \sin 5ax \cdot dx$
2. $\int \tan (3x + 1) \cdot dx$
3. $\int \cos 3x \cdot dx$
4. $\int (\tan 3\theta - 1)^2 \cdot d\theta$
5. $\int \sin (5x + 3) \, dx$
6. $\int \cos (3x - 1) \, dx$
7. $\int \tan 4x \cdot dx$
8. $\int \cot 3x \cdot dx$
9. $\int \sec 3x \cdot dx$
10. $\int \tan^2 x \cdot dx$
11. $\int \cot^2 x \cdot dx$
12. $\int \sqrt{\sin x} \cdot \cos x \cdot dx$
13. $\int \dfrac{\sin x \cdot \cos x \cdot dx}{3 + \sin^2 x}$
14. $\int \operatorname{cosec} 3x \cdot dx$
15. $\int \sec 3x \cdot \tan 3x \cdot dx$
16. $\int \cos^2 3x \cdot \sin 3x \cdot dx$
17. $\int \operatorname{cosec} 3x \cdot \cot 3x \cdot dx$
18. $\int (3 \sin x - 3)^2 \cdot dx$
19. $\int \sin^3 x \cdot \cos^2 x \cdot dx$
HINT.—$\int (1 - \cos^2 x) \cos^2 x \cdot \sin x \cdot dx$
20. $\int \sec^3 x \cdot \tan x \cdot dx$

CHAPTER XXIII

INTEGRATION BY INSPECTION

451. What is the integral of a function that can be separated by inspection into two simple factors the second of which is the derivative of the first?

It is one-half the square of the first factor.

$$\int (u) \cdot du \quad = \quad \frac{u^2}{2} + C$$

(factors, first and second)   (one-half the square of the first factor)

452. If $y = \int (x^4 + 5x)(4x^3 + 5)\, dx$, what is the integral?

Let $\qquad u = x^4 + 5x = $ first factor

and $du = (4x^3 + 5) \cdot dx = $ second factor = derivative of first factor

$\therefore \dfrac{u^2}{2} + C = \dfrac{(x^4 + 5x)^2}{2} + C = $ the integral = one-half the square of the first factor

453. If $y = \int \sin x \cdot \cos x \cdot dx$, what is the integral?

Let $u = \sin x = $ first factor

and

$\qquad du = \cos x \cdot dx = $ second factor = derivative of first factor

$\therefore \dfrac{u^2}{2} + C = \dfrac{\sin^2 x}{2} + C = $ the integral = one-half the square of the first factor

454. What is a more general form of the above arrangement of $\int u \cdot du$?

A more general form is $y = \displaystyle\int u^n \cdot du = \dfrac{u^{n+1}}{n+1} + C$

(factors, first and second)

455. If $y = \int x^3(a^2 + x^4)^{\frac{3}{4}} \cdot dx$, what is the integral?

Let $\quad (a^2 + x^4) = u = $ the first factor

and $\quad\quad \frac{4}{4}x^3 \cdot dx = \frac{1}{4} du = $ the second factor

Then $y = \dfrac{1}{4} \displaystyle\int (a^2 + x^4)^{\frac{3}{4}} \cdot 4x^3 \cdot dx = \dfrac{1}{4} \int u^{\frac{3}{4}} \cdot du$

$$= \frac{1}{4} \cdot \frac{u^{\frac{3}{4}+1}}{\frac{3}{4}+1} = \frac{u^{\frac{7}{4}}}{11}$$

$\therefore y = \dfrac{(a^2 + x^4)^{\frac{7}{4}}}{11} + C = $ the integral

456. If $\dfrac{dy}{dx} = \dfrac{x^3 + 1}{x}$, what is the integral by inspection?

Inspection indicates division by x first.

$$y = \int \frac{(x^3 + 1)}{x} dx = \int \left(x^2 + \frac{1}{x} \right) dx$$

$$= \int x^2 \cdot dx + \int \frac{1}{x} \cdot dx \text{ (transformed)}$$

$$= \frac{x^3}{3} + \log_e x + C = \text{the integral}$$

457. If $\dfrac{dy}{dx} = x^2 \sqrt{1 - x^3}$, what is the integral?

Inspection indicates the form $\int u^n \cdot du$, as the exponents of x differ by 1.

$$y = \int x^2(1 - x^3)^{\frac{1}{2}} dx = -\frac{1}{3} \cdot \int (1 - x^3)^{\frac{1}{2}} \cdot -3x^2 \cdot dx$$

Then $\quad\quad y = -\frac{1}{3} \cdot \frac{2}{3}(1 - x^3)^{\frac{3}{2}}$

$\quad\quad\quad \therefore y = -\frac{2}{9}(1 - x^3)^{\frac{3}{2}} + C = $ the integral

458. If $\dfrac{dy}{dx} = x^2$, what is the integral?

Inspection indicates the form $\int u^n \cdot du$.

$$y = \int x^2 \cdot dx = \frac{1}{3}\int 3x^2 \cdot dx = \frac{1}{3} \cdot x^3 + C = \text{the integral}$$

459. What are some simple examples of integrating by inspection of the form $y = \int u^n \cdot du$?

Examples

(a) $\quad y = \int x^{\frac{1}{4}} \cdot dx = \dfrac{x^{\frac{1}{4}+1}}{\frac{1}{4}+1} = \dfrac{4}{5} \cdot x^{\frac{5}{4}} + C$

(b) $\quad y = \int x^{-2.41} \cdot dx = -\dfrac{x^{-1.41}}{1.41} + C$

460. What are some examples of the algebraic special case of $\int u^n \cdot du$?

Form $y = \int (ax^m + b)^n x^{m-1} \cdot dx = \dfrac{1}{ma} \dfrac{(ax^m + b)^{n+1}}{n+1} + C$

Examples

(a) $y = \int (x^3 + 42)^{\frac{1}{2}} x^2 \cdot dx = \dfrac{1}{3 \cdot 1} \cdot (x^3 + 42)^{\frac{3}{2}} \cdot \dfrac{2}{3}$

$\qquad\qquad = \frac{2}{9}(x^3 + 42)^{\frac{3}{2}} + C$

(b) $y = \int (4x^3 + 7)^4 \cdot x^2 \cdot dx = \dfrac{1}{3 \cdot 4} \cdot \dfrac{(4x^3 + 7)^5}{5}$

$\qquad\qquad = \dfrac{(4x^3 + 7)^5}{60} + C$

461. What are some examples of the trigonometric special case of $\int u^n \cdot du$?

Form $y = \int \sin^n x \cdot \cos x \cdot dx = \dfrac{1}{n+1} \sin^{n+1} x + C$

Examples

(a) $y = \int \sin^3 x \cdot \cos x \cdot dx = \frac{1}{4} \cdot \sin^4 x + C$

(b) $y = \int \dfrac{\cos x}{\sin^2 x} \cdot dx = \int \sin^{-2} x \cdot \cos x \, dx$

$\qquad = \dfrac{1}{-2+1} \cdot \sin^{-2+1} x = -\sin^{-1} x = -\dfrac{1}{\sin x}$

Form $y = \int \cos^n x \cdot \sin x \cdot dx = -\dfrac{1}{n+1} \cos^{n+1} x + C$

Examples

(a) $y = \int \cos^6 x \cdot \sin x \cdot dx = -\frac{1}{7} \cos^7 x + C$

(b) $y = \int \dfrac{\sin x \cdot dx}{\cos^2 x} = \int \cos^{-2} x \cdot \sin x \cdot dx$

$$= +\frac{1}{1} \cos^{-1} x = \frac{1}{\cos x}$$

462. What is the integral when by mere inspection the numerator is seen to be the derivative of the denominator?

When the numerator = the derivative of the denominator, the integral equals the *log of the denominator*.

or
$$y = \int \frac{du}{u} = \log_e u + C$$

463. If $y = \int \dfrac{\epsilon^x}{\epsilon^x + 3} \cdot dx$, what is the integral?

Derivative of the denominator $(\epsilon^x + 3) = \epsilon^x =$ the numerator.

$$\therefore y = \int \frac{\epsilon^x \cdot dx}{\epsilon^x + 3} = \log_e (\epsilon^x + 3) + C = \text{log of denominator}$$

464. If $y = \int \dfrac{3x^2}{4 + x^3} \cdot dx$, what is the integral?

The numerator $= 3x^2 =$ the derivative of the denominator $(4 + x^3)$.

$$\therefore y = \int \frac{3x^2}{4 + x^3} \cdot dx = \log_e (4 + x^3) + C = \text{log of}$$
$$\text{denominator}$$

465. If $y = \int \dfrac{4x^3 + 8}{x^4 + 8x} \cdot dx$, what is the integral?

The numerator $4x^3 + 8 =$ the derivative of $x^4 + 8x$ or denominator.

$$\therefore y = \int \frac{4x^3 + 8}{x^4 + 8x} \cdot dx = \log_\epsilon (x^4 + 8x) + C = \text{log of}$$
$$\text{denominator}$$

466. If $y = \int \dfrac{x^{m-1}\, dx}{ax^m + b},$ **what is the integral?**

To make the numerator the derivative of the denominator, multiply the expression by $\dfrac{ma}{ma}.$

$$\therefore y = \frac{1}{ma} \log_\epsilon (ax^m + b) + C$$

467. If $y = \int \cot ax \cdot dx,$ **what is the integral?**

$$y = \int \frac{\cos ax}{\sin ax} \cdot dx = \frac{1}{a} \log_\epsilon \sin ax + C$$

To make the numerator the derivative of the denominator multiply by $\dfrac{a}{a}.$

468. If $y = \int \tan ax \cdot dx,$ **what is the integral?**

$$y = \int \frac{\sin ax}{\cos ax} \cdot dx = -\frac{1}{a} \log_\epsilon \cos ax + C$$

469. If $y = \int \sec ax \cdot dx,$ **what is the integral?**

It can be readily shown that

$$\sec ax = \frac{\sec^2 ax + \sec ax \cdot \tan ax}{\sec ax + \tan ax}$$

Then $\qquad y = \displaystyle\int \frac{\sec^2 ax + \sec ax \cdot \tan ax}{\sec ax + \tan ax} \cdot dx$

Multiply by $\dfrac{a}{a}$ to make the numerator the derivative of the denominator.

$$\therefore y = \frac{1}{a} \log_\epsilon (\sec ax + \tan ax) + C$$

470. If $y = \int \text{cosec } ax \cdot dx$, what is the integral?

It can be readily shown that

$$\text{cosec } ax = \frac{\text{cosec}^2 ax - \text{cosec } ax \cdot \cot ax}{\text{cosec } ax - \cot ax}$$

Then
$$y = \int \frac{\text{cosec}^2 ax - \text{cosec } ax \cdot \cot ax}{\text{cosec } ax - \cot ax} \cdot dx$$

$$\therefore y = +\frac{1}{a} \log_\epsilon (\text{cosec } ax - \cot ax) + C$$

Multiply by $\dfrac{a}{a}$ to make the numerator the **derivative** of the denominator.

PROBLEMS

Integrate the following by inspection:

1. $\int (x^5 + 9x)(5x^4 + 9)\, dx$
2. $\int \tan x \cdot \sec^2 x \cdot dx$
3. $\int x^2(a^2 + x^3)^{\frac{3}{2}} \cdot dx$
4. $\int x^{\frac{7}{2}} \cdot dx$
5. $\int x^{-3.62} \cdot dx$
6. $\int (x^4 + 6)^{\frac{3}{2}} \cdot x^3 \cdot dx$
7. $\int (3x^2 + 4)^3 \cdot x \cdot dx$
8. $\int \sin^4 x \cdot \cos x \cdot dx$
9. $\int \dfrac{\cos x}{\sin^3 x} \cdot dx$
10. $\int \cos^5 x \cdot \sin x \cdot dx$
11. $\int \dfrac{\sin x \cdot dx}{\cos^3 x}$
12. $\int \dfrac{\epsilon^x}{\epsilon^x + 1} \cdot dx$
13. $\int \dfrac{4x^3}{7 + x^4} \cdot dx$
14. $\int \dfrac{3x^2 - 2}{x^3 - 2x} \cdot dx$
15. $\int \dfrac{x^4}{5x^5 + 2} \cdot dx$
16. $\int \cot 3\theta \cdot d\theta$
17. $\int \tan 4\theta \cdot d\theta$
18. $\int \sec 3\theta \cdot d\theta$
19. $\int \text{cosec } 4\theta \cdot d\theta$
20. $\int x^3(1 - x^4)^{\frac{1}{4}} \cdot dx$

CHAPTER XXIV

INTEGRATION BY SUBSTITUTION

471. What is integration by substitution?

Before integrating, the integrand is reduced to one of the fundamental forms by means of the substitution of a new variable.

472. If $\dfrac{dy}{dx} = \sqrt{5 + x}$, what is the integral?

$$dy = \sqrt{5 + x} \cdot dx$$

Then
$$y = \int \sqrt{5 + x} \cdot dx$$

Let
$$u = (5 + x)$$

Then
$$du = dx$$

and, substituting these in the original function,

$$y = \int u^{\frac{1}{2}} \cdot du = \tfrac{2}{3}u^{\frac{3}{2}} = \tfrac{2}{3}(5 + x)^{\frac{3}{2}} = \text{the integral}$$

473. If $\dfrac{dy}{dx} = \dfrac{2}{\epsilon^{2x} + \epsilon^{-2x}}$, what is the integral?

$$y = 2 \int \frac{dx}{\epsilon^{2x} + \epsilon^{-2x}}$$

Let
$$u = \epsilon^{2x}$$

Then
$$\frac{du}{dx} = 2\epsilon^{2x}$$

$$dx = \frac{du}{2\epsilon^{2x}}$$

and, substituting these in the original function,

$$y = 2 \int \frac{dx}{\epsilon^{2x} + \epsilon^{-2x}} = 2 \int \frac{du}{2\epsilon^{2x}(\epsilon^{2x} + \epsilon^{-2x})}$$

or
$$y = \frac{2}{2} \int \frac{du}{u\left(u + \dfrac{1}{u}\right)} = \int \frac{du}{u^2 + 1}$$

$$\therefore \; y = \text{arc tan } u = \text{arc tan } \epsilon^{2x} = \text{the integral}$$

474. If $\dfrac{dy}{dx} = (a + bx)^n$, what is the integral when $n = 1$, $n = 2$, and $n = -2$?

$$y = \int (a + bx)^n \cdot dx$$

Let
$$u = (a + bx)$$

Then
$$x = \frac{u - a}{b} \quad \text{and} \quad \frac{dx}{du} = \frac{1}{b}$$

and, substituting these in the original function

$$y = \int (a + bx)^n \cdot dx = \int u^n \cdot \frac{du}{b} = \frac{1}{b} \int u^n \cdot du = \frac{u^{n+1}}{b(n + 1)}$$

or $y = \dfrac{(a + bx)^{n+1}}{b(n + 1)} + C = $ the integral in the general form

For $n = 1$,

$$\int (a + bx)^1 \, dx = \frac{(a + bx)^{1+1}}{b(1 + 1)} = \frac{(a + bx)^2}{2b} + C$$

For $n = 2$,

$$\int (a + bx)^2 \, dx = \frac{(a + bx)^{2+1}}{b(2 + 1)} = \frac{(a + bx)^3}{3b} + C$$

For $n = -2$,

$$\int \frac{dx}{(a + bx)^2} = \int (a + bx)^{-2} \cdot dx = \frac{(a + bx)^{-2+1}}{b(-2 + 1)}$$

$$= -\frac{1}{b(a + bx)} + C$$

475. If $\dfrac{dy}{dx} = (a - bx)^n$, what is the integral when $n = 2$?

$$y = \int (a - bx)^n \cdot dx$$

Let
$$(a - bx) = u$$

Then
$$x = -\frac{(u - a)}{b} \quad \text{and} \quad \frac{dx}{du} = -\frac{1}{b}$$

Substituting these in the original function

$$y = \int (a - bx)^n \cdot dx = \int u^n \cdot -\frac{1}{b} \cdot du$$

$$= -\frac{1}{b} \int u^n \cdot du = -\frac{1u^{n+1}}{b(n+1)}$$

$$\therefore y = -\frac{(a-bx)^{n+1}}{b(n+1)} + C = \text{the integral in the general}$$

form

For $n = 2$,

$$y = \int (a - bx)^2 \cdot dx = -\frac{(a-bx)^3}{3b} + C$$

476. If $\dfrac{dy}{dx} = (a + bx)^{-1}$, what is the integral?

This is an exceptional case where $n = -1$.

$$y = \int \frac{dx}{a+bx}$$

Let $$u = (a + bx)$$

Then $$x = \frac{u-a}{b} \quad \text{and} \quad \frac{dx}{du} = \frac{1}{b}$$

Substituting in the original function,

$$\therefore y = \int \frac{dx}{a+bx} = \frac{1}{b} \int \frac{du}{u} = \frac{1}{b} \log_e u = \frac{1}{b} \log_e (a + bx)$$

$$+ C = \text{the integral}$$

477. If $\dfrac{dy}{dx} = \epsilon^{nx}$, what is the integral?

$$y = \int \epsilon^{nx} \cdot dx$$

Let $$u = nx$$

Then $$x = \frac{u}{n} \quad \text{and} \quad \frac{dx}{du} = \frac{1}{n},$$

Substituting,

$$y = \int \epsilon^{nx} \cdot dx = \int \epsilon^u \cdot \frac{1}{n} \cdot du = \frac{1}{n} \int \epsilon^u \cdot du = \frac{1}{n} \epsilon^u = \frac{1}{n} \epsilon^{nx}$$

For $n = 4$,

$$y = \int \epsilon^{4x} \cdot dx = \tfrac{1}{4}\epsilon^{4x} + C = \text{the integral}$$

478. If $\dfrac{dy}{dx} = (x^3 - 4x^2)(3x^2 - 8x)$, what is the integral?

$$y = \int (x^3 - 4x^2)(3x^2 - 8x) \cdot dx$$

Let $\qquad\qquad u = x^3 - 4x^2$

Then $\quad \dfrac{du}{dx} = 3x^2 - 8x \qquad$ and $\qquad \dfrac{dx}{du} = \dfrac{1}{3x^2 - 8x}$

Then $y = \displaystyle\int (x^3 - 4x^2)(3x^2 - 8x)\, dx$

$$= \int \frac{(x^3 - 4x^2)(3x^2 - 8x)}{(3x^2 - 8x)}\, du$$

$$= \int u \cdot du = \frac{u^2}{2}$$

$$\therefore \; y = \frac{(x^3 - 4x^2)^2}{2}$$

479. If $\dfrac{dy}{dx} = \epsilon^x \cdot \cos \epsilon^x$, what is the integral?

$$y = \int \epsilon^x \cdot \cos \epsilon^x \cdot dx$$

Let $\qquad\qquad u = \epsilon^x$

Then $\qquad\qquad \log_\epsilon u = x \log_\epsilon \epsilon = x$

and $\qquad\qquad \dfrac{dx}{du} = \dfrac{1}{u} = \dfrac{1}{\epsilon^x}$

Then $y = \displaystyle\int \epsilon^x \cdot \cos \epsilon^x \cdot dx = \int \epsilon^x \cdot \cos \epsilon^x \cdot \dfrac{du}{\epsilon^x}$

$$= \int \cos \epsilon^x \cdot du$$

$$= \int \cos u \cdot du$$

$$\therefore \; y = \sin \epsilon^x + C = \text{the integral}$$

480. What is the integral if $y = \int x \sqrt{2x + 1} \cdot dx$?

Let $\qquad\qquad u = \sqrt{2x + 1}$

Substituting these in the original function

$$y = \int (a - bx)^n \cdot dx = \int u^n \cdot -\frac{1}{b} \cdot du$$

$$= -\frac{1}{b} \int u^n \cdot du = -\frac{1u^{n+1}}{b(n+1)}$$

$$\therefore y = -\frac{(a-bx)^{n+1}}{b(n+1)} + C = \text{the integral in the general}$$

form

For $n = 2$,

$$y = \int (a - bx)^2 \cdot dx = -\frac{(a - bx)^3}{3b} + C$$

476. If $\dfrac{dy}{dx} = (a + bx)^{-1}$, **what is the integral?**

This is an exceptional case where $n = -1$.

$$y = \int \frac{dx}{a + bx}$$

Let $\qquad u = (a + bx)$

Then $\qquad x = \dfrac{u - a}{b} \qquad$ and $\qquad \dfrac{dx}{du} = \dfrac{1}{b}$

Substituting in the original function,

$$\therefore y = \int \frac{dx}{a + bx} = \frac{1}{b} \int \frac{du}{u} = \frac{1}{b} \log_\epsilon u = \frac{1}{b} \log_\epsilon (a + bx)$$

$$+ C = \text{the integral}$$

477. If $\dfrac{dy}{dx} = \epsilon^{nx}$, **what is the integral?**

$$y = \int \epsilon^{nx} \cdot dx$$

Let $\qquad u = nx$

Then $\qquad x = \dfrac{u}{n} \qquad$ and $\qquad \dfrac{dx}{du} = \dfrac{1}{n}$,

Substituting,

$$y = \int \epsilon^{nx} \cdot dx = \int \epsilon^u \cdot \frac{1}{n} \cdot du = \frac{1}{n} \int \epsilon^u \cdot du = \frac{1}{n} \epsilon^u = \frac{1}{n} \epsilon^{nx}$$

For $n = 4$,

$$y = \int \epsilon^{4x} \cdot dx = \tfrac{1}{4}\epsilon^{4x} + C = \text{the integral}$$

478. If $\dfrac{dy}{dx} = (x^3 - 4x^2)(3x^2 - 8x)$, what is the integral?

$$y = \int (x^3 - 4x^2)(3x^2 - 8x) \cdot dx$$

Let $\qquad u = x^3 - 4x^2$

Then $\qquad \dfrac{du}{dx} = 3x^2 - 8x \qquad$ and $\qquad \dfrac{dx}{du} = \dfrac{1}{3x^2 - 8x}$

Then $y = \displaystyle\int (x^3 - 4x^2)(3x^2 - 8x)\, dx$

$$= \int \dfrac{(x^3 - 4x^2)(3x^2 - 8x)}{(3x^2 - 8x)}\, du$$

$$= \int u \cdot du = \dfrac{u^2}{2}$$

$$\therefore y = \dfrac{(x^3 - 4x^2)^2}{2}$$

479. If $\dfrac{dy}{dx} = \epsilon^x \cdot \cos \epsilon^x$, what is the integral?

$$y = \int \epsilon^x \cdot \cos \epsilon^x \cdot dx$$

Let $\qquad u = \epsilon^x$

Then $\qquad \log_\epsilon u = x \log_\epsilon \epsilon = x$

and $\qquad \dfrac{dx}{du} = \dfrac{1}{u} = \dfrac{1}{\epsilon^x}$

Then $y = \displaystyle\int \epsilon^x \cdot \cos \epsilon^x \cdot dx = \int \epsilon^x \cdot \cos \epsilon^x \cdot \dfrac{du}{\epsilon^x}$

$$= \int \cos \epsilon^x \cdot du$$

$$= \int \cos u \cdot du$$

$$\therefore y = \sin \epsilon^x + C = \text{the integral}$$

480. What is the integral if $y = \int x \sqrt{2x + 1} \cdot dx$?

Let $\qquad u = \sqrt{2x + 1}$

Then $u^2 = 2x + 1$ or $x = \dfrac{u^2 - 1}{2}$

and $dx = u \cdot du$

Then $\displaystyle\int x \sqrt{2x + 1} \cdot dx = \int \left(\dfrac{u^2 - 1}{2}\right) u \cdot u \cdot du$

$$= \int \dfrac{u^4 - u^2}{2} \cdot du$$

$$= \dfrac{1}{2} \int u^4 \cdot du - \dfrac{1}{2} \int u^2 \cdot du$$

$$= \dfrac{1}{2} \cdot \dfrac{u^5}{5} - \dfrac{1}{2} \cdot \dfrac{u^3}{3}$$

$$= \dfrac{u^5}{10} - \dfrac{u^3}{6} = \dfrac{(2x + 1)^{\frac{5}{2}}}{10}$$

$$- \dfrac{(2x + 1)^{\frac{3}{2}}}{6}$$

$$= (2x + 1)^{\frac{3}{2}} \left(\dfrac{2x + 1}{10} - \dfrac{1}{6}\right)$$

$$= (2x + 1)^{\frac{3}{2}} \left(\dfrac{6x + 3 - 5}{30}\right)$$

$$= \tfrac{1}{15}(2x + 1)^{\frac{3}{2}}(3x - 1)$$

481. What is $\displaystyle\int \dfrac{dx}{x^{\frac{1}{2}} - x^{\frac{1}{4}}}$?

The common multiple of the denominators of $\frac{1}{2}$ and $\frac{1}{4}$ is 4.   Then substitute u^4 for x, and get

$$x^{\frac{1}{2}} = u^2, \qquad x^{\frac{1}{4}} = u, \qquad \text{and} \qquad dx = 4u^3 \cdot du$$

$$\therefore \int \dfrac{dx}{x^{\frac{1}{2}} - x^{\frac{1}{4}}} = \int \dfrac{4u^3 \cdot du}{u^2 - u} = 4 \int \dfrac{u^2 \cdot du}{u - 1}$$

$$= 4 \int \left(u + 1 + \dfrac{1}{u - 1}\right) du$$

$$= 4 \int u \cdot du + 4 \int du + 4 \int \dfrac{du}{u - 1}$$

$$= 4 \cdot \dfrac{u^2}{2} + 4u + 4 \log_e (u - 1)$$

$$= 2x^{\frac{1}{2}} + 4x^{\frac{1}{4}} + 4 \log_e (x^{\frac{1}{4}} - 1)$$

482. What is $\int \dfrac{dx}{x\sqrt{3x^2 + 2x - 1}}$?

Substitute $\dfrac{1}{u}$ for x.

Then $\qquad x = \dfrac{1}{u} \qquad$ and $\qquad dx = -\dfrac{du}{u^2}$

and $\quad \displaystyle\int \dfrac{dx}{x\sqrt{3x^2 + 2x - 1}} = \int \dfrac{-\dfrac{du}{u^2}}{\dfrac{1}{u}\sqrt{\dfrac{3}{u^2} + \dfrac{2}{u} - 1}}$

$$= -\int \dfrac{\dfrac{du}{u^2}}{\dfrac{1}{u^2}\sqrt{3 + 2u - u^2}}$$

$$= -\int \dfrac{du}{\sqrt{3 + 2u - u^2}}$$

Now complete the square.

$$\int \dfrac{dx}{x\sqrt{3x^2 + 2x - 1}} = -\int \dfrac{du}{\sqrt{4 - (u^2 - 2u + 1)}}$$

$$= -\int \dfrac{du}{\sqrt{(2)^2 - (u - 1)^2}}$$

By formula (V),

$$\therefore \int \dfrac{dx}{x\sqrt{3x^2 + 2x - 1}} = \sin^{-1}\dfrac{(u - 1)}{2} = \sin^{-1}\dfrac{(1 - x)}{2x}$$

483. In integrating expressions containing radicals

$$\sqrt{a^2 - u^2} \qquad \text{or} \qquad \sqrt{u^2 \pm a^2},$$

what trigonometric substitutions are generally made?

a. When $\sqrt{a^2 - u^2}$ occurs, let $u = a\sin\theta$.

b. When $\sqrt{a^2 + u^2}$ occurs, let $u = a\tan\theta$.

c. When $\sqrt{u^2 - a^2}$ occurs, let $u = a\sec\theta$.

These substitutions eliminate the radical sign in each case, as

(a) $\sqrt{a^2 - a^2 \sin^2 \theta} = a \sqrt{1 - \sin^2 \theta} = a \cos \theta$

(b) $\sqrt{a^2 + a^2 \tan^2 \theta} = a \sqrt{1 + \tan^2 \theta} = a \sec \theta$

(c) $\sqrt{a^2 \sec^2 \theta - a^2} = a \sqrt{\sec^2 \theta - 1} = a \tan \theta$

484. What is $\displaystyle\int \frac{dx}{(a^2 - x^2)^{\frac{3}{2}}}$?

Let $x = a \sin \theta$

Then $dx = a \cos \theta \cdot d\theta$

Then $\displaystyle\int \frac{dx}{(a^2 - x^2)^{\frac{3}{2}}} = \int \frac{a \cos \theta \cdot d\theta}{a^3 \cos^3 \theta} = \frac{1}{a^2} \int \frac{d\theta}{\cos^2 \theta}$

$$= \frac{1}{a^2} \int \sec^2 \theta \cdot d\theta = \frac{\tan \theta}{a^2}$$

$$= \frac{x}{a^2 \sqrt{a^2 - x^2}} + C$$

485. What is $\displaystyle\int \frac{dx}{x \sqrt{x^2 + 4}}$?

Let $x = u$.   Then $dx = du$, and let $a = 2$.

Then $\displaystyle\int \frac{dx}{x \sqrt{x^2 + 4}} = \int \frac{du}{u \sqrt{u^2 + a^2}}$

Now let $u = a \tan \theta$.   Then $du = a \sec^2 \theta \cdot d\theta$.

and $\displaystyle\int \frac{du}{u \sqrt{u^2 + a^2}} = \int \frac{a \sec^2 \theta \cdot d\theta}{a \tan \theta \cdot a \sec \theta} = \frac{1}{a} \int \frac{\sec \theta \cdot d\theta}{\tan \theta}$

$$= \frac{1}{a} \int \frac{d\theta}{\sin \theta} = \frac{1}{a} \int \operatorname{cosec} \theta \cdot d\theta$$

$$= \frac{1}{a} \log_\epsilon (\operatorname{cosec} \theta - \cot \theta)$$

$$= \frac{1}{a} \log_\epsilon \left(\frac{\sqrt{u^2 + a^2}}{u} - \frac{a}{u} \right)$$

QUES. 484.

QUES. 485.

$$= \frac{1}{a} \log_\epsilon \frac{\sqrt{u^2 + a^2} - 2}{u}$$

$$\therefore \int \frac{dx}{x \sqrt{x^2 + 4}} = \frac{1}{2} \log_\epsilon \frac{\sqrt{x^2 + 4} - 2}{x} + C$$

PROBLEMS

Find the integrals of the following by substitution:

1. $\int \frac{x \cdot dx}{\sqrt{3 - 4x}}$.   Let $u = 3 - 4x$.

2. $\int \frac{x \cdot dx}{\sqrt[3]{1 + x}}$.   Let $u^3 = 1 + x$.

3. $\int \frac{x \cdot dx}{1 + x^{\frac{1}{2}}}$.   Let $u^4 = x$.

4. $\int \frac{dx}{x^{\frac{2}{3}} + x^{\frac{1}{2}}}$.   Let $x = u^6$.

5. $\int \frac{dx}{x \sqrt{3x + 5}}$.   Let $3x + 5 = u^2$.

6. $\int \frac{dx}{x \sqrt{x^2 - 2x + 5}}$.   Let $x = \frac{1}{u}$.

7. $\int \sqrt{4 - x^2} \cdot dx$.   Let $x = 2 \sin \theta$.

8. $\int \frac{\sqrt{4 + x^2}}{x} \cdot dx$.   Let $x = 2 \tan \theta$.

9. $\int \frac{\sqrt{x^2 - 4}}{x^2} \cdot dx$.   Let $x = 2 \sec \theta$.

10. $\int \frac{(x - 4)^{\frac{1}{2}} \cdot dx}{(x - 4) + 6}$.   Let $(x - 4)^{\frac{1}{2}} = u$.

11. $\int \frac{dx}{1 - \sqrt[3]{x}}$.   Let $x = u^3$.

12. $\int \frac{dx}{x + 3 + \sqrt{x + 4}}$.   Let $x + 4 = u^2$.

13. $\int \frac{\sqrt{\epsilon^x} \cdot dx}{\epsilon^x + 3}$.   Let $\epsilon^x = u^2$.

14. $\int \frac{dx}{x \sqrt{4x^2 + 9}}$.   Let $2x = u$ and $a = 3$; then let $u = a \tan \theta$.

15. $\int \frac{dx}{(a^2 - x^2)^{\frac{3}{2}}}$

16. $\int \frac{dx}{3x \sqrt{9x^2 + 16}}$

17. $\int \frac{dx}{x^3 \sqrt{x^2 - 9}}$

CHAPTER XXV

INTEGRATION OF TRIGONOMETRIC FUNCTIONS BY TRANSFORMATION AND REDUCTION

486. How would you integrate $\int \sin^m \theta \cdot \cos^n \theta \cdot d\theta$ when m is a *positive odd integer* no matter what n is?

Let $\sin^m \theta = \sin^{m-1} \theta \cdot \sin \theta$, making $m - 1$ an even power. Then substitute $\sin^2 \theta = 1 - \cos^2 \theta$ to transform the function to the form $\int u^n \cdot du$.

487. What is $\int \sin^3 \theta \cdot \cos^2 \theta \cdot d\theta$?

$$\int \sin^3 \theta \cdot \cos^2 \theta \, d\theta = \int \sin^2 \theta \cdot \sin \theta \cdot \cos^2 \theta \cdot d\theta$$

Now $\qquad\qquad \sin^2 \theta = 1 - \cos^2 \theta$

$$\therefore \int \sin^3 \theta \cdot \cos^2 \theta \cdot d\theta = \int (1 - \cos^2 \theta) \sin \theta \cdot \cos^2 \theta \cdot d\theta$$
$$= \int \cos^2 \theta \cdot \sin \theta \cdot d\theta$$
$$- \int \cos^4 \theta \cdot \sin \theta \cdot d\theta$$
$$= \int (\cos \theta)^2 \cdot \sin \theta \cdot d\theta$$
$$- \int (\cos \theta)^4 \cdot \sin \theta \cdot d\theta$$
$$= -\frac{\cos^3 \theta}{3} + \frac{\cos^5 \theta}{5} + C$$

488. How would you integrate $\int \sin^m \theta \cdot \cos^n \theta \cdot d\theta$ when n is a *positive odd integer*?

Let $\cos^n \theta = \cos^{n-1} \theta \cdot \cos \theta$, making $n - 1$ even. Then substitute $\cos^2 \theta = 1 - \sin^2 \theta$, and transform to the form $\int u^n \cdot du$.

489. What is $\int \sin^2 \theta \cos^3 \theta \cdot d\theta$?

$$\int \sin^2 \theta \cdot \cos^3 \theta \, d\theta = \int \sin^2 \theta \cdot \cos^2 \theta \cdot \cos \theta \cdot d\theta$$

Now $\qquad\qquad \cos^2 \theta = 1 - \sin^2 \theta$

Then $\int \sin^2 \theta \cdot \cos^3 \theta \, d\theta = \int \sin^2 \theta \, (1 - \sin^2 \theta) \cos \theta \, d\theta$

$$= \int \sin^2 \theta \cdot \cos \theta \, d\theta$$

$$- \int \sin^4 \theta \cdot \cos \theta \cdot d\theta$$

$$= \frac{\sin^3 \theta}{3} - \frac{\sin^5 \theta}{5} + C$$

490. How would you integrate $\int \tan^n \theta \cdot d\theta$ when n is an integer?

Use the relation

$$\tan^n \theta = \tan^{n-2} \theta \cdot \tan^2 \theta = \tan^{n-2} \theta(\sec^2 \theta - 1)$$

491. What is $\int \tan^5 \theta \cdot d\theta$?

$\int \tan^5 \theta \cdot d\theta = \int \tan^3 \theta \cdot \tan^2 \theta \cdot d\theta$

$$= \int \tan^2 \theta \cdot \tan \theta \cdot \tan^2 \theta \cdot d\theta$$

$$= \int \tan \theta(\sec^2 \theta - 1)^2 \cdot d\theta = \int \tan \theta \sec^4 \theta \cdot d\theta$$

$$- 2\int \sec^2 \theta \cdot \tan \theta \cdot d\theta + \int \tan \theta \cdot d\theta$$

$$= \int \sec^3 \theta \cdot \sec \theta \cdot \tan \theta \cdot d\theta$$

$$- 2\int \sec \theta \cdot \sec \theta \cdot \tan \theta \cdot d\theta + \log_e \sec \theta$$

$$= \frac{\sec^4 \theta}{4} - \sec^2 \theta + \log_e \sec \theta$$

492. How would you integrate $\int \cot^n \theta \cdot d\theta$ when n is an integer?

Use the relation

$$\cot^n \theta = \cot^{n-2} \theta \cdot \cot^2 \theta = \cot^{n-2} \theta(\operatorname{cosec}^2 \theta - 1)$$

493. What is $\int \cot^3 2\theta \cdot d\theta$?

Let $2\theta = u$. Then $\theta = \dfrac{u}{2}$ and $d\theta = \tfrac{1}{2} \cdot du$.

Then $\int \cot^3 2\theta \cdot d\theta = \tfrac{1}{2}\int \cot^3 u \cdot du = \tfrac{1}{2}\int \cot^2 u \cdot \cot u \cdot du$

$$= \tfrac{1}{2}\int \cot u(\operatorname{cosec}^2 u - 1) \, du$$

$$= \tfrac{1}{2}\int \operatorname{cosec} u \cdot \operatorname{cosec} u \cdot \cot u \cdot du$$

$$- \tfrac{1}{2}\int \cot u \cdot du$$

$$= -\tfrac{1}{4} \operatorname{cosec}^2 u - \tfrac{1}{2} \log_e \sin u + C$$

$$= -\tfrac{1}{4} \operatorname{cosec}^2 2\theta - \tfrac{1}{2} \log_e \sin 2\theta + C$$

494. How would you integrate $\int \sec^n \theta \cdot d\theta$ when n is a *positive even integer?*

Use the relation

$$\sec^n \theta = \sec^{n-2} \theta \cdot \sec^2 \theta = (\tan^2 \theta + 1)^{\frac{n-2}{2}} \cdot \sec^2 \theta$$

495. What is $\int \sec^4 2\theta \cdot d\theta$?

Let $2\theta = u$. Then $\theta = \dfrac{u}{2}$ and $d\theta = \tfrac{1}{2} \cdot du$.

$$\begin{aligned}
\text{Then } \int \sec^4 2\theta \cdot d\theta &= \tfrac{1}{2}\int \sec^4 u \cdot du \\
&= \tfrac{1}{2}\int (\tan^2 u + 1)^{\frac{4-2}{2}} \cdot \sec^2 u \cdot du \\
&= \tfrac{1}{2}\int (\tan^2 u + 1) \cdot \sec^2 u \cdot du \\
&= \tfrac{1}{2}\int \tan^2 u \cdot \sec^2 u \cdot du + \tfrac{1}{2}\int \sec^2 u \cdot du \\
&= \frac{1}{2} \cdot \frac{\tan^3 u}{3} + \frac{1}{2} \tan u
\end{aligned}$$

$\therefore \int \sec^4 2\theta \cdot d\theta = \tfrac{1}{6} \tan^3 2\theta + \tfrac{1}{2} \tan 2\theta + C$

496. How would you integrate $\int \operatorname{cosec}^n \theta \cdot d\theta$ when n is a *positive even integer?*

Use the relation

$$\operatorname{cosec}^n \theta = \operatorname{cosec}^{n-2} \theta \cdot \operatorname{cosec}^2 \theta = (\cot^2 \theta + 1)^{\frac{n-2}{2}} \operatorname{cosec}^2 \theta$$

497. What is $\int \operatorname{cosec}^4 2\theta \cdot d\theta$?

Let $2\theta = u$. Then $\theta = \dfrac{u}{2}$ and $d\theta = \dfrac{1}{2} \cdot du$.

$$\begin{aligned}
\text{Then } \int \operatorname{cosec}^4 2\theta \cdot d\theta &= \tfrac{1}{2}\int \operatorname{cosec}^4 u \cdot du \\
&= \tfrac{1}{2}\int (\cot^2 u + 1)^{\frac{4-2}{2}} \cdot \operatorname{cosec}^2 u \cdot du \\
&= \tfrac{1}{2}\int (\cot^2 u + 1) \operatorname{cosec}^2 u \cdot du \\
&= \tfrac{1}{2}\int \cot^2 u \cdot \operatorname{cosec}^2 u \cdot du \\
&\qquad + \tfrac{1}{2}\int \operatorname{cosec}^2 u \cdot du \\
&= -\frac{1}{2} \cdot \frac{\cot^3 u}{3} - \frac{1}{2} \cot u \\
&= -\frac{1}{6} \cot^3 u - \frac{1}{2} \cot u
\end{aligned}$$

$\therefore \int \operatorname{cosec}^4 2\theta \cdot d\theta = -\tfrac{1}{6} \cot^3 2\theta - \tfrac{1}{2} \cot 2\theta$

498. How would you integrate $\int \tan^m \theta \cdot \sec^n \theta \cdot d\theta$ when **n is a *positive even integer?***

Use the relation

$$\sec^n \theta = \sec^{n-2} \theta \cdot \sec^2 \theta = (\tan^2 \theta + 1)^{\frac{n-2}{2}} \cdot \sec^2 \theta$$

499. What is $\int \tan^8 \theta \cdot \sec^6 \theta \cdot d\theta$?

$$\int \tan^8 \theta \cdot \sec^6 \theta \cdot d\theta = \int \tan^8 \theta \cdot \sec^{6-2} \theta \cdot \sec^2 \theta \cdot d\theta$$
$$= \int \tan^8 \theta \cdot \sec^2 \theta (\tan^2 \theta + 1)^{\frac{6-2}{2}} d\theta$$
$$= \int \tan^8 \theta (\tan^2 \theta + 1)^2 \sec^2 \theta \cdot d\theta$$
$$= \int \tan^8 \theta \sec^2 \theta (\tan^4 \theta + 2 \tan^2 \theta + 1)\, d\theta$$
$$= \int \tan^{12} \theta \cdot \sec^2 \theta \cdot d\theta$$
$$+ 2\int \tan^{10} \theta \cdot \sec^2 \theta \cdot d\theta$$
$$+ \int \tan^8 \theta \cdot \sec^2 \theta \cdot d\theta$$
$$= \frac{\tan^{13} \theta}{13} + \frac{2}{11} \tan^{11} \theta + \frac{\tan^9 \theta}{9} + C$$

500. How would you integrate $\int \cot^m \theta \cdot \operatorname{cosec}^n \theta \cdot d\theta$ when **n is a *positive even integer?***

Use the relation

$$\operatorname{cosec}^n \theta = \operatorname{cosec}^{n-2} \theta \operatorname{cosec}^2 \theta = (\cot^2 \theta + 1)^{\frac{n-2}{2}} \cdot \operatorname{cosec}^2 \theta$$

501. What is $\int \cot^8 \theta \cdot \operatorname{cosec}^6 \theta \cdot d\theta$?

$$\int \cot^8 \theta \cdot \operatorname{cosec}^6 \theta \cdot d\theta = \int \cot^8 \theta \cdot \operatorname{cosec}^{6-2} \theta \cdot \operatorname{cosec}^2 \theta \cdot d\theta$$
$$= \int \cot^8 \theta (\cot^2 \theta + 1)^{\frac{6-2}{2}} \cdot \operatorname{cosec}^2 \theta \cdot d\theta$$
$$= \int \cot^8 \theta (\cot^2 \theta + 1)^2 \operatorname{cosec}^2 \theta \cdot d\theta$$
$$= \int \cot^8 \theta (\cot^4 \theta + 2 \cot^2 \theta + 1) \operatorname{cosec}^2 \theta \cdot d\theta$$
$$= \int \cot^{12} \theta \operatorname{cosec}^2 \theta \cdot d\theta$$
$$+ 2\int \cot^{10} \theta \cdot \operatorname{cosec}^2 \theta \cdot d\theta$$
$$+ \int \cot^8 \theta \cdot \operatorname{cosec}^2 \theta \cdot d\theta$$
$$= -\frac{\cot^{13} \theta}{13} - \frac{2}{11} \cot^{11} \theta$$
$$- \frac{\cot^9 \theta}{9} + C$$

502. How would you integrate $\int \tan^m \theta \cdot \sec^n \theta \cdot d\theta$ when m is an *odd integer*?

Reduce m and n by 1, and substitute $(\sec^2 \theta - 1)$ for $\tan^2 \theta$.

503. What is $\int \tan^7 \theta \sec^5 \theta \cdot d\theta$?

$$
\begin{aligned}
\int \tan^7 \theta \cdot \sec^5 \theta \cdot d\theta &= \int \tan^6 \theta \cdot \sec^4 \theta \cdot \sec \theta \cdot \tan \theta \cdot d\theta \\
&= \int (\sec^2 \theta - 1)^3 \sec^4 \theta \cdot \sec \theta \cdot \tan \theta \cdot d\theta \\
&= \int (\sec^6 \theta - 3 \sec^4 \theta + 3 \sec^2 \theta \\
&\qquad\qquad - 1) \sec^4 \theta \cdot \sec \theta \cdot \tan \theta \cdot d\theta \\
&= \int (\sec^{10} \theta - 3 \sec^8 \theta + 3 \sec^6 \theta \\
&\qquad\qquad - \sec^4 \theta) \sec \theta \cdot \tan \theta \cdot d\theta \\
&= \int \sec^{10} \theta \cdot \sec \theta \cdot \tan \theta \cdot d\theta \\
&\qquad - 3\int \sec^8 \theta \cdot \sec \theta \cdot \tan \theta \cdot d\theta \\
&\qquad + 3\int \sec^6 \theta \cdot \sec \theta \cdot \tan \theta \cdot d\theta \\
&\qquad - \int \sec^4 \theta \cdot \sec \theta \cdot \tan \theta \cdot d\theta \\
&= \frac{\sec^{11} \theta}{11} - \frac{3 \sec^9 \theta}{9} + \frac{3}{7} \sec^7 \theta \\
&\qquad\qquad\qquad\qquad - \frac{\sec^5 \theta}{5}
\end{aligned}
$$

504. How would you integrate $\int \sin^m \theta \cdot \cos^n \theta \cdot d\theta$ when m and n are both *positive even integers*?

Use the following relations, and transform the function to an expression involving sines and cosines of multiple angles:

$$
\begin{aligned}
\sin \theta \cdot \cos \theta &= \tfrac{1}{2} \sin 2\theta \\
\sin^2 \theta &= \tfrac{1}{2} - \tfrac{1}{2} \cos 2\theta \\
\cos^2 \theta &= \tfrac{1}{2} + \tfrac{1}{2} \cos 2\theta
\end{aligned}
$$

505. What is $\int \cos^2 \theta \cdot d\theta$?

$$
\begin{aligned}
\int \cos^2 \theta \cdot d\theta &= \int (\tfrac{1}{2} + \tfrac{1}{2} \cos 2\theta) \, d\theta \\
&= \tfrac{1}{2}\int d\theta + \tfrac{1}{2}\int \cos 2\theta \cdot d\theta \\
&= \frac{\theta}{2} + \frac{1}{2} \cdot \frac{1}{2} \sin 2\theta \\
\therefore \int \cos^2 \theta \cdot d\theta &= \frac{\theta}{2} + \frac{1}{4} \sin 2\theta + C
\end{aligned}
$$

506. What is $\int \sin^4 \theta \cdot \cos^4 \theta \cdot d\theta$?

Now $\qquad\qquad \sin 2\theta = 2 \sin \theta \cdot \cos \theta$

and $\qquad\qquad (\sin 2\theta)^4 = 16 \sin^4 \theta \cdot \cos^4 \theta$

Then $\qquad \int \sin^4 \theta \cdot \cos^4 \theta \cdot d\theta = \int \frac{\sin^4 2\theta}{16} \cdot d\theta$

Now $\quad \sin^2 2\theta = \frac{1}{2} - \frac{1}{2} \cos 4\theta = \frac{1}{2}(1 - \cos 4\theta)$

Then $\frac{1}{16}\int \sin^4 2\theta \cdot d\theta$

$$= \frac{1}{64}\int (1 - \cos 4\theta)^2 \cdot d\theta$$

$$= \frac{1}{64}\int (1 - 2 \cos 4\theta + \cos^2 4\theta)\, d\theta$$

$$= \frac{1}{64}(\int d\theta - 2\int \cos 4\theta \cdot d\theta + \int \cos^2 4\theta \cdot d\theta)$$

Now $\qquad\qquad \cos^2 4\theta = \frac{1}{2} + \frac{1}{2} \cos 8\theta$

$\frac{1}{16}\int \sin^4 2\theta \cdot d\theta = \frac{1}{64}[\theta - 2 \cdot \frac{1}{4} \sin 4\theta + \frac{1}{2}\int (1 + \cos 8\theta)\, d\theta]$

$$= \frac{1}{64}(\theta - \frac{1}{2} \sin 4\theta + \frac{1}{2}\int d\theta + \frac{1}{2}\int \cos 8\theta \cdot d\theta)$$

$$= \frac{1}{64}\left(\theta - \tfrac{1}{2} \sin 4\theta + \frac{\theta}{2} + \frac{1}{2} \cdot \frac{1}{8} \cdot \sin 8\theta\right)$$

$$\therefore \int \sin^4 \theta \cdot \cos^4 \theta \cdot d\theta = \frac{3\theta}{128} - \frac{\sin 4\theta}{128} + \frac{\sin 8\theta}{1024} + C$$

507. How would you integrate expressions of the form $\int \sin m\theta \cdot \cos n\theta \cdot d\theta$, $\int \sin m\theta \cdot \sin n\theta \cdot d\theta$, or $\int \cos m\theta \cdot \cos n\theta \cdot d\theta$ when m is not equal to n?

From trigonometry, the sum of the sines of two angles is

$$\sin A + \sin B = 2 \sin \left(\frac{A + B}{2}\right) \cos \left(\frac{A - B}{2}\right)$$

Now, if $\dfrac{A + B}{2} = m\theta$ and $\dfrac{A - B}{2} = n\theta$, it can be shown that

$$\sin m\theta \cdot \cos n\theta = \tfrac{1}{2} \sin (m + n)\theta + \tfrac{1}{2} \sin (m - n)\theta$$

$$\therefore \int \sin m\theta \cdot \cos n\theta \cdot d\theta$$

$$= \tfrac{1}{2}\int \sin (m + n)\theta \cdot d\theta + \tfrac{1}{2}\int \sin (m - n)\theta \cdot d\theta$$

$$= - \frac{\cos (m + n)\theta}{2(m + n)} - \frac{\cos (m - n)\theta}{2(m - n)} + C$$

Similarly,

$\int \sin m\theta \cdot \sin n\theta \cdot d\theta$

$$= -\frac{\sin (m+n)\theta}{2(m+n)} + \frac{\sin (m-n)\theta}{2(m-n)} + C$$

$\int \cos m\theta \cdot \cos n\theta \cdot d\theta$

$$= \frac{\sin (m+n)\theta}{2(m+n)} + \frac{\sin (m-n)\theta}{2(m-n)} + C$$

508. What is $\int \sin 2\theta \cdot \cos 4\theta \cdot d\theta$?

$$\int \sin m\theta \cdot \cos n\theta \cdot d\theta = -\frac{\cos (m+n)\theta}{2(m+n)} - \frac{\cos (m-n)\theta}{2(m-n)}$$

Here $m = 2$, and $n = 4$.

$$\therefore \int \sin 2\theta \cdot \cos 4\theta \cdot d\theta = -\frac{\cos (2+4)\theta}{2(2+4)} - \frac{\cos (2-4)\theta}{2(2-4)}$$

$$= -\frac{\cos 6\theta}{12} + \frac{\cos (-2\theta)}{4}$$

$$= \frac{\cos 2\theta}{4} - \frac{\cos 6\theta}{12} + C$$

509. What is $\int \sin 3\theta \cdot \sin 2\theta \cdot d\theta$?

$$\int \sin m\theta \cdot \sin n\theta \cdot d\theta = -\frac{\sin (m+n)\theta}{2(m+n)} + \frac{\sin (m-n)\theta}{2(m-n)} + C$$

Here $m = 3$, and $n = 2$.

$$\therefore \int \sin 3\theta \cdot \sin 2\theta \cdot d\theta = -\frac{\sin (3+2)\theta}{2(3+2)} + \frac{\sin (3-2)\theta}{2(3-2)}$$

$$= -\frac{\sin 5\theta}{10} + \frac{\sin \theta}{2} + C$$

510. What is $\int \cos 4\theta \cdot \cos 3\theta \cdot d\theta$?

$$\int \cos m\theta \cdot \cos n\theta \cdot d\theta = \frac{\sin (m+n)\theta}{2(m+n)} + \frac{\sin (m-n)\theta}{2(m-n)}$$

Here $m = 4$, and $n = 3$.

$$\therefore \int \cos 4\theta \cdot \cos 3\theta \cdot d\theta = \frac{\sin (4 + 3)\theta}{2(4 + 3)} + \frac{\sin (4 - 3)\theta}{2(4 - 3)}$$

$$= \frac{\sin 7\theta}{14} + \frac{\sin \theta}{2} + C$$

PROBLEMS

Integrate the following by transformation and reduction:

1. $\int \sin^5 \theta \cdot \cos^2 \theta \cdot d\theta$

2. $\int \sin^2 \theta \cdot \cos^5 \theta \cdot d\theta$

3. $\int \tan^3 \theta \cdot d\theta$

4. $\int \cot^4 2\theta \cdot d\theta$

5. $\int \sec^6 2\theta \cdot d\theta$

6. $\int \operatorname{cosec}^3 2\theta \cdot d\theta$

7. $\int \tan^4 \theta \cdot \sec^4 \theta \cdot d\theta$

8. $\int \cot^4 \theta \cdot \operatorname{cosec}^4 \theta \cdot d\theta$

9. $\int \tan^5 \theta \cdot \sec^3 \theta \cdot d\theta$

10. $\int \sin^6 \theta \cdot \cos^4 \theta \cdot d\theta$

11. $\int \sin 3\theta \cdot \cos 5\theta \cdot d\theta$

12. $\int \sin 6\theta \cdot \sin 4\theta \cdot d\theta$

13. $\int \cos 3\theta \cdot \cos 2\theta \cdot d\theta$

14. $\int \cos^4 \theta \cdot d\theta$

15. $\int (2 - \sin \theta)^2 \cdot d\theta$

16. $\int \sin^2 \frac{\theta}{2} \cos^2 \frac{\theta}{2} \cdot d\theta$

Chapter XXVI

INTEGRATION BY PARTS

511. What is the formula for integration by parts?

$$\int u \cdot dv = u \cdot v - \int v \cdot du$$

The function under the integral sign is broken up into two parts, u and dv; and if $\int v \cdot du$ can be found, then $\int u \cdot dv$ can also be found. The choice should be so made that $\int v \cdot du$ will be simpler or not any more difficult than the original integral $\int u \cdot dv$. Choose the most complicated part as part of dv.

If u and v are functions of the same independent variable, then

$$d(u \cdot v) = u \cdot dv + v \cdot du = \text{differentiation of a product}$$

or

$$u \cdot dv = d(u \cdot v) - v \cdot du$$
$$\therefore \int u \cdot dv = u \cdot v - \int v \cdot du = \text{the formula}$$

512. When is the formula for integration by parts most useful?

In integrating products of two or more functions, in integrals of logarithmic, exponential, and inverse trigonometric functions.

513. If $\dfrac{dy}{dx} = x^2 \log_e x$, how would you integrate by parts?

$$dy = x^2 \log_e x \cdot dx$$

Now, let
$$u = \log_e x$$

Then
$$du = \frac{1}{x} \cdot dx$$

And let
$$dv = x^2 \cdot dx$$

Then
$$v = \frac{x^3}{3}$$

Now,
$$\int u \cdot dv = u \cdot v - \int v \cdot du$$

$$\therefore y = \int \overbrace{\log_\epsilon x}^{u} \cdot \overbrace{x^2 \cdot dx}^{dv} = \int u \cdot dv = \frac{x^3}{3} \log_\epsilon x - \int \frac{x^3}{3} \cdot \frac{1}{x} \cdot dx$$

or $y = \dfrac{x^3}{3} \log_\epsilon x - \dfrac{x^3}{9} + C = $ the integral

514. If $\dfrac{dy}{dx} = \sin^2 x$, how would you integrate by parts?

$$dy = \sin^2 x \cdot dx$$

Let
$$u = \sin x$$
Then
$$du = \cos x \cdot dx$$
Let
$$dv = \sin x \cdot dx$$
Then
$$v = - \cos x$$

$$\therefore y = \int \overbrace{\sin x}^{u} \cdot \overbrace{\sin x \cdot dx}^{dv} = \int u \cdot dv$$
$$= - \sin x \cdot \cos x + \int \cos x \cdot \cos x \cdot dx$$
$$y = \int \sin^2 x \cdot dx = - \sin x \cdot \cos x + \int \cos^2 x \cdot dx \quad (1)$$

But
$$\cos^2 x = 1 - \sin^2 x$$

$$\therefore \int \cos^2 x \cdot dx = \int dx - \int \sin^2 x \cdot dx = x - \int \sin^2 x \cdot dx$$

Substituting in Eq. (1),

$$y = \int \sin^2 x \cdot dx = - \sin x \cdot \cos x + x - \int \sin^2 x \cdot dx$$

transposing the last term.

$$y = 2\int \sin^2 x \cdot dx = x - \sin x \cdot \cos x$$
$$\therefore y = \int \sin^2 x \cdot dx = \frac{x}{2} - \frac{\sin x \cdot \cos x}{2} + C = \text{the integral}$$

515. If $\dfrac{dy}{dx} = x^2 \sin x$, how would you integrate by parts?

$$dy = x^2 \sin x \cdot dx$$
$$\int u \cdot dv = u \cdot v - \int v \cdot du$$

Let
$$u = x^2$$

Then $\qquad du = 2x \cdot dx$

Let $\qquad dv = \sin x \cdot dx$

Then $\qquad v = -\cos x$

$$\therefore y = \int \overset{u}{\overbrace{x^2}} \overset{dv}{\overbrace{\sin x \cdot dx}} = \int u \cdot dv = -x^2 \cos x + \int 2x \cdot \cos x \cdot dx$$

$$y = -x^2 \cos x + 2\int x \cdot \cos x \cdot dx$$

To integrate $\qquad \int x \cdot \cos x \cdot dx$

Let $\qquad u = x$

Then $\qquad du = dx$

Let $\qquad dv = \cos x \cdot dx$

Then $\qquad v = \sin x$

By parts,

$$uv - \int v \cdot du = x \sin x - \int \sin x \cdot dx = x \sin x + \cos x$$

Finally, $y = 2x \sin x + 2 \cos x - x^2 \cos x = $ the integral

516. If $\dfrac{dy}{dx} = \cos^2 x$, **what is the integral by parts?**

$$dy = \cos^2 x \cdot dx$$
$$\int u\, dv = uv - \int v \cdot du$$

Let $\qquad u = \cos x$

Then $\qquad du = -\sin x \cdot dx$

Let $\qquad dv = \cos x \cdot dx$

Then $\qquad v = \sin x$

$$y = \int \overset{u}{\overbrace{\cos x}} \cdot \overset{dv}{\overbrace{\cos x \cdot dx}} = \int u \cdot dv = \sin x \cdot \cos x - \int - \sin^2 x \cdot dx$$

$$= \int \cos^2 x \cdot dx = \frac{2 \cos x \cdot \sin x}{2} + \int (1 - \cos^2 x)\, dx$$

$$= \int \cos^2 x \cdot dx = \frac{\sin 2x}{2} + \int dx - \int \cos^2 x \cdot dx,$$

transposing the last term

$$= 2\int \cos^2 x \cdot dx = \frac{\sin 2x}{2} + x$$

$$\therefore y = \int \cos^2 x \cdot dx = \frac{\sin 2x}{4} + \frac{x}{2} + C = \text{ the integral}$$

517. What is $\int x\epsilon^{2x} \cdot dx$?

$$\text{Let} \qquad u = x \qquad \text{and} \qquad dv = \epsilon^{2x} \cdot dx$$
$$\text{Then} \qquad du = dx \qquad \text{and} \qquad v = \int\epsilon^{2x} \cdot dx = \tfrac{1}{2}\epsilon^{2x}$$
$$\text{Now} \qquad \int u \cdot dv = uv - \int v \cdot du$$

$$\int \overset{u}{x} \cdot \overbrace{\epsilon^{2x} \cdot dx}^{dv} = x \cdot \tfrac{1}{2}\epsilon^{2x} - \int \tfrac{1}{2}\epsilon^{2x} \cdot dx = \tfrac{1}{2}x\epsilon^{2x} - \tfrac{1}{4}\epsilon^{2x}$$
$$= \tfrac{1}{4}\epsilon^{2x}(2x - 1)$$

Show that, if another choice had been made for u and dv, as $u = \epsilon^{2x}$ and $dv = x \cdot dx$, a more difficult integral to evaluate than the original would result.

518. What is the integration of $\int\epsilon^{ax} \cdot \sin 3x \cdot dx$?

$$\text{Let} \qquad u = \epsilon^{ax} \qquad \text{and} \qquad dv = \sin 3x \cdot dx$$

Then

$$du = a\epsilon^{ax} \cdot dx \qquad \text{and} \qquad v = \int \sin 3x \cdot dx = -\tfrac{1}{3} \cos 3x$$
$$\int u \cdot dv = uv - \int v \cdot du$$
$$\int\epsilon^{ax} \cdot \sin 3x \cdot dx = \epsilon^{ax} \cdot -\tfrac{1}{3} \cos 3x - \int -\tfrac{1}{3} \cos 3x \cdot a\epsilon^{ax} \cdot dx$$
$$= -\frac{1}{3} \epsilon^{ax} \cdot \cos 3x + \frac{a}{3} \int \epsilon^{ax} \cdot \cos 3x \cdot dx$$

$$(1)$$

Now apply integration by parts to $\int\epsilon^{ax} \cdot \cos 3x \cdot dx$.

$$\text{Let} \qquad\qquad\qquad u = \epsilon^{ax}$$
$$\text{and} \qquad\qquad\qquad dv = \cos 3x \cdot dx$$
$$\text{Then } du = a\epsilon^{ax} \cdot dx \qquad \text{and} \qquad v = \int \cos 3x \cdot dx = \tfrac{1}{3} \sin 3x$$
$$\int u \cdot dv = uv - \int v \cdot du$$
$$\text{Then } \int\epsilon^{ax} \cdot \cos 3x \cdot dx = \epsilon^{ax} \cdot \tfrac{1}{3} \sin 3x - \int \tfrac{1}{3} \sin 3x \cdot a\epsilon^{ax} \cdot dx$$

$$= \frac{1}{3} \epsilon^{ax} \cdot \sin 3x - \frac{a}{3} \int \epsilon^{ax} \cdot \sin 3x \cdot dx$$

Substitute this result in Eq. (1).

$$\int \epsilon^{ax} \cdot \sin 3x \, dx = -\frac{1}{3} \epsilon^{ax} \cdot \cos 3x + \frac{a}{3} \left(\frac{1}{3} \epsilon^{ax} \cdot \sin 3x \right.$$
$$\left. -\frac{a}{3} \int \epsilon^{ax} \cdot \sin 3x \cdot dx \right)$$
$$= -\frac{1}{3} \epsilon^{ax} \cdot \cos 3x + \frac{a}{9} \cdot \epsilon^{ax} \cdot \sin 3x$$
$$-\frac{a^2}{9} \int \epsilon^{ax} \cdot \sin 3x \cdot dx$$

Transpose the integral in the right member to the left side, and unite, getting

$$\left(1 + \frac{a^2}{9} \right) \int \epsilon^{ax} \cdot \sin 3x \cdot dx = -\frac{1}{3} \epsilon^{ax} \cdot \cos 3x$$
$$+ \frac{a}{9} \epsilon^{ax} \cdot \sin 3x$$

Now divide by $\left(1 + \dfrac{a^2}{9} \right)$ and simplify.

$$\int \epsilon^{ax} \cdot \sin 3x \cdot dx = \frac{\epsilon^{ax}(a \sin 3x - 3 \cos 3x)}{9 + a^2} + C$$

Although $\int v \cdot du$ was no simpler than the original $\int u \cdot dv$, an integral like the original finally appeared and was transposed and simplified. Work out the integral, choosing $dv = \epsilon^{ax} \cdot dx$ and $u = \sin 3x$, and compare.

519. What is the integration of $\int x \cdot \cos x \cdot dx$?

Let $\qquad u = x \qquad$ and $\qquad dv = \cos x \cdot dx$
Then $\quad du = dx \qquad$ and $\qquad v = \int \cos x \cdot dx = \sin x$
$\int u \cdot dv = uv - \int v \cdot du$ becomes

$$\int x \cdot \cos x \cdot dx = x \cdot \sin x - \int \sin x \cdot dx$$
$$= x \sin x + \cos x + C$$

520. What is $\int \sec^3 x \cdot dx$?

Let $\quad u = \sec x \quad$ and $\quad dv = \sec^2 x \cdot dx$
Then $du = \sec x \cdot \tan x \cdot dx \quad$ and $\quad v = \int \sec^2 x \cdot dx$
$$= \tan x$$

Now $\int u \cdot dv = uv - \int v \cdot du$ becomes

$\int \sec x \cdot \sec^2 x \cdot dx = \sec x \cdot \tan x$
$$- \int \tan x \cdot \sec x \cdot \tan x \cdot dx$$
$$= \sec x \cdot \tan x - \int \tan^2 x \cdot \sec x \cdot dx$$
But $\qquad\qquad \tan^2 x = \sec^2 x - 1$

Substitute this to get

$\int \sec^3 x \cdot dx = \sec x \cdot \tan x - \int \sec^3 x \cdot dx + \int \sec x \cdot dx$
$$= \sec x \cdot \tan x - \int \sec^3 x \cdot dx + \log_\epsilon (\sec x$$
$$+ \tan x)$$

Now transform and simplify, obtaining

$$2\int \sec^3 x \cdot dx = \sec x \cdot \tan x + \log_\epsilon (\sec x + \tan x)$$
or $\quad \displaystyle\int \sec^3 x \cdot dx = \frac{\sec x \cdot \tan x + \log_\epsilon (\sec x + \tan x)}{2}$

PROBLEMS

Integrate the following by parts:

1. $\int x^3 \epsilon^{ax} \cdot dx$
2. $\int x \log_\epsilon x \cdot dx$
3. $\int x^2 \cdot \cos x \cdot dx$
4. $\int \epsilon^x \cdot \sin x \cdot dx$
5. $\int \epsilon^x \cdot \cos x \cdot dx$
6. $\int \epsilon^{ax} \cdot \cos 4x \cdot dx$
7. $\int \epsilon^x \cdot \sin 4x \cdot dx$
8. $\int \sin^{-1} x \cdot dx$
9. $\int \cos^{-1} x \cdot dx$
10. $\int \tan^{-1} x \cdot dx$
11. $\int x^2 \sin 2x \cdot dx$
12. $\int x^2 \sin x \cdot \cos x \cdot dx$
13. $\int \epsilon^{ax} \cdot \cos 3x \cdot dx$
14. $\displaystyle\int x \sin \frac{x}{2} \cdot dx$
15. $\int x \cdot \cos 3x \cdot dx$
16. $\int x \cdot a^x \cdot dx$
17. $\displaystyle\int \frac{\log_\epsilon x \cdot dx}{(1 + x)^2}$
18. $\int \tan^{-1} x \cdot dx$
19. $\displaystyle\int \frac{x \epsilon^x \cdot dx}{(1 + x)^2}$
20. $\int x \epsilon^{-x} \cdot dx$

Chapter XXVII

INTEGRATION OF FRACTIONS

521. What is a rational fraction in x?

A fraction having the numerator and denominator polynomials in x. The variable is not affected with negative or fractional exponents.

522. How would you integrate when the degree of the numerator is equal to or greater than the degree of the denominator?

First, the fraction is reduced by dividing the numerator by the denominator.

Example

If $y = \displaystyle\int \frac{x^4 + 6x^3}{x^2 + 3x + 1} \cdot dx$, dividing out first, we get

$$\frac{x^4 + 6x^3}{x^2 + 3x + 1} = x^2 + 3x - 10 + \frac{27x + 10}{x^2 + 3x + 1}$$

$$\therefore y = \int \frac{x^4 + 6x^3}{x^2 + 3x + 1} \cdot dx = \int x^2 \cdot dx + 3 \int x \cdot dx$$

$$- 10 \int dx + \int \frac{27x + 10}{x^2 + 3x + 1} \cdot dx$$

$$= \frac{x^3}{3} + \frac{3x^2}{2} - 10x + \int \frac{27x + 10}{x^2 + 3x + 1} \cdot dx$$

523. How would you integrate the expression

$$\int \frac{3x + 6}{x^3 + 2x^2 - 3x} \cdot dx$$

by the method of proper fractions?

$$\frac{3x + 6}{x^3 + 2x^2 - 3x} = \frac{3x + 6}{x(x - 1)(x + 3)} = \frac{A}{x} + \frac{B}{(x - 1)} + \frac{C}{x + 3}$$

Clearing fractions,

$$
\begin{aligned}
3x + 6 &= A(x - 1)(x + 3) + Bx(x + 3) + Cx(x - 1) \\
&= A(x^2 + 2x - 3) + Bx^2 + 3Bx + Cx^2 - Cx \\
&= x^2(A + B + C) + x(2A + 3B - C) - 3A
\end{aligned}
$$

Equating coefficients of like powers,

x^2,	$A + B + C = 0$	(1)
x,	$2A + 3B - C = 3$	(2)
Constants,	$-3A = 6$	(3)
	$\therefore A = -2$	

Put the value of A in Eq. (1).

$$B + C = 2 \quad \text{or} \quad C = 2 - B$$

Put the value of A in Eq. (2).

$$
\begin{aligned}
-4 + 3B - C = 3 \quad &\text{or} \quad 3B - C = 7 \\
\text{or,} \quad 3B - (2 - B) = 7 \quad &\text{or} \quad 4B = 9 \\
\therefore B &= \tfrac{9}{4} \\
\therefore C = 2 - B = 2 - \tfrac{9}{4} &= -\tfrac{1}{4}
\end{aligned}
$$

$$\therefore \frac{3x + 6}{x^3 + 2x^2 - 3x} = -\frac{2}{x} + \frac{9}{4(x - 1)} - \frac{1}{4(x + 3)}$$

Finally, $y = \displaystyle\int \frac{(3x + 6)\, dx}{x^3 + 2x^2 - 3x} = -2 \int \frac{dx}{x} + \frac{9}{4} \int \frac{dx}{x - 1}$

$$- \frac{1}{4} \int \frac{dx}{x + 3}$$

$$= -2 \log_\epsilon x + \tfrac{9}{4} \log_\epsilon (x - 1)$$

$$- \tfrac{1}{4} \log_\epsilon (x + 3) + C$$

$$\therefore y = \log_\epsilon \frac{C(x - 1)^{\frac{9}{4}}}{x^2(x + 3)^{\frac{1}{4}}}$$

524. How would you integrate a proper rational fraction whose denominator is a single factor repeated, and whose numerator is of degree not less than the degree of that factor, as $\int \dfrac{4x + 2}{(x + 3)^2} \cdot dx$?

First by division, using the factor to the first power as a divisor.

$$\frac{4x + 2}{x + 3} = 4 - \frac{10}{x + 3}$$

Then

$$\frac{4x + 2}{(x + 3)^2} = \frac{4}{x + 3} - \frac{10}{(x + 3)^2}$$

and

$$\int \frac{(4x + 2) \cdot dx}{(x + 3)^2} = \int \frac{4 \cdot dx}{(x + 3)} - \int \frac{10 \cdot dx}{(x + 3)^2}$$

$$= 4 \int \frac{dx}{x + 3} - 10 \int (x + 3)^{-2} \cdot dx$$

$$= 4 \log_e (x + 3) - \frac{10 \cdot (x + 3)^{-2+1}}{-2 + 1}$$

$$= 4 \log_e (x + 3) + \frac{10}{x + 3} + C$$

525. How would you integrate a proper rational fraction that has a denominator consisting only of factors of the first or second degree none of which are repeated, as $\int \dfrac{4x - 2}{(x + 2)(x - 4)} \cdot dx$?

The fraction can be separated into just as many proper partial fractions as there are factors, and each fraction will have one of the factors as a denominator.

$$\frac{4x - 2}{(x + 2)(x - 4)} = \frac{A}{x + 2} + \frac{B}{x - 4} \qquad (1)$$

Clearing of fractions,

$$4x - 2 = A(x - 4) + B(x + 2) \qquad (2)$$

This equation will be satisfied by all values of x. Choose $x = 4$ and $x = -2$ to reduce the factors to zero.

For $x = 4$ in Eq. (2),

$$4 \cdot 4 - 2 = A(4 - 4) + B(4 + 2)$$

or
$$14 = 6B$$
$$\therefore B = \tfrac{7}{3}$$

For $x = -2$ in Eq. (2),

$$4 \cdot (-2) - 2 = A(-2 - 4) + B(-2 + 2)$$
or
$$-10 = -6A$$
$$\therefore A = \tfrac{5}{3}$$

Now substitute the values of A and B in Eq. (1).

$$\frac{4x - 2}{(x + 2)(x - 4)} = \frac{5}{3(x + 2)} + \frac{7}{3(x - 4)}$$

Then $\displaystyle\int \frac{4x - 2}{(x + 2)(x - 4)} \cdot dx = \frac{5}{3} \int \frac{dx}{(x + 2)}$

$$+ \frac{7}{3} \int \frac{dx}{(x - 4)}$$

$$= \tfrac{5}{3} \log_e (x + 2)$$
$$+ \tfrac{7}{3} \log_e (x - 4)$$
$$= \log_e (x + 2)^{\frac{5}{3}}(x - 4)^{\frac{7}{3}} + C$$

526. What is the integral of $\displaystyle\int \frac{x^3 + 5x^2 + 2x - 4}{(x^4 - 1)} \cdot dx$?

The denominator $(x^4 - 1)$ can be factored into

$$(x - 1)(x + 1)(x^2 + 1).$$

Then $\dfrac{x^3 + 5x^2 + 2x - 4}{(x^4 - 1)} = \dfrac{A}{x - 1} + \dfrac{B}{x + 1} + \dfrac{Cx + D}{x^2 + 1}$

When the denominator contains a term in x^2, an x term must appear in the numerator in addition to a constant.

Now, clearing fractions, we get

$$x^3 + 5x^2 + 2x - 4 = A(x + 1)(x^2 + 1)$$
$$+ B(x - 1)(x^2 + 1) + (Cx + D)(x - 1)(x + 1)$$

Let $x = 1$.

Then $1 + 5 + 2 - 4 = A(1 + 1)(1 + 1) + B \cdot 0$
$$+ (C + D) \cdot 0$$

or $$4 = 4A$$
$$\therefore A = 1$$

Let $x = -1$.

Then $-1 + 5 - 2 - 4 = A \cdot 0 + B(-1 - 1)(1 + 1)$
$$+ [C \cdot (-1) + D] \cdot 0$$

or $$-2 = -4B$$
$$\therefore B = \tfrac{1}{2}$$

Let $x = 0$.

Then $0 + 0 + 0 - 4 = A(0 + 1)(0 + 1)$
$$+ B(0 - 1)(0 + 1) + (0 + D)(0 - 1)(0 + 1)$$

or $$-4 = A - B - D = 1 - \tfrac{1}{2} - D$$
$$\therefore D = 4\tfrac{1}{2}$$

Let $x = 2$.

Then $8 + 20 + 4 - 4 = A(2 + 1)(4 + 1)$
$$+ B(2 - 1)(4 + 1) + (C \cdot 2 + D)(2 - 1)(2 + 1)$$

or $$28 = 15A + 5B + 6C + 3D$$
$$28 = 15 \cdot 1 + 5 \cdot \tfrac{1}{2} + 6C + 13\tfrac{1}{2}$$
$$-3 = 6C$$
$$\therefore C = -\tfrac{1}{2}$$

Then $\dfrac{x^3 + 5x^2 + 2x - 4}{(x^4 - 1)} = \dfrac{1}{(x - 1)} + \dfrac{1}{2(x + 1)}$
$$+ \dfrac{-\tfrac{1}{2} \cdot x + 4\tfrac{1}{2}}{x^2 + 1}$$

$$= \dfrac{1}{x - 1} + \dfrac{1}{2(x + 1)} + \dfrac{9 - x}{2(x^2 + 1)}$$

Then $\displaystyle\int \dfrac{x^3 + 5x^2 + 2x - 4}{(x^4 - 1)} \cdot dx = \int \dfrac{dx}{x - 1} + \dfrac{1}{2} \int \dfrac{dx}{x + 1}$
$$+ \dfrac{9}{2} \int \dfrac{dx}{(x^2 + 1)} - \dfrac{1}{2} \int \dfrac{x \cdot dx}{(x^2 + 1)}$$

$\displaystyle\therefore \int \dfrac{x^3 + 5x^2 + 2x - 4}{(x^4 - 1)} \cdot dx = \log_e (x - 1)$
$$+ \dfrac{1}{2} \log_e (x + 1) + \dfrac{9}{2} \tan^{-1} x - \dfrac{1}{4} \log_e (x^2 + 1) + C$$

527. How would you integrate a proper rational fraction that has a denominator consisting of factors of the first and second degree one or more of which are repeated?

Separate the fraction into as many partial fractions as there are different factors in the denominator. Each not-repeated factor will be a denominator, and each repeated factor affected with an exponent that is the number of times it is repeated will be a denominator. Then the partial fractions having repeated factors as denominators may be further separated by actual division.

528. What is the integral of $\int \dfrac{x^3 + 5x^2 + 2x - 4}{x(x^2 + 4)^2} \cdot dx$?

$$\frac{x^3 + 5x^2 + 2x - 4}{x(x^2 + 4)^2} = \frac{A}{x} + \frac{Bx^3 + Cx^2 + Dx + E}{(x^2 + 4)^2} \quad (1)$$

Always provide a numerator of one less degree than the denominator.

Clearing fractions,

$$x^3 + 5x^2 + 2x - 4 = A(x^2 + 4)^2 + x(Bx^3 + Cx^2 + Dx + E)$$

$$x^3 + 5x^2 + 2x - 4 = Ax^4 + 8Ax^2 + 16A + Bx^4 + Cx^3 + Dx^2 + Ex \quad (2)$$

Now equate coefficients of like powers of x.

Coefficients of x^4, $0 = A + B \qquad \therefore B = +\frac{1}{4}$
(see value of A below)

Coefficients of x^3, $\ 1 = C \qquad \therefore C = 1$

Coefficients of x^2, $5 = 8A + D, \qquad 5 = 8 \cdot (-\frac{1}{4}) + D$
$\therefore D = 7$ (see value of A below)

Coefficients of x, $\quad 2 = E \qquad \therefore E = 2$

Coefficients of x^0, $-4 = 16A \qquad \therefore A = -\frac{1}{4}$

Now substitute these in Eq. (1).

$$\frac{x^3 + 5x^2 + 2x - 4}{x(x^2 + 4)^2} = \frac{-\frac{1}{4}}{x} + \frac{\frac{1}{4}x^3 + x^2 + 7x + 2}{(x^2 + 4)^2}$$

Factor out $\frac{1}{4}$ of the right-hand member,

or $\qquad \frac{1}{4}\left[-\frac{1}{x} + \frac{x^3 + 4x^2 + 28x + 8}{(x^2 + 4)^2} \right]$

Divide out the fraction.

$$\frac{x^3 + 4x^2 + 28x + 8}{x^2 + 4} = x + 4 + \frac{24x - 8}{x^2 + 4}$$

$$\therefore \frac{x^3 + 4x^2 + 28x + 8}{(x^2 + 4)^2} = \frac{x + 4}{(x^2 + 4)} + \frac{24x - 8}{(x^2 + 4)^2}$$

Then $\dfrac{x^3 + 5x^2 + 2x - 4}{x(x^2 + 4)^2} = \dfrac{1}{4}\left[-\dfrac{1}{x} + \dfrac{x + 4}{(x^2 + 4)} \right.$

$$\left. + \frac{24x - 8}{(x^2 + 4)^2} \right]$$

$$\int \frac{x^3 + 5x^2 + 2x - 4}{x(x^2 + 4)^2} \cdot dx$$

$$= -\frac{1}{4} \int \frac{dx}{x} + \frac{1}{4} \int \frac{x \cdot dx}{x^2 + 4} + \int \frac{dx}{x^2 + 4} + \frac{1}{4} \int \frac{24x \cdot dx}{(x^2 + 4)^2}$$

$$- 2 \int \frac{dx}{(x^2 + 4)^2}$$

$$= -\frac{1}{4} \log_e x + \frac{1}{8} \log_e (x^2 + 4) + \frac{1}{2} \tan^{-1} \frac{x}{2}$$

$$+ 3 \int (x^2 + 4)^{-2} \cdot 2x \cdot dx - 2 \int \frac{dx}{(x^2 + 4)^2}$$

$$= \log_e \frac{(x^2 + 4)^{\frac{1}{8}}}{x^{\frac{1}{4}}} + \frac{1}{2} \tan^{-1} \frac{x}{2} - \frac{3}{(x^2 + 4)} - 2 \int \frac{dx}{(x^2 + 4)^2}$$

(The last term is not easily integrated by simple methods.)

529. How would you integrate $\displaystyle\int \frac{x^2 + 4x}{(x - 2)^2(x^2 + 4)} \cdot dx$**?**

$$\frac{x^2 + 4x}{(x - 2)^2(x^2 + 4)} = \frac{Ax + B}{(x - 2)^2} + \frac{Cx + D}{(x^2 + 4)} \qquad (1)$$

Clearing fractions,

$$x^2 + 4x = (Ax + B)(x^2 + 4) + (Cx + D)(x - 2)^2 \qquad (2)$$

For $x = 2$,

$$4 + 8 = (2A + B)(4 + 4) + (2C + D) \cdot 0$$
$$12 = 16A + 8B \quad \text{or} \quad 3 = 4A + 2B \qquad (3)$$

For $x = 0$,

$$0 + 0 = (0 + B)(0 + 4) + (0 + D)(0 - 2)^2$$
$$0 = 4B + 4D \quad \text{or} \quad B = -D$$

Equating coefficients of x^3 of Eq. (2),

$$0 = A + C \quad \text{or} \quad A = -C$$

Equating coefficients of x of Eq. (2),

$$4 = 4A + 4C - 4D$$
$$4 = 4A - 4A - 4D \quad \text{or} \quad D = -1 \quad \text{and} \quad \therefore B = 1$$

Now, from Eq. (3),

$$3 = 4A + 2 \cdot 1 \quad \text{or} \quad A = \tfrac{1}{4} \quad \text{and} \quad \therefore C = -\tfrac{1}{4}$$

Then substitute these values in Eq. (1).

$$\frac{x^2 + 4x}{(x - 2)^2(x^2 + 4)} = \frac{\tfrac{1}{4}x + 1}{(x - 2)^2} + \frac{-\tfrac{1}{4}x - 1}{(x^2 + 4)}$$

$$= \frac{1}{4}\left[\frac{x + 4}{(x - 2)^2}\right] - \frac{1}{4}\left[\frac{x + 4}{(x^2 + 4)}\right]$$

$$= \frac{1}{4}\left[\frac{x + 4}{(x - 2)^2} - \frac{x + 4}{(x^2 + 4)}\right]$$

By division,

$$\frac{x + 4}{x - 2} = 1 + \frac{6}{x - 2}$$

Then

$$\frac{x + 4}{(x - 2)^2} = \frac{1}{(x - 2)} + \frac{6}{(x - 2)^2}$$

$$\therefore \frac{x^2 + 4x}{(x - 2)^2(x^2 + 4)} = \frac{1}{4}\left[\frac{1}{(x - 2)}\right.$$

$$\left. + \frac{6}{(x - 2)^2} - \frac{x + 4}{(x^2 + 4)}\right]$$

$$\int \frac{x^2 + 4x}{(x-2)^2(x^2+4)} \cdot dx = \frac{1}{4} \int \frac{dx}{x-2} + \frac{3}{2} \int (x-2)^{-2} \cdot dx$$

$$- \frac{1}{4} \int \frac{x \cdot dx}{x^2+4} - \int \frac{dx}{x^2+4}$$

$$= \frac{1}{4} \log_e (x-2) - \frac{3}{2(x-2)}$$

$$- \frac{1}{8} \log_e (x^2+4) - \frac{1}{2} \tan^{-1} \frac{x}{2}$$

PROBLEMS

Integrate the following by the methods for fractions:

1. $\int \frac{4x+2}{(x+1)^2} \cdot dx$,

2. $\int \frac{3x^4 + 2x^3 - 3x + 1}{(x^2+1)^3} \cdot dx$

3. $\int \frac{4x-2}{(x-2)(x+2)} \cdot dx$

4. $\int \frac{x^3 + 5x^2 - 2x + 6}{x^4 - 1} \cdot dx$

5. $\int \frac{x^3 + x^2 + 4x - 3}{x(x^2+2)^2} \cdot dx$

6. $\int \frac{x^2 + x}{(x-2)^2(x^2+3)} \cdot dx$

7. $\int \frac{x-5}{(x-3)^2} \cdot dx$

8. $\int \frac{x^2 + 4x - 2}{(x^2+1)^2} \cdot dx$

9. $\int \frac{4x^2 - 3x - 2}{x(x^2-4)} \cdot dx$

10. $\int \frac{x^4}{x^3+1} \cdot dx$

11. $\int \frac{dx}{x^2(x^2+3)}$

12. $\int \frac{dx}{(x-2)^2(x^2+2)^2}$

CHAPTER XXVIII

PLANE AREAS BY INTEGRATION
DEFINITE INTEGRAL
DIFFERENTIAL OF AN AREA
LIMIT OF A SUM

530. How can we find the area under any curve when an equation giving the height y in terms of the horizontal distance x is known?

CB is a fixed ordinate. DM moves to the right.

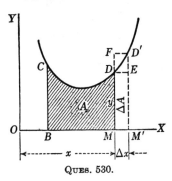

QUES. 530.

$$DM = y, \qquad D'M' = y + \Delta y$$

Area A will vary with x. When x is increased by Δx, A is increased by ΔA and

Rectangle $DEM'M < \Delta A$
$\qquad < $ rectangle $FD'M'M$

or

$$y \cdot \Delta x < \Delta A < (y + \Delta y) \cdot \Delta x$$

Dividing by Δx, we get

$$y < \frac{\Delta A}{\Delta x} < (y + \Delta y)$$

As Δx approaches zero, $(y + \Delta y)$ approaches y as a limit, and $\frac{\Delta A}{\Delta x}$ has a value between y and an expression that has y as a limiting value. Therefore, $\frac{\Delta A}{\Delta x}$ approaches y as a limit as Δx approaches zero.

Now, $\dfrac{\Delta A}{\Delta x}$ is the average rate of increase of A for the interval Δx. And the instantaneous rate of increase at any point is

$$\lim_{\Delta x \to 0} \left[\frac{\Delta A}{\Delta x} \right] = \frac{dA}{dx} = y = f(x) \quad [y \text{ and } f(x) \text{ are the same.}]$$

Therefore the rate at which the area is increasing at any point is equal to the ordinate at that point, and

$$dA = y \cdot dx \qquad \text{or} \qquad A = \int y \cdot dx = \int f(x) \cdot dx$$

If we can integrate $y \cdot dx$, we have a method of finding the area. This is called the *differential of the area method* and applies to falling as well as to rising curves.

531. What is meant by a definite integral?

Integration between limits, *i.e.*, between an upper limit b and a lower limit a, results in a definite integral. The integral then has a definite value.

$\int_a^b y \cdot dx$ or $\int_a^b f(x) \cdot dx$ is the symbol for a definite integral and is read "the integral from a to b of $y \cdot dx$."

532. How can we find a definite area under a curve?

Integrate the differential of the area under the curve and substitute in this, first the upper limit and then the lower limit for the variable, and subtract the last result from the first. The constant of integration disappears in subtracting and it is not necessary to bring it in.

In symbols,

$$\text{Area } A = \int_a^b y \cdot dx = F(b) - F(a)$$

This procedure applies only when the lines bound an area; *i.e.*, the curve does not rise or fall to infinity and does not cross the x axis, and the limits a and b are finite.

533. What is the value of the definite integral

$$\int_{x_1}^{x_2} (a + bx^2)\, dx?$$

The indefinite integral is $ax + \dfrac{bx^3}{3} + C$.

$$\left[ax + \frac{bx^3}{3} + C \right]_{x_1}^{x_2}$$ means "substitute the limits."

Substitute the superior limit

$$ax_2 + \frac{bx_2^3}{3} + C \tag{1}$$

Substitute the inferior limit

$$ax_1 + \frac{bx_1^3}{3} + C \tag{2}$$

Subtract Eq. (2) from Eq. (1), getting

$$a(x_2 - x_1) + \frac{b}{3}(x_2^3 - x_1^3) = \text{definite integral}$$

The added constant C always disappears in the definite integral.

534. What is the area under the curve $y = 2x^2$ between the fixed ordinates $x = 2$ and $x = 6$?

By the differential of the area method, $A = \int y \cdot dx$ becomes $A = \int_2^6 2x^2 \cdot dx$, since $y = 2x^2$.

$$A = 2\left[\frac{x^3}{3} \right]_2^6 = \frac{2}{3}\left[x^3 \right]_2^6 = \frac{2}{3}(6^3 - 2^3) = \frac{2}{3} \cdot 208 = 138\frac{2}{3}$$

535. What is the area of the curve $y = \dfrac{a}{x + b}$ between $x = x_2$ and $x = x_1$?

$$A = \int_{x_1}^{x_2} y \cdot dx = \int_{x_1}^{x_2} \frac{a}{x + b} \cdot dx$$

$$A = a \left[\log_\epsilon (x + b) + C \right]_{x_1}^{x_2}$$
$$= a[\log_\epsilon (x_2 + b) + C - \log_\epsilon (x_1 + b) - C]$$
$$= a \log_\epsilon \frac{x_2 + b}{x_1 + b}$$

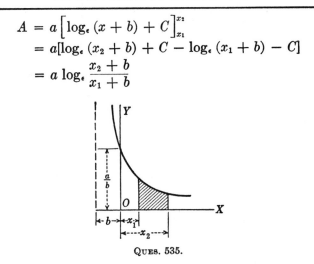

QUES. 535.

536. How do we determine the area when the curve is below the x axis?

Set the function equal to zero, and determine the points where the curve crosses the x axis. Integrate the function or differential between these determined limits.

Then the area is the negative of the integral.

$$A = -\int y \cdot dx = -\int f(x) \cdot dx$$

537. How do we determine the area when the curve is both above and below the x axis?

Each of the portions above and below the x axis is computed separately by applying the appropriate limits, and then these are added together to get the total area.

538. What is the area between the curve $y = x^2 - 4x + 3$ and the x axis?

Let $y = 0$, and solve for the points of intersection with the x axis.

$$0 = x^2 - 4x + 3 = (x - 1)(x - 3)$$
$$\therefore x = 1 \quad \text{and} \quad x = 3$$

These are the end values of x.

Then $A = -\int_1^3 (x^2 - 4x + 3)\, dx$ (An area below the
x axis is the negative of the integral.)

and $A = -\left[\dfrac{x^3}{3} - 2x^2 + 3x\right]_1^3$

$= -[(\tfrac{1}{3}\cdot 3^3 - 2\cdot 3^2 + 3\cdot 3)$
$\qquad\qquad - (\tfrac{1}{3}\cdot 1^3 - 2\cdot 1^2 + 3\cdot 1)]$

$= -(9 - 18 + 9 - \tfrac{1}{3} + 2 - 3) = 1\tfrac{1}{3}$ square units
for the area below the x axis

QUES. 538.

QUES. 539.

**539. What is the area between the curve $y = \cos x$ and
the x axis from $x = \dfrac{\pi}{4}$ to $x = \dfrac{3}{2}\pi$.**

$A_1 = \text{area} = \int_{\frac{\pi}{4}}^{\frac{\pi}{2}} y\cdot dx = \int_{\frac{\pi}{4}}^{\frac{\pi}{2}} \cos x\cdot dx$

$= \left[\sin x\right]_{\frac{\pi}{4}}^{\frac{\pi}{2}} = \left(\sin\frac{\pi}{2} - \sin\frac{\pi}{4}\right) = \left(1 - \frac{\sqrt{2}}{2}\right)$

$A_2 = -\int_{\frac{\pi}{2}}^{\frac{3\pi}{2}} \cos x\cdot dx = -\left[\sin x\right]_{\frac{\pi}{2}}^{\frac{3\pi}{2}}$

$= -\left(\sin\frac{3}{2}\pi - \sin\frac{\pi}{2}\right) = -(-1 - 1) = 2$

Then $A = A_1 + A_2 = \left(1 - \frac{\sqrt{2}}{2}\right) + 2 = 3 - \frac{\sqrt{2}}{2}$

$= 2.293$ square units

540. What is the expression for finding the area between a curve, the y axis, and two abscissas?

$$A = \int_{y_1}^{y_2} x \cdot dy$$

This expression is obtained in a similar manner to that of the $\int y \cdot dx$ method.

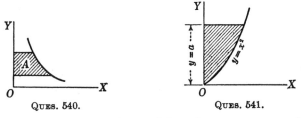

QUES. 540.   QUES. 541.

541. What is the area formed by the curve $y = x^2$, the y axis, and the abscissas $y = 0$ and $y = a$?

$$A = \int_{y_1}^{y_2} x \cdot dy = \int_0^a y^{\frac{1}{2}} \cdot dy = \left[\frac{y^{\frac{3}{2}}}{\frac{3}{2}}\right]_0^a$$

$$\therefore A = \tfrac{2}{3} a^{\frac{3}{2}}$$

542. What is the area formed by the curve $y^2 = x$, the x axis, and the ordinates at $x = 0$ and $x = a$?

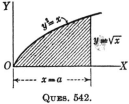

QUES. 542.

$$A = \int_{x_1}^{x_2} y \cdot dx = \int_0^a x^{\frac{1}{2}} \cdot dx$$

$$\therefore A = \left[x^{\frac{3}{2}} \cdot \tfrac{2}{3}\right]_0^a = \tfrac{2}{3} a^{\frac{3}{2}}$$

543. What is the effect of interchanging the limits of integration?

Interchanging the limits is equivalent to changing the sign of the definite integral.

Example

If $y_1 = \int_a^b f(x) \, dx$, then after integrating between a and b, with $b = $ upper limit, the result is

$$y_1 = F(b) - F(a)$$

Now if $y_2 = \int_b^a f(x)\, dx$ then after integrating between b and a with $a =$ upper limit, the result is

$$y_2 = F(a) - F(b)$$

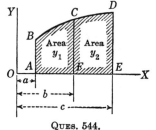

QUES. 544.

and this is *the first integral y_1 with sign changed.*

544. What is the effect of decomposing the limits of integration?

Decomposing the limits is equivalent to adding the integrals of the separate parts.

Example

If $y_1 = \int_a^b f(x)\, dx = F(b) - F(a) = \text{area } y_1$ (1)

and $y_2 = \int_b^c f(x)\, dx = F(c) - F(b) = \text{area } y_2$ (2)

then adding the above expressions (1) and (2) results in

$$y_3 = y_1 + y_2 = F(c) - F(a) = \text{the total area } ABDE$$
$$\therefore\ y_3 = \int f(x) \cdot dx = \int_a^b f(x) \cdot dx + \int_b^c f(x) \cdot dx = \text{the}$$
sum of the integrals of the separate parts

545. How can we find an area when the equations of the curve are given in parametric form?

If $x = f(t)$ and $y = F(t)$ are the parametric equations of the curve with t as the parameter, then $dx = f'(t)\, dt$ and area $= \int_a^b y \cdot dx$ becomes

$$\int_{t_1}^{t_2} F(t) \cdot f'(t) \cdot dt$$

where $t = t_1$ when $x = a$,
         $t = t_2$ when $x = b$.

546. What is the area bounded by one arch of the cycloid $x = a(\theta - \sin \theta)$, $y = a(1 - \cos \theta)$, and the x axis?

$$\frac{dx}{d\theta} = a(1 - \cos \theta) \quad \text{or} \quad dx = a(1 - \cos \theta)\, d\theta$$

$$\text{Area} = \int_0^{2\pi} a(1 - \cos \theta) \cdot a(1 - \cos \theta)\, d\theta$$

$$= a^2 \int_0^{2\pi} (1 - \cos \theta)^2\, d\theta$$

$$= a^2 \int_0^{2\pi} (1 - 2 \cos \theta + \cos^2 \theta)\, d\theta$$

$$= a^2 \left[\theta - 2 \sin \theta + \frac{\theta}{2} + \frac{1}{4} \sin 2\theta \right]_0^{2\pi}$$

$$\therefore \text{Area} = a^2(2\pi - 0 + \pi + 0) = 3\pi a^2 \text{ square units}$$

547. What are polar coordinates?

Polar coordinates of a point in a plane locate the point by giving its distance from a given fixed point and its angular direction from a given fixed line in the plane. The fixed point is O, called the *pole* or *origin*. The fixed line is OX, or *polar axis*. $OP = \rho = $ radius vector of P. Angle $\theta = $ vectorial angle of P. Counterclockwise angles are positive.

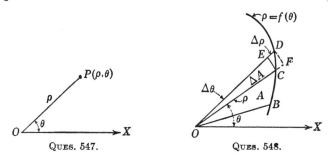

QUES. 547.   QUES. 548.

548. How do we find an area in polar coordinates by the differential of the area method?

As θ increases by $\Delta\theta$, area $BOC(= A)$ increases by ΔA (which is COD).

Area of sector COE of the circle is

$\frac{1}{2}$ the radius times arc $CE = \frac{1}{2}\rho \cdot \rho \cdot \Delta\theta = \frac{1}{2}\rho^2 \cdot \Delta\theta$

Area of sector FOD of the circle is

$$\frac{1}{2}(\rho + \Delta\rho)(\rho + \Delta\rho)\,\Delta\theta$$

Then, by comparing sectors,

$$\frac{1}{2}\rho^2 \cdot \Delta\theta < \Delta A < \frac{1}{2}(\rho + \Delta\rho)^2\,\Delta\theta$$

Divide by $\Delta\theta$.

$$\frac{\rho^2}{2} < \frac{\Delta A}{\Delta\theta} < \frac{(\rho + \Delta\rho)^2}{2}$$

Now, as $\Delta\theta$ approaches zero, $\dfrac{(\rho + \Delta\rho)^2}{2}$ approaches $\dfrac{\rho^2}{2}$ and $\dfrac{\Delta A}{\Delta\theta}$ has a value that is between $\dfrac{\rho^2}{2}$ and a quantity that approaches $\dfrac{\rho^2}{2}$ as a limit.   Therefore, $\dfrac{\Delta A}{\Delta\theta}$ approaches $\dfrac{\rho^2}{2}$ as $\Delta\theta$ approaches zero and

$$\lim_{\Delta\theta \to 0}\left[\frac{\Delta A}{\Delta\theta}\right] = \frac{dA}{d\theta} = \frac{\rho^2}{2} = \frac{1}{2}\,[f(\theta)]^2$$

Finally, $dA = \dfrac{\rho^2}{2} \cdot d\theta$ and $A = \dfrac{1}{2}\displaystyle\int \rho^2 \cdot d\theta$

$$= \frac{1}{2}\int f(\theta)^2 \cdot d\theta$$

549. What is the area enclosed by the curve

$$\rho = a(1 - \cos\theta)^{\frac{1}{2}}?$$

$A = \frac{1}{2}\int\rho^2 \cdot d\theta$ and $\rho = a(1 - \cos\theta)^{\frac{1}{2}}$

This is the equation of a curve whose limits are 0 and 2π.

Then $A = \dfrac{a^2}{2}\displaystyle\int_0^{2\pi}(1 - \cos\theta)\,d\theta = \dfrac{a^2}{2}\left[\theta - \sin\theta\right]_0^{2\pi}$

$$= \frac{a^2}{2}\,(2\pi - 0 - 0 + 0) = \frac{a^2 \cdot 2\pi}{2} = \pi a^2$$

550. What is the area of a sector, containing 2 radians, of a circle whose radius is ρ?

$$A = \frac{1}{2} \int_0^2 \rho^2 \cdot d\theta = \frac{\rho^2}{2} \int_0^2 d\theta = \frac{\rho^2}{2} \Big[\theta \Big]_0^2 = \frac{2\rho^2}{2} = \rho^2$$

sq. radians

QUES. 549.   QUES. 551.

551. What is the area bounded by the curve $\rho = \sin \theta$?

The curve is generated while θ varies from 0 to π radians.

$$A = \tfrac{1}{2} \int_0^\pi \rho^2 \cdot d\theta = \tfrac{1}{2} \int_0^\pi \sin^2 \theta \cdot d\theta$$

$$\sin^2 \theta = 1 - \cos^2 \theta$$

$$A = \tfrac{1}{2} \int_0^\pi (1 - \cos^2 \theta)\, d\theta = \tfrac{1}{2} \Big[\theta - \int \cos^2 \theta \cdot d\theta \Big]_0^\pi$$

$$\cos^2 \theta = \tfrac{1}{2} + \tfrac{1}{2} \cos 2\theta$$

Then
$$- \int_0^\pi \cos^2 \theta \cdot d\theta = - \tfrac{1}{2} \int_0^\pi (1 + \cos 2\theta)\, d\theta$$

$$= - \tfrac{1}{2} \Big[\theta + \tfrac{1}{2} \sin 2\theta \Big]_0^\pi$$

$$= - \tfrac{1}{2} \Big[\theta + \sin \theta \cdot \cos \theta \Big]_0^\pi$$

$$\therefore\ A = \frac{1}{2} \Big[\theta - \frac{1}{2} \Big(\theta + \sin \theta \cdot \cos \theta \Big) \Big]_0^\pi$$

$$= \frac{1}{2} \Big[\theta - \frac{\theta}{2} - \frac{\sin \theta \cdot \cos \theta}{2} \Big]_0^\pi$$

$$= \Big[\frac{\theta}{2} - \frac{\theta}{4} - \frac{\sin \theta \cdot \cos \theta}{4} \Big]_0^\pi = \Big[\frac{\theta}{4} - \frac{\sin \theta \cdot \cos \theta}{4} \Big]_0^\pi$$

$$= \frac{\pi}{4} = 0.7854 \text{ square units}$$

The curve is a circle with a radius $= \tfrac{1}{2}$.

552. What is meant by the fundamental theorem of the integral calculus?

The fundamental theorem considers integration as a summation. That is, the desired quantity is obtained as the limit of the sum when the number of its parts is increased indefinitely. This is also known as the method of summation.

In symbols, $$\lim_{\substack{n \to \infty \\ \Delta x \to 0}} \sum_{x=a}^{x=b} f(x)\,\Delta x = \int_a^b f(x)\,dx$$

This is read "The limit of the sum of the parts $f(x) \cdot \Delta x$ within the interval from $x = a$ to $x = b$ when the number of parts, n, becomes infinite as Δx approaches zero is equal to the integral of $f(x)\,dx$ between the limits a and b."

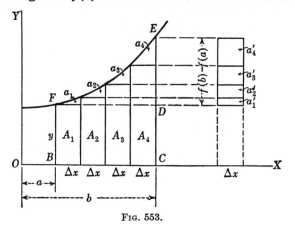

Fig. 553.

553. How is the method of summation applied in finding an area under a curve?

Area $FBCE =$ the sum of the rectangles

$$(A_1 + A_2 + A_3 + A_4)$$

plus the sum of the leftovers $(a_1 + a_2 + a_3 + a_4)$, or

$$\text{Area } FBCE = \sum_{x=a}^{x=b} A_n + \sum_{x=a}^{x=b} a_n$$

where $n =$ the number of divisions. The Greek letter Σ (sigma) indicates the summation of the elementary areas.

Let the rectangle to the right of the figure of height $= DE$ and base $= \Delta x$ be the sum of rectangles

$$(a_1' + a_2' + a_3' + a_4').$$

Evidently, $\displaystyle\sum_{x=a}^{x=b} a_n < \sum_{x=a}^{x=b} a_n'(= DE \cdot \Delta x)$

Now, as the number of rectangles becomes infinite, Δx approaches zero as a limit and $DE \cdot \Delta x$ approaches zero as a limit. And

$$\text{Area } FBCE = \lim_{n \to \infty} \sum_{x=a}^{x=b} A_n$$

Since the area of each elementary rectangle is

$$y \cdot \Delta x = f(x)\, \Delta x$$

then,

$$\text{Area } FBCE = \lim_{n \to \infty} \sum_{x=a}^{x=b} y \cdot \Delta x = \int_a^b y \cdot dx *$$

This is read, "The area $FBCE$ equals the limit of the sum, as n becomes infinite, of $y \cdot \Delta x$ from $x = a$ to $x = b$, equals the definite integral $\int_a^b y \cdot dx$."

554. What is $\displaystyle\sum_{x=1}^{x=1.6} 3x^2 \cdot \Delta x$, when $\Delta x = 0.1$?

When $\Delta x = 0.1$,

$$\sum_{x=1}^{x=1.6} 3x^2 \cdot \Delta x = 3(1^2 + 1.1^2 + 1.2^2 + 1.3^2 + 1.4^2 + 1.5^2) \cdot 0.1$$
$$= 2.865$$

---

* It is important to note that it is the limit of the sum of the rectangles, and not the sum itself which is our only concern, and this limit is a precise quantity no part of which is discarded or neglected.

555. What is the limit $\displaystyle\lim_{n\to\infty}\sum_{x=1}^{x=1.6} 3x^2 \cdot \Delta x$?

$$\lim_{n\to\infty}\sum_{x=1}^{x=1.6} 3x^2 \cdot \Delta x = \int_1^{1.6} 3x^2 \cdot dx = 3\int_1^{1.6} x^2 \cdot dx$$

$$= \Big[x^3\Big]_1^{1.6} = 3.096$$

556. What would be the area of a right triangle of base 15 inches and height 3 inches by the summation method?

$$\lim_{n\to\infty}\sum_{x=0}^{x=15} y \cdot \Delta x = \int_0^{15} y \cdot dx = \text{total area}$$

The equation of the sloping line is $y = \dfrac{x}{5}$.

$$\therefore A = \int_0^{15}\frac{x}{5}\cdot dx = \frac{1}{5}\Big[\frac{x^2}{2}\Big]_0^{15} = \frac{1}{5}\cdot\frac{225}{2} = 22.5 \text{ sq. in.}$$

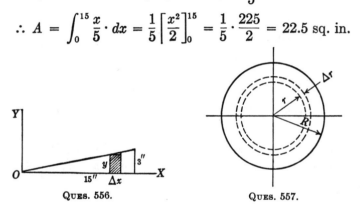

QUES. 556.   QUES. 557.

557. What is the area of a circle of radius R as found by area summation?

Assume elementary circular zones of areas radiating from the center where $r = 0$ to $r = R$. One such circular zone has a breadth $= \Delta r$ and a length $= 2\pi r$ and therefore an area $2\pi r \cdot \Delta r$.

Then $\displaystyle\lim_{n\to\infty}\sum_0^R 2\pi r \cdot \Delta r = \int_0^R 2\pi r \cdot dr = \text{the total area} = A$

or $A = 2\pi \displaystyle\int_0^R r \cdot dr = 2\pi\Big[\frac{r^2}{2}\Big]_0^R = \pi R^2 = \text{the area of the}$

whole circle

558. What is the area of the ellipse $\dfrac{x^2}{16} + \dfrac{y^2}{4} = 1$ by area summation?

Consider the first quadrant only, and multiply by 4 for the total area.

When $y = 0$, then $x = 4$ in the first quadrant. Therefore, the limits are $x = 0$ and $x = 4$. Solve for y.

$$x^2 + 4y^2 = 16, \qquad 4y^2 = 16 - x^2, \qquad y = \pm \tfrac{1}{2}\sqrt{16 - x^2}$$

The area of an elementary strip is $y \cdot \Delta x$. Then

$$\lim_{n \to \infty} \sum_{x=0}^{x=4} y \cdot \Delta x = \int_0^4 y \cdot dx = \frac{A}{4} \text{ for one quadrant}$$

or $$\frac{A}{4} = \int_0^4 \frac{1}{2} \sqrt{16 - x^2} \cdot dx$$

and $$A = 2 \int_0^4 \sqrt{16 - x^2} \cdot dx$$

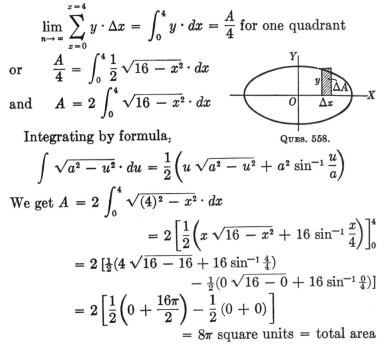

QUES. 558.

Integrating by formula,

$$\int \sqrt{a^2 - u^2} \cdot du = \frac{1}{2}\left(u \sqrt{a^2 - u^2} + a^2 \sin^{-1}\frac{u}{a}\right)$$

We get $A = 2 \displaystyle\int_0^4 \sqrt{(4)^2 - x^2} \cdot dx$

$$= 2\left[\frac{1}{2}\left(x\sqrt{16 - x^2} + 16 \sin^{-1}\frac{x}{4}\right)\right]_0^4$$

$$= 2\left[\tfrac{1}{2}(4\sqrt{16 - 16} + 16 \sin^{-1}\tfrac{4}{4})\right.$$
$$\left.- \tfrac{1}{2}(0\sqrt{16 - 0} + 16 \sin^{-1}\tfrac{0}{4})\right]$$

$$= 2\left[\frac{1}{2}\left(0 + \frac{16\pi}{2}\right) - \frac{1}{2}(0 + 0)\right]$$

$$= 8\pi \text{ square units} = \text{total area}$$

559. What is the area included between the parabolas $y^2 = 4x$ and $x^2 = 4y$ by area summation?

Solve the equations simultaneously for the coordinates of the points of intersection.

$$y = 2\sqrt{x} = \frac{x^2}{4}$$

or $\qquad 4x = \dfrac{x^4}{16},\qquad 64x = x^4,\qquad x(64 - x^3) = 0$

giving $x = 0$ and $x = 4$.

The height of the elementary strip goes from the lower curve $x^2 = 4y$ where $y = \dfrac{x^2}{4}$ to the upper curve $y^2 = 4x$ where $y = 2\sqrt{x}$, giving a net height that is the difference of the two values of y, or

$$y(\text{for } \Delta A) = 2\sqrt{x} - \frac{x^2}{4}$$

$$\lim_{n \to \infty} \sum_{x=0}^{x=4} y \cdot \Delta x = \int_0^4 y \cdot dx = \int_0^4 \left(2\sqrt{x} - \frac{x^2}{4}\right) dx$$

$$= \text{total area between the curves}$$

$$A = 2\int_0^4 x^{\frac{1}{2}} \cdot dx - \frac{1}{4}\int_0^4 x^2 \cdot dx = 2\left[\frac{2}{3} \cdot x^{\frac{3}{2}}\right]_0^4 - \left[\frac{1}{4} \cdot \frac{x^3}{3}\right]_0^4$$

$$= \tfrac{4}{3}\sqrt{4^3} - \tfrac{1}{12} \cdot 4^3 = \tfrac{4}{3} \cdot 8 - \tfrac{64}{12} = \tfrac{16}{3} \text{ square units}$$

QUES. 559.

QUES. 560.

560. How is the method of summation applied to an area in polar coordinates?

Let $\rho = f(\theta)$ be the polar equation of the curve. The area is required between the curve and two radius vectors at α and β. The required area is the limit of the sum of the circular sectors.

$$\tfrac{1}{2}\rho_1{}^2 \cdot \Delta\theta_1 + \tfrac{1}{2}\rho_2{}^2 \cdot \Delta\theta_2 + \cdots + \tfrac{1}{2}\rho_n{}^2 \cdot \Delta\theta_n = \sum_1^n \tfrac{1}{2}\rho_i{}^2 \cdot \Delta\theta_i$$

= the sum of the areas of the elementary sectors

Then $\lim\limits_{n \to \infty} \sum\limits_1^n \tfrac{1}{2}\rho_i{}^2 \cdot \Delta\theta_i = \int_\alpha^\beta \tfrac{1}{2}\rho^2 \cdot d\theta$ = the total area

$$\text{Area} = \tfrac{1}{2} \int_\alpha^\beta \rho^2 \cdot d\theta$$

561. What is the entire area of the four-leaved rose $\rho = a \sin 2\theta$?

The limits for the area of one leaf are 0 and $\dfrac{\pi}{2}$. Since the other leaves are alike, find the area of one leaf, and multiply by 4.

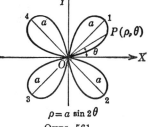

$\rho = a \sin 2\theta$
QUES. 561.

Area $= \tfrac{1}{2} \int_\alpha^\beta \rho^2 \cdot d\theta$ of one leaf

Then $4 \cdot \tfrac{1}{2} \int_0^{\frac{\pi}{2}} a^2 \sin^2 2\theta \cdot d\theta$ = area of 4 leaves

Now $\sin^2 2\theta = \tfrac{1}{2} - \tfrac{1}{2} \cos 4\theta$

Then

$$A = 2 \int_0^{\frac{\pi}{2}} a^2(\tfrac{1}{2} - \tfrac{1}{2}\cos 4\theta)\, d\theta = a^2 \int_0^{\frac{\pi}{2}} (1 - \cos 4\theta)\, d\theta$$

$$= a^2 \left[\theta - \frac{1}{4} \sin 4\theta \right]_0^{\frac{\pi}{2}} = a^2 \left[\frac{\pi}{2} - 0 - 0 + 0 \right] = \frac{\pi a^2}{2}$$

= the area of the 4 leaves

PROBLEMS

1. What is the value of the area $\int y \cdot dx$ when $y = 3a + 2bx^3$ from $x_1 = 1$ to $x_2 = 3$, when $a = 1$ and $b = 2$?

2. Find by integration the area of a right triangle whose base $x = 24$ in. and whose height $y = 8$ in., and find the area included between $x = 7$ and $x = 19$.

3. What is the area under the curve $y = \dfrac{2a}{2x + b}$ from $x = 1$ to $x = 2$, when $a = 1$ and $b = 1$?

4. What is the area between the curve $y = x^2 - 6x + 8$ and the x axis?

5. What is the area between the curve $y = \sin x$ and the x axis from $x = \dfrac{\pi}{6}$ to $x = \dfrac{3\pi}{2}$?

6. What is the area included between $y^2 = 2x$ and $x^2 = 2y$?

7. What is the area formed by the curve $y = 2x^2$ and the y axis and limited by the abscissas at $y = 0$ and $y = 3$?

8. What is the area formed by the curve $y^2 = 3x$, the x axis, and the ordinates at $x = 0$ and $x = 3$?

9. What is the area bounded by the cardioid $\rho = a(1 - \cos\theta)$, taking twice the area after integrating from 0 to π?

10. What is the area bounded by the curve $\rho = \cos\theta$?

11. Evaluate $\displaystyle\int_{\frac{\pi}{3}}^{\frac{\pi}{2}} \sin^2\theta \cdot \cos\theta \cdot d\theta$.

12. What is the area of the ellipse $\dfrac{x^2}{36} + \dfrac{y^2}{16} = 1$?

13. Evaluate $\displaystyle\int_0^1 2x \cdot \epsilon^{x^2} \cdot dx$.

14. Evaluate $\displaystyle\int_0^1 \dfrac{x \cdot dx}{4 + x^2}$.

15. What is the area bounded by the lines $y = 3x$, $y = 15 - 3x$, and the x axis?

16. What is the area bounded by the curve $\rho = 2\cos\theta$ between $\theta = 0$ and $\dfrac{\pi}{2}$?

17. What is the area generated by one revolution of $\rho = 2\theta$, from $\theta = 0$?

18. Find the area bounded by the curve $\rho = 3\sin\theta$.

19. What is the area bounded by the circle $x^2 + y^2 = 25$, the x axis, and $x = -3$ and 4?

20. Find the area bounded by the circle $\rho = a\cos\theta$ and the lines $\theta = 0$ and $\theta = 60°$.

21. Find the area bounded by the parabola $\rho(1 + \cos\theta) = a$ and the lines $\theta = 0$ and $\theta = 120°$.

22. What is the area enclosed by $\rho = \sin^2\dfrac{\theta}{2}$?

23. What is the area bounded by $\rho^2 = a^2 \sin 4\theta$?

24. Find the area enclosed by $\rho = 2 - \cos\theta$.

CHAPTER XXIX

MEAN VALUE

562. What is meant by average, or mean, value?

It is the average ordinate of the curve for the considered x distance. It is that height which when multiplied by the base = the area under the curve.

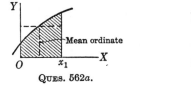

QUES. 562a.

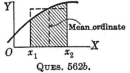

QUES. 562b.

Therefore the mean $y = \dfrac{1}{x_1} \displaystyle\int_0^{x_1} y \cdot dx$ = the mean ordinate of any curve from $x = 0$ to $x = x_1$.

And the mean $y = \dfrac{1}{x_2 - x_1} \displaystyle\int_{x_1}^{x_2} y \cdot dx$ = the mean ordinate of any curve from $x = x_1$ to $x = x_2$.

Then, the mean value is the area of the curve divided by the abscissa between the limits.

563. When does the area under a curve represent the distance traveled?

When the ordinates = velocity and the abscissas = time elapsed.

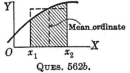

FIG. 563.

Then s = distance = the area under the curve

or s = the mean velocity $\times$ time elapsed

564. How would you express the distance formula by area summation?

243

s = distance = $\int v \cdot dt$ for a velocity-time graph

This adds up the elementary areas under the curve.

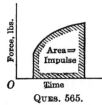

QUES. 565.

565. When does the area represent impulse?

When the ordinates = force and the abscissas = time.

Then I = impulse = the area under the curve = the mean force × time elapsed

566. How would you express the impulse formula by area summation?

I = impulse = $\int F \cdot dt$ for a force-time graph

567. When does the area represent velocity?

When the ordinates = acceleration and the abscissas = time.

Then v = velocity = area under the curve = the mean acceleration × time elapsed

568. How would you express the velocity formula by area summation?

v = velocity = $\int \alpha \cdot dt$ for acceleration-time graph

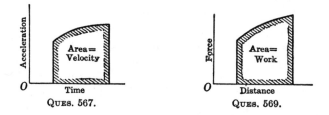

QUES. 567.   QUES. 569.

569. When does the area represent work?

When the ordinates = force and the abscissas = distance.

Then W = work = area under the curve = the mean force × distance traveled

570. How would you express the work formula by area summation?

$$W = \text{work} = \int F \cdot ds \text{ for force-distance graph}$$

571. When does the area represent volume?

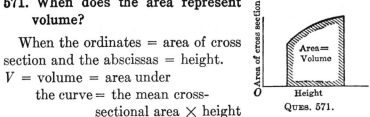

QUES. 571.

When the ordinates = area of cross section and the abscissas = height.
V = volume = area under
    the curve = the mean cross-
        sectional area $\times$ height

572. How would you express the volume formula by area summation?

$$V = \text{volume} = \int A_s \cdot dx \text{ for cross section–height graph}$$

573. What is the work done in stretching a spring from a length of 10 inches to 14 inches if a weight of 10 pounds will stretch it 1 inch and it is 8 inches long when no force is applied?

$$F = ks$$

or force varies as the elongation in this case.

Now find k by substituting the given quantities.

$$\therefore \ 10 = k \cdot 1 \qquad \text{or} \qquad k = 10$$

and $\qquad\qquad\qquad F = 10s$

Then $W = \text{work} = \int_{2}^{6} 10s \cdot ds = \left[5s^2\right]_{2}^{6} = 5(6^2 - 2^2)$

$$= 160 \text{ in.-lb. } (s \text{ is in inches})$$

574. What is the average, or mean, value of the ordinates of the first quadrant of the circle $x^2 + y^2 = r^2$?

$$\bar{y}_x = \frac{1}{x_1} \int_{0}^{x_1} y \cdot dx = \frac{1}{r} \int_{0}^{r} \sqrt{r^2 - x^2} \cdot dx$$

By formula (VII),

$$\bar{y}_x = \frac{1}{r}\left[\frac{x}{2}\sqrt{r^2 - x^2} + \frac{r^2}{2}\sin^{-1}\frac{x}{r}\right]_0^r$$

$$= \frac{1}{r}\left(\frac{r}{2}\sqrt{r^2 - r^2} + \frac{r^2}{2}\sin^{-1}\frac{r}{r} - 0 - \frac{r^2}{2}\sin^{-1}\frac{0}{r}\right)$$

$$= \frac{1}{r}\cdot\frac{\pi r^2}{4} = \frac{\pi r}{4} = 0.785r = \text{mean value}$$

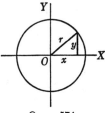

QUES. 574.

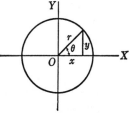

QUES. 575.

575. What is the mean value of the ordinates of the first quadrant of the circle $y = r \sin \theta$?

$$\bar{y}_\theta = \frac{1}{\frac{\pi}{2} - 0}\int_0^{\frac{\pi}{2}} y \cdot d\theta = \frac{2}{\pi}\int_0^{\frac{\pi}{2}} r \sin \theta \cdot d\theta$$

$$= \frac{2r}{\pi}\left[-\cos \theta\right]_0^{\frac{\pi}{2}} = \frac{2r}{\pi}\left(-\cos\frac{\pi}{2} + \cos 0\right) = \frac{2r}{\pi} = 0.637r$$

This shows that the average, or mean, value of a given function y will depend upon the choice of the independent variable. This is indicated by means of a subscript after the mean value, as $\bar{y}_\theta$ indicates that the mean value is to be taken with respect to θ.

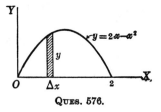

QUES. 576.

576. What is the mean value or mean ordinate of the positive part of the curve $y = 2x - x^2$?

First determine the length of the base to fix the limits of integration for the area by setting y equal to zero, or

$$0 = 2x - x^2 = x(2 - x)$$

Then $\qquad x_1 = 0 \qquad$ and $\qquad x_2 = 2$

Now $\bar{y}_x = \dfrac{1}{x_2 - x_1} \displaystyle\int y \cdot dx = \dfrac{1}{2 - 0} \displaystyle\int_0^2 (2x - x^2) \, dx$

$\qquad = \displaystyle\int_0^2 \left(x - \dfrac{x^2}{2}\right) dx = \left[\dfrac{x^2}{2} - \dfrac{x^3}{6}\right]_0^2 = \dfrac{4}{2} - \dfrac{8}{6} = \dfrac{2}{3}$

$\qquad\qquad\qquad\qquad\qquad\qquad\qquad\qquad = $ the mean ordinate

577. What is meant by the quadratic mean of a variable quantity?

It is the square root of the mean of the *squares* of all the values between the limits considered.

Example

If y is the function considered and if the limits of x are $x = 0$ and $x = x_1$, then

$$\sqrt{\dfrac{1}{x_1} \int_0^{x_1} y^2 \cdot dx} = \text{the quadratic mean}$$

The quadratic mean is particularly useful in the study of alternating currents.

578. What is meant by form factor?

Form factor is the ratio of the quadratic to the arithmetic

mean, or $\dfrac{\sqrt{\dfrac{1}{x_1} \displaystyle\int_0^{x_1} y^2 \cdot dx}}{\dfrac{1}{x_1} \displaystyle\int_0^{x_1} y \cdot dx},$

assuming that the limits are from 0 to x_1.

579. What is the quadratic mean and what is the form factor of the function $y = 5x$?

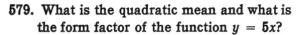

QUES. 579.

$$\int_0^l y^2 \cdot dx = \int_0^l 25x^2 \cdot dx = 25 \int_0^l x^2 \cdot dx = 25 \cdot \left[\dfrac{x^3}{3}\right]_0^l$$

Therefore, the integral is $\dfrac{25l^3}{3}$. Now divide by l, or

$\dfrac{25l^3}{3} \cdot \dfrac{1}{l} = \dfrac{25l^2}{3}$. Now take the square root, or $\dfrac{5l}{\sqrt{3}} = $ the

quadratic mean and the arithmetic mean is

$$\bar{y}_x = \frac{1}{x_1} \int_0^{x_1} y \cdot dx = \frac{1}{l} \int_0^l 5x \cdot dx = \frac{5}{l} \int_0^l x \cdot dx$$
$$= \frac{5}{l} \cdot \frac{l^2}{2} = \frac{5l}{2}$$

Therefore, the form factor is $\dfrac{\dfrac{5l}{\sqrt{3}}}{\dfrac{5l}{2}} = \dfrac{5l}{\sqrt{3}} \cdot \dfrac{2}{5l} = \dfrac{2}{\sqrt{3}} = 1.155$

NOTE.—The form factor does not depend upon the x limit.

580. What is the quadratic mean of $y = x^{2a}$?

The integral is

$$\int_{x=0}^{x=l} x^{4a} \cdot dx = \left[\frac{x^{4a+1}}{4a+1}\right]_0^l = \frac{l^{4a+1}}{4a+1}$$

Now divide by l.   Then

$$\frac{l^{4a+1}}{4a+1} \cdot \frac{1}{l} = \frac{l^{4a}}{4a+1}$$

Now take the square root, and

$$\therefore \text{ Quadratic mean } = \sqrt{\frac{l^{4a}}{4a+1}} = l^{2a}\sqrt{\frac{1}{4a+1}}.$$

581. What is the quadratic mean of $y = a^{\frac{x}{4}}$?

The integral is

$$\int_{x=0}^{x=l} (a^{\frac{x}{4}})^2 \cdot dx = \int_0^l a^{\frac{x}{2}} \cdot dx = \left[\frac{2a^{\frac{x}{2}}}{\log_e a} \right]_0^l$$

$$= \left(\frac{2a^{\frac{l}{2}}}{\log_e a} - \frac{2a^{\frac{0}{4}}}{\log_e a} \right) = \frac{2}{\log_e a} (a^{\frac{l}{2}} - 1)$$

$$\therefore \text{Quadratic mean} = \sqrt{\frac{2}{\log_e a} (a^{\frac{l}{2}} - 1)}$$

PROBLEMS

1. What is the mean ordinate of $y = 5x - 2x^2$?

2. What are the quadratic mean and the form factor of $y = 7x + 3$?

3. What is the quadratic mean of $y = 3x^{4a}$, from 0 to l, and when $l = 2$ and $a = 1$?

4. What is the quadratic mean of $y = 2a^{\frac{x}{3}}$, from $x = 0$ to $x = l$?

5. Find the area under the curve $y = 2x^2 + 2x - 3$, between $x = 0$ and $x = 5$, and the mean ordinate between these limits.

6. Find the area under the curve $y = \sin^2 \theta$, between 0 and π radians, and the mean ordinate.

7. Find the mean ordinate of $y = \frac{2}{x}$, between $x = 1$ and $x = 2$.

8. What are the arithmetical mean, quadratic mean, and the form factor of $y = \sin x$ between 0 and π radians?

9. Find the arithmetical and quadratic means of $y = x^2 + 2x + 1$, from $x = 0$ to $x = 2$.

10. What is the work done in stretching a spring from a length of 8 to 12 in. if a weight of 5 lb. will stretch it 1 in. and it is 10 in. long when no force is applied?

CHAPTER XXX

LENGTH OF AN ARC

582. How do we find the length of an arc of a curve?

Required to find the length of arc MC. Assume $y = f(x)$ to be a continuous single-valued function with no cusp points between $x = a$ and $x = b$. Let $s =$ the distance the tracing point has moved when it has reached any point, say, P. At this instant, y is

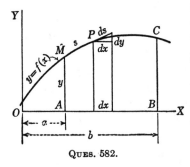

QUES. 582.

changing at the same rate it would if the point were moving along the tangent line at P.

Now, if the change in y is to become and remain uniform with respect to x, the point *must move along the tangent.* Then, the change in s, if this change becomes and remains uniform, is ds, corresponding to dx and dy. Therefore, from the right triangle,

$$(ds)^2 = (dx)^2 + (dy)^2 \qquad (1)$$

Dividing by $(dx)^2$,

$$\frac{(ds)^2}{(dx)^2} = 1 + \frac{(dy)^2}{(dx)^2}$$

Taking the square root,

$$\frac{ds}{dx} = \sqrt{1 + \left(\frac{dy}{dx}\right)^2}$$

or

$$ds = \sqrt{1 + \left(\frac{dy}{dx}\right)^2} \cdot dx$$

And $s = \displaystyle\int_a^b \sqrt{1 + \left(\frac{dy}{dx}\right)^2} \cdot dx$

= the length of the arc between $x = a$ and $x = b$

Now if $x = f(y)$ is a single-valued and continuous function for the limits $y = c$ and $y = d$, then dividing Eq. (1) by $(dy)^2$ we get

$$\frac{(ds)^2}{(dy)^2} = \frac{(dx)^2}{(dy)^2} + 1$$

and $\dfrac{ds}{dy} = \sqrt{1 + \left(\dfrac{dx}{dy}\right)^2}$ or $ds = \sqrt{1 + \left(\dfrac{dx}{dy}\right)^2} \cdot dy$

And $s = \displaystyle\int_c^d \sqrt{1 + \left(\dfrac{dx}{dy}\right)^2} \cdot dy$

     = the length of the arc between $y = c$ and $y = d$

583. When is the formula for the length of an arc between the y limits used?

In the event that the integral between the x limits leaves a doubt as to the exact portion of the curve whose length is desired, the integral between the y limits is used.

Arcs AB, CD, EF, AD, AF, and EB all have the same x limits but different y limits.   Use the y integral here.

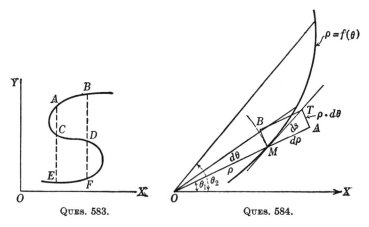

QUES. 583.   QUES. 584.

584. How do we find the length of an arc of a curve in polar coordinates?

Let s = the length of the curve generated by the moving point.   Now, if at any point, as M, the changes in ρ and s

become and remain uniform with respect to θ, then the differential of the arc ds is along the tangent line MT, the differential of the radius vector $d\rho$ is along the radius vector OM, and the differential of the motion perpendicular to the radius vector is $\rho \cdot d\theta$ along MB tangent to the circle having OM for a radius.

Therefore, in the right triangle MBT,

$$ds = \sqrt{(\rho \cdot d\theta)^2 + (d\rho)^2}$$

Dividing by $d\theta$,

$$\frac{ds}{d\theta} = \sqrt{\rho^2 + \left(\frac{d\rho}{d\theta}\right)^2} \quad \text{or} \quad ds = \sqrt{\rho^2 + \left(\frac{d\rho}{d\theta}\right)^2} \cdot d\theta$$

$$\therefore \ s = \int_{\theta_1}^{\theta_2} \sqrt{\rho^2 + \left(\frac{d\rho}{d\theta}\right)^2} \cdot d\theta$$

$$= \text{the length of arc from } \theta_1 \text{ to } \theta_2$$

585. What is the length of the curve $y = x^2$ from $x = 2$ to $x = 5$?

$$\text{Length } s = \int_2^5 \sqrt{1 + \left(\frac{dy}{dx}\right)^2} \cdot dx$$

If

$$y = x^2,$$

then

$$\frac{dy}{dx} = 2x$$

and

$$\left(\frac{dy}{dx}\right)^2 = 4x^2$$

$$\therefore \ s = \int_2^5 \sqrt{1 + 4x^2} \cdot dx$$

Now integrate by parts.

Let

$$u = (1 + 4x^2)^{\frac{1}{2}}$$

Now, if

$$(1 + 4x^2) = w,$$

then

$$dw = 8x \cdot dx$$

and

$$u = w^{\frac{1}{2}}$$

and

$$\frac{du}{dw} = \frac{1}{2} w^{-\frac{1}{2}}$$

or

$$du = \tfrac{1}{2} w^{-\frac{1}{2}} \cdot dw$$

$$= \frac{1}{2(1 + 4x^2)^{\frac{1}{2}}} \cdot 8x \cdot dx$$

$$= \frac{4x}{(1 + 4x^2)^{\frac{1}{2}}} \cdot dx$$

Then $\dfrac{du}{dx} = \dfrac{4x}{(1 + 4x^2)^{\frac{1}{2}}}$

Also, let $dv = dx$

Then $v = x$

The formula for integration by parts is

$$\int u \cdot dv = uv - \int v \cdot du$$

or

$$\int_2^5 (1 + 4x^2)^{\frac{1}{2}} \cdot dx = x(1 + 4x^2)^{\frac{1}{2}} - \int \frac{4x^2}{(1 + 4x^2)^{\frac{1}{2}}} \cdot dx \quad (1)$$

Also, in another form,

$$\int_2^5 (1 + 4x^2)^{\frac{1}{2}} \cdot dx = \int \frac{(1 + 4x^2)\, dx}{(1 + 4x^2)^{\frac{1}{2}}} = \int \frac{dx}{(1 + 4x^2)^{\frac{1}{2}}}$$
$$+ 4 \int \frac{x^2 \cdot dx}{(1 + 4x^2)^{\frac{1}{2}}} \quad (2)$$

Adding the preceding two equations (1) and (2) eliminates
$4 \displaystyle\int \frac{x^2 \cdot dx}{(1 + 4x^2)^{\frac{1}{2}}},$

getting

$$2 \int_2^5 (1 + 4x^2)^{\frac{1}{2}} \cdot dx = x(1 + 4x^2)^{\frac{1}{2}} + \int \frac{dx}{(1 + 4x^2)^{\frac{1}{2}}}$$

or $\displaystyle\int (1 + 4x^2)^{\frac{1}{2}} \cdot dx = \frac{x}{2}(1 + 4x^2)^{\frac{1}{2}} + \int \frac{dx}{(\frac{1}{4} + x^2)^{\frac{1}{2}}} \quad (3)$

Now integrate $\displaystyle\int \frac{dx}{(\frac{1}{4} + x^2)^{\frac{1}{2}}},$ which is of the form of

$$\int \frac{dx}{(a^2 + x^2)^{\frac{1}{2}}} = \log_\epsilon (x + \sqrt{a^2 + x^2})$$

Here $a^2 = \frac{1}{4}$, and

$$\therefore \int \frac{dx}{(\frac{1}{4} + x^2)^{\frac{1}{2}}} = \log_\epsilon (x + \sqrt{\tfrac{1}{4} + x^2})$$

Substitute this in Eq. (3).

$$\therefore \int_2^5 (1 + 4x^2)^{\frac{1}{2}} \cdot dx = \left[\frac{x}{2} (1 + 4x^2)^{\frac{1}{2}} + \log_\epsilon (x + \sqrt{\tfrac{1}{4} + x^2}) \right]_2^5$$

$$= [\tfrac{5}{2}(1 + 4 \cdot 25)^{\frac{1}{2}} + \log_\epsilon (5 + \sqrt{\tfrac{1}{4} + 25})$$
$$- \tfrac{2}{2}(1 + 4 \cdot 4)^{\frac{1}{2}} - \log_\epsilon (2 + \sqrt{\tfrac{1}{4} + 4})$$
$$= \tfrac{5}{2} \sqrt{101} + \log_\epsilon 10.025 - \sqrt{17}$$
$$- \log_\epsilon 4.062$$
$$= \tfrac{5}{2} \cdot 10.05 + 2.31 - 4.12 - 1.40$$
$$= 25.12 + 2.31 - 4.12 - 1.40$$

$$s = \int_2^5 (1 + 4x^2)^{\frac{1}{2}} \cdot dx = 27.43 - 5.52 = 21.91 \text{ units of arc}$$

586. What is the length of the curve $\rho = (1 - \cos \theta)$ from $\theta = 0$ to $\theta = \pi$?

$$s = \int_0^\pi \left[\rho^2 + \left(\frac{d\rho}{d\theta} \right)^2 \right]^{\frac{1}{2}} \cdot d\theta$$

If

$$p = (1 - \cos \theta)$$

then

$$\frac{dp}{d\theta} = \sin \theta$$

And $s = \displaystyle\int_0^\pi [(1 - \cos \theta)^2 + \sin^2 \theta]^{\frac{1}{2}} \cdot d\theta$

$$= \int_0^\pi (1 - 2 \cos \theta + \cos^2 \theta + \sin^2 \theta)^{\frac{1}{2}} \cdot d\theta$$

$$= \int_0^\pi (2 - 2 \cos \theta)^{\frac{1}{2}} \cdot d\theta$$

$$= \int_0^\pi \sqrt{2} (1 - \cos \theta)^{\frac{1}{2}} \cdot \frac{\sqrt{2}}{\sqrt{2}} \cdot d\theta$$

$$= \int_0^\pi \sqrt{2} \sqrt{2} \left(\frac{1 - \cos \theta}{2} \right)^{\frac{1}{2}} \cdot d\theta$$

$$= 2 \int_0^\pi \sin \frac{\theta}{2} \cdot d\theta \cdot \frac{2}{2}$$

because $\sin\dfrac{\theta}{2} = \left(\dfrac{1-\cos\theta}{2}\right)^{\frac{1}{2}}$

$$s = 4\int_0^\pi \sin\frac{\theta}{2}\cdot\frac{d\theta}{2} = \left[4\cdot-\cos\frac{\theta}{2}\right]_0^\pi$$

$$= \left[-4\cos\frac{\theta}{2}\right]_0^\pi = -4\cos\frac{\pi}{2} + 4\cos\frac{0}{2}$$

$$= -0 + 4 = 4 = \text{length of curve}$$

587. What is the length of the circumference of a circle?

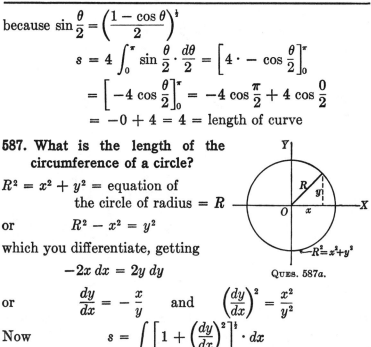

QUES. 587a.

$R^2 = x^2 + y^2 =$ equation of the circle of radius $= R$

or $\qquad R^2 - x^2 = y^2$

which you differentiate, getting

$$-2x\,dx = 2y\,dy$$

or $\qquad \dfrac{dy}{dx} = -\dfrac{x}{y}$ and $\left(\dfrac{dy}{dx}\right)^2 = \dfrac{x^2}{y^2}$

Now $\qquad s = \int\left[1 + \left(\dfrac{dy}{dx}\right)^2\right]^{\frac{1}{2}}\cdot dx$

First get the length of a quadrant of the circle with limits from $x = 0$ to $x = R$. Then

$$\frac{s}{4} = \int_0^R\left(1 + \frac{x^2}{y^2}\right)^{\frac{1}{2}}\cdot dx = \int_0^R\left(1 + \frac{x^2}{R^2-x^2}\right)^{\frac{1}{2}}\cdot dx$$

$$= \int_0^R\frac{R}{(R^2-x^2)^{\frac{1}{2}}}\cdot dx$$

To integrate this function, let us work backward and assume that

$$y = R\text{ arc }\sin\left(\frac{x}{R}\right)$$

Then $\qquad \dfrac{x}{R} = \sin\dfrac{y}{R}$ and $\dfrac{dx}{dy} = \cos\dfrac{y}{R}$

or $\qquad \dfrac{dy}{dx} = \dfrac{1}{\cos\dfrac{y}{R}}$

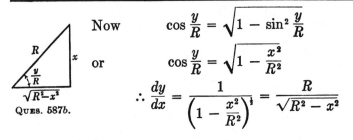

Now $\qquad \cos \dfrac{y}{R} = \sqrt{1 - \sin^2 \dfrac{y}{R}}$

or $\qquad \cos \dfrac{y}{R} = \sqrt{1 - \dfrac{x^2}{R^2}}$

$\therefore \dfrac{dy}{dx} = \dfrac{1}{\left(1 - \dfrac{x^2}{R^2}\right)^{\frac{1}{2}}} = \dfrac{R}{\sqrt{R^2 - x^2}}$

Ques. 587b.

which is the expression within the integral sign above.

$$\therefore \frac{s}{4} = \int_0^R \frac{R}{(R^2 - x^2)^{\frac{1}{2}}} \cdot dx = \left[R \text{ arc sin} \left(\frac{x}{R} \right) \right]_0^R$$

$$= \left(R \text{ arc sin} \frac{R}{R} - R \text{ arc sin} \frac{0}{R} \right)$$

$$= \frac{\pi}{2} \cdot R - 0 = \frac{\pi R}{2}$$

$\therefore s = 2\pi R = $ the length of the circumference of a circle

588. What is the length of the arc of $\rho = \epsilon^{2\theta}$ between $\theta = 0$ and $\theta = 2$ radians?

If $\qquad\qquad \rho = \epsilon^{2\theta}$

then $\qquad\qquad \dfrac{d\rho}{d\theta} = 2\epsilon^{2\theta} = 2\rho$

Now $s = \displaystyle\int \left[\rho^2 + \left(\frac{d\rho}{d\theta} \right)^2 \right]^{\frac{1}{2}} \cdot d\theta$

or $\quad s = \displaystyle\int_0^2 [\rho^2 + (2\rho)^2]^{\frac{1}{2}} \cdot d\theta = \int_0^2 (5\rho^2)^{\frac{1}{2}} \cdot d\theta$

$\qquad = \sqrt{5} \displaystyle\int_0^2 \rho \cdot d\theta = \sqrt{5} \int_0^2 \epsilon^{2\theta} \cdot d\theta$

$$= \sqrt{5} \int_0^2 \epsilon^{2\theta} \cdot \tfrac{2}{2} \cdot d\theta$$

$$= \frac{\sqrt{5}}{2} \int_0^2 \epsilon^{2\theta} \cdot 2 \, d\theta = \frac{\sqrt{5}}{2} \left[\epsilon^{2\theta} \right]_0^2$$

$$= \frac{2.236}{2} (\epsilon^4 - \epsilon^0) = 1.12(54.598 - 1)$$

$$= 1.12 \cdot 53.598$$

$\therefore s = 60.03$ in.

589. What is the length of arc of the catenary curve

$$y = \frac{a}{2} \left(\varepsilon^{\frac{x}{a}} + \varepsilon^{-\frac{x}{a}} \right) \text{ from } x = 0$$

to $x = 2a$?

If

$$y = \frac{a}{2} \epsilon^{\frac{x}{a}} + \frac{a}{2} \epsilon^{-\frac{x}{a}}$$

then

$$\frac{dy}{dx} = \frac{1}{2} \epsilon^{\frac{x}{a}} - \frac{1}{2} \epsilon^{-\frac{x}{a}}$$

And $s = \displaystyle\int_0^{2a} \sqrt{1 + \left(\frac{dy}{dx} \right)^2} \cdot dx$

QUES. 589.

$$= \int_0^{2a} \sqrt{1 + \frac{1}{4} \left(\epsilon^{\frac{x}{a}} - \epsilon^{-\frac{x}{a}} \right)^2} \cdot dx$$

$$= \int_0^{2a} \sqrt{\frac{4 + \epsilon^{\frac{2x}{a}} - 2\epsilon^{\frac{x}{a} - \frac{x}{a}} + \epsilon^{-\frac{2x}{a}}}{4}} \cdot dx$$

$$= \tfrac{1}{2} \int_0^{2a} \sqrt{4 + \epsilon^{\frac{2x}{a}} - 2 + \epsilon^{-\frac{2x}{a}}} \cdot dx$$

because

$$\epsilon^{\frac{x}{a} - \frac{x}{a}} = \epsilon^0 = 1$$

Thus

$$s = \tfrac{1}{2} \int_0^{2a} \sqrt{2 + \epsilon^{\frac{2x}{a}} + \epsilon^{-\frac{2x}{a}}} \cdot dx$$

Now replace 2 by $2 \cdot \epsilon^0 = 2 \cdot \epsilon^{\frac{x}{a} - \frac{x}{a}}$, or

$$s = \tfrac{1}{2} \int_0^{2a} \sqrt{\epsilon^{\frac{2x}{a}} + 2\epsilon^{\frac{x}{a} - \frac{x}{a}} + \epsilon^{-\frac{2x}{a}}} \cdot dx$$

$$= \tfrac{1}{2} \int_0^{2a} \sqrt{\left(\epsilon^{\frac{x}{a}} + \epsilon^{-\frac{x}{a}} \right)^2} \cdot dx$$

$$= \tfrac{1}{2} \int_0^{2a} \left(\epsilon^{\frac{x}{a}} + \epsilon^{-\frac{x}{a}} \right) \cdot dx = \tfrac{1}{2} \int_0^{2a} \epsilon^{\frac{x}{a}} \cdot dx + \tfrac{1}{2} \int_0^{2a} \epsilon^{-\frac{x}{a}} \cdot dx$$

$$= \frac{a}{2} \left[\epsilon^{\frac{x}{a}} - \epsilon^{-\frac{x}{a}} \right]_0^{2a} = \frac{a}{2} \left(\epsilon^{\frac{2a}{a}} - \epsilon^{-\frac{2a}{a}} - \epsilon^0 + \epsilon^{-\frac{0}{a}} \right)$$

$$= \frac{a}{2} \left(\epsilon^2 - \epsilon^{-2} - 1 + 1 \right) = \frac{a}{2} \left(\epsilon^2 - \frac{1}{\epsilon^2} \right) = \text{length of the arc}$$

PROBLEMS

1. Find the length of one arch of the cycloid $x = a(\theta - \sin \theta)$, $y = a(1 - \cos \theta)$.

2. Find the length of the circle $x^2 + y^2 = 25$.

3. Find the length of the four-cusped hypocycloid $x^{\frac{2}{3}} + y^{\frac{2}{3}} = a^{\frac{2}{3}}$.

4. Find the length of the cables between the supports of a suspension bridge if the cables take the form of a parabola $x^2 = 2py$ and if the distance between the tops of the supports is 1,500 ft. and the lowest point of the cable is 150 ft. below this level.

5. Find the length of the spiral of Archimedes $\rho = a\theta$ between $\theta = 0$ and 2π radians.

6. Find the length of the curve $\rho = 3a \sin \theta$ from 0 to π radians.

7. Find the length of the curve $\rho = a \cos \theta$ from 0 to π radians.

8. Find the perimeter of the cardioid $\rho = a(1 + \cos \theta)$.

9. Find the length of the curve whose equation is $y^3 = x^2$ between (1, 1) and (4, 2).

10. Find the length of the parabola $x^2 = 6y$ from (0, 0) to (4, $\frac{8}{3}$).

11. Find the length of the curve $y = x^2$ from $x = 3$ to $x = 6$.

CHAPTER XXXI

AREAS OF SURFACES—VOLUMES
BY SINGLE INTEGRATION

590. What is a surface of revolution?

It is a surface considered as generated by a straight
line or a curve revolving about an axis.

Example

A right circular cone may be considered as generated by a line OA revolving about OX as an axis. Also, it
may be thought of as generated by a
variable circle starting at a point O and
continually getting larger as it moves along to the right.

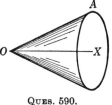

QUES. 590.

591. Under what conditions is the differential of an area of a surface of revolution most readily obtained?

When the surface of revolution is thought of as generated
by a circle moving perpendicular to the axis of revolution
and when the equation of the curve guiding the circle
is known, then the differential of the area is readily obtained.

592. How do we find the area of a surface of revolution?

Consider the surface as generated by a circle starting
at A and moving toward the right guided by the curve
$y = f(x)$. Required the area of the surface from $x = a$ to
$x = b$.

The center of the circle moves along the x axis uniformly.
The rate at which the surface is being generated at any
point, as C, is equal to the length of the circle times the rate
at which C is moving at that instant.

259

But point C is moving in a direction along the tangent to the curve $y = f(x)$ at such a rate that it moves a distance $CC' = ds$ while x increases by dx (or B reaches D). Since the rate of change of the surface S must become and remain constant while x changes by dx, the circumference of the circle must be considered as constant. Therefore, the change in S is that of a cylinder with radius $CB = y$ and $CC' = ds$ for an altitude.

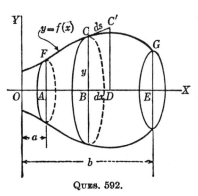

QUES. 592.

$$\therefore dS = 2\pi y \cdot ds \qquad \text{or} \qquad S = 2\pi \int_a^b y \cdot ds$$

As the length of an arc of a curve $ds = \sqrt{1 + \left(\dfrac{dy}{dx}\right)^2} \cdot dx$, then

$$S = 2\pi \int y \sqrt{1 + \left(\frac{dy}{dx}\right)^2} \cdot dx$$

593. Under what conditions may the above formula be used?

It may be used to find any surface of revolution generated by a curve $y = f(x)$ revolving about the x axis when $f(x)$ is above the x axis and is single valued and continuous and has no cusp points within the limits considered. For a curve below the x axis, use $-y$.

594. What is the area generated by revolving the parabola $y^2 = 4x$ about the x axis and between $x = 0$ and $x = 5$?

$$y^2 = 4x \qquad \text{or} \qquad y = 2x^{\frac{1}{2}}$$

and

$$\frac{dy}{dx} = 2 \cdot \frac{1}{2} x^{\frac{1}{2}-1} = x^{-\frac{1}{2}} = \frac{1}{\sqrt{x}}$$

$$S = \text{surface} = 2\pi \int_a^b y \left[1 + \left(\frac{dy}{dx} \right)^2 \right]^{\frac{1}{2}} \cdot dx$$

$$= 2\pi \int_0^5 2x^{\frac{1}{2}} \cdot \left(1 + \frac{1}{x} \right)^{\frac{1}{2}} \cdot dx = 4\pi \int_0^5 (x+1)^{\frac{1}{2}} \cdot dx$$

$$= 4\pi \left[(x+1)^{\frac{1}{2}+1} \cdot \frac{2}{3} \right]_0^5 = \frac{8\pi}{3} \left[(x+1)^{\frac{3}{2}} \right]_0^5$$

$$= 36.53\pi \text{ square units}$$

595. What is the area of the surface of revolution formed by revolving the circle $r^2 = x^2 + y^2$ about the x axis?

Given $x^2 + y^2 = r^2$ or $y = \pm (r^2 - x^2)^{\frac{1}{2}}$

$$\frac{dy}{dx} = \frac{1}{2}(r^2 - x^2)^{-\frac{1}{2}} \cdot -2x = -\frac{x}{(r^2 - x^2)^{\frac{1}{2}}}$$

$$S = \text{area} = 2\pi \int_a^b y \left[1 + \left(\frac{dy}{dx} \right)^2 \right]^{\frac{1}{2}} \cdot dx$$

Choose the plus value of y for the upper half of the circle, and note that the area is twice that generated when x varies from 0 to r.

$$\therefore S = 2 \cdot 2\pi \int_0^r (r^2 - x^2)^{\frac{1}{2}} \left[\left(1 + \frac{x^2}{(r^2 - x^2)} \right) \right]^{\frac{1}{2}} \cdot dx$$

$$= 4\pi \int_0^r (r^2 - x^2 + x^2)^{\frac{1}{2}} \cdot dx = 4\pi \int_0^r r \cdot dx$$

$$= 4\pi r \left[x \right]_0^r = 4\pi r^2 \text{ square units}$$

QUES. 595.

QUES. 596.

596. What is the area of the surface formed by revolving an arc of the cubical parabola $x^3 = 4y$ about the x axis between $x = 0$ and $x = 2$?

$$y = \frac{x^3}{4}, \qquad \frac{dy}{dx} = \frac{3x^2}{4}$$

$$S = 2\pi \int_a^b y \left[1 + \left(\frac{dy}{dx} \right)^2 \right]^{\frac{1}{2}} \cdot dx$$

= formula for surface of revolution

$$\therefore \ S = 2\pi \int_0^2 \frac{x^3}{4} \left(1 + \frac{9x^4}{16} \right)^{\frac{1}{2}} \cdot dx$$

$$= \frac{2\pi}{16} \int_0^2 x^3 (16 + 9x^4)^{\frac{1}{2}} \cdot dx$$

$$= \frac{\pi}{8 \cdot 36} \int_0^2 (16 + 9x^4)^{\frac{1}{2}} \cdot 36x^3 \cdot dx$$

$$= \frac{\pi}{288} \left[(16 + 9x^4)^{\frac{3}{2}} \cdot \frac{2}{3} \right]_0^2 = \frac{\pi}{432} \left[(16 + 9x^4)^{\frac{3}{2}} \right]_0^2$$

$$= \frac{\pi}{432} \left[(16 + 9 \cdot 16)^{\frac{3}{2}} - (16)^{\frac{3}{2}} \right] = \frac{\pi}{432} (2{,}023.86 - 64)$$

= 14.25 square units

QUES. 597.

597. What is the area of the lateral surface of a right circular cone with radius of base = r and altitude = h?

By similar triangles,

$$x{:}y = h{:}r \qquad \text{or} \qquad yh = xr$$

Then $\quad y = \dfrac{r}{h} x \quad$ and $\quad \dfrac{dy}{dx} = \dfrac{r}{h}$

$$S = 2\pi \int_a^b y \left[1 + \left(\frac{dy}{dx} \right)^2 \right]^{\frac{1}{2}} \cdot dx$$

$$= 2\pi \int_0^h \frac{r}{h} x \left(1 + \frac{r^2}{h^2} \right)^{\frac{1}{2}} \cdot dx = \frac{2\pi r}{h^2} \int_0^h (h^2 + r^2)^{\frac{1}{2}} x \cdot dx$$

$$= \frac{2\pi r}{h^2} (h^2 + r^2)^{\frac{1}{2}} \int_0^h x \cdot dx = \frac{2\pi r}{h^2} (h^2 + r^2)^{\frac{1}{2}} \left[\frac{x^2}{2} \right]_0^h$$

$$= \pi r (h^2 + r^2)^{\frac{1}{2}}$$

The slant height of the cone $= s = OA = (h^2 + r^2)^{\frac{1}{2}}$.

$\therefore \ S = \pi r s =$ area of the lateral surface of the cone

598. What is a volume of revolution?

It is a volume considered as generated by revolving an area about an axis.

Example

When a right triangular area is revolved about one of the sides, not the hypotenuse, a right circular cone is formed. And, by revolving the area of a circle about a diameter as an axis, a sphere is formed.

599. How can we find a volume of revolution by means of differentials?

Think of the volume as being generated by the area of a variable circle perpendicular to the x axis. At any point, as at M, the differential of the volume is the volume of a right circular cylinder of radius $MB = y$ and height $dx = BC$, because at M the rate of change of the volume becomes and remains constant while x increases by dx. At M the circle travels along the tangent to M and the center of the circle B travels at a uniform rate to C.

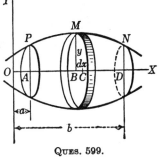

QUES. 599.

$$\therefore \; dV = \pi y^2 \cdot dx$$

or $$V = \pi \int_a^b y^2 \cdot dx = \text{volume of revolution}$$

600. How may a volume of revolution be found by the fundamental theorem, or summation, method?

The formula is the same as for differentials,

$$V = \pi \int_a^b y^2 \cdot dx,$$

but the consideration is that of the limit of a sum instead of an instantaneous rate of change at a point.

The volume is thought of as being generated by revolving area $ABCD$ about the x axis, dividing BC into n parts at Δx intervals, forming rectangles.

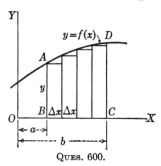

The volume of a cylinder generated by any rectangle is

$$\pi y^2 \cdot \Delta x.$$

And summing up all the rectangles between a and b, noting that $\Delta x \to 0$ as $n \to \infty$,

Ques. 600.

$$V = \lim_{\Delta x \to 0} \sum_a^{o} \pi y^2 \cdot \Delta x = \pi \int_a^b y^2 \cdot dx = \text{volume of revolution}$$

601. How do we find the volume of a solid that is not a solid of revolution either by the differential of the volume method or by the method of summation?

By summation: Cut slices by parallel planes Δx apart. Let the variable area of the section of the solid forming the base of any one of the slices be equal to A.   Then

$$\Delta V = A \cdot \Delta x = \text{volume of one of the slices}$$

The exact volume is equal to the limit of the sum of these slices as the number increases without limit, or

$$V = \lim_{n \to \infty} \sum A \cdot \Delta x$$

For the actual value of this summation between limits for x it is necessary to express A in terms of x, and therefore $V = \int_a^b A \cdot dx$.

By differentials: The same formula results; the only difference is the point of view.   Here the solid is thought of as generated by a variable area A.   At any instant, the differential of the volume is $A \cdot dx$, and the volume is $V = \int_a^b A \cdot dx$.   Here, also, A must be expressed as a function of x.

602. What is the volume generated by revolving the area of the upper half of the circle $x^2 + y^2 = r^2$ about the x axis?

$y = + \sqrt{r^2 - x^2}$ is single valued for the upper half of the circle. Consider the volume generated by a circle starting at O and moving toward the right to A. The total volume is twice that.

Now, $dV = \pi y^2 \cdot dx =$ the differential of the volume when the moving circle has reached any point, say, B. And the volume $= \pi \int y^2 \cdot dx$. For half the volume of the sphere the limits are $x = 0$ to $x = r$.

$$\therefore V = 2\pi \int_0^r y^2 \cdot dx = 2\pi \int_0^r (r^2 - x^2) \cdot dx$$

$$= 2\pi \left[r^2 \cdot x - \frac{x^3}{3} \right]_0^r$$

$$= 2\pi \left(r^3 - \frac{r^3}{3} \right) = \frac{4}{3}\pi r^3 = \text{total volume of the sphere}$$

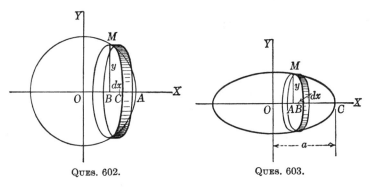

QUES. 602.   QUES. 603.

603. What is the volume generated by revolving the ellipse $\frac{x^2}{a^2} + \frac{y^2}{b^2} = 1$ about the x axis?

Consider one-half the volume of the ellipse generated by a circle starting at O and moving toward the right to C. The total volume is twice that.

Now, $dV = \pi y^2 \cdot dx$ and $V = \pi \int y^2 \cdot dx$ for one-half the ellipse. But $y^2 = \dfrac{b^2(a^2 - x^2)}{a^2}$, and the limits for the half ellipse are 0 and a.

Then

$$\frac{V}{2} = \pi \int_0^a y^2 \cdot dx = \pi \int_0^a \frac{b^2}{a^2}(a^2 - x^2) \cdot dx$$

$$V = \frac{2\pi b^2}{a^2}\left[\int_0^a (a^2 - x^2) \cdot dx\right]$$

$$= \frac{2\pi b^2}{a^2}\left(\int_0^a a^2 \cdot dx - \int_0^a x^2 \cdot dx\right)$$

$$= \frac{2\pi b^2}{a^2}\left[a^2 x - \frac{x^3}{3}\right]_0^a = \frac{2\pi b^2}{a^2}\left(a^3 - \frac{a^3}{3}\right)$$

$$= \frac{2\pi b^2}{a^2} \cdot \frac{2a^3}{3} = \frac{4}{3}\pi ab^2 = \text{the volume of the ellipsoid}$$

604. What is the volume of a right circular cone with radius of base $= r$ and altitude $= h$?

From similar triangles OAM and OBC,

$$x{:}y = h{:}r \quad \text{and} \quad y = \frac{rx}{h}$$

$$V = \pi \int_0^h y^2 \cdot dx = \pi \int_0^h \frac{r^2 x^2}{h^2} \cdot dx$$

$$= \frac{\pi r^2}{h^2} \int_0^h x^2 \cdot dx = \frac{\pi r^2}{h^2}\left[\frac{x^3}{3}\right]_0^h$$

$$= \tfrac{1}{3}\pi r^2 h = \text{the volume of the cone}$$

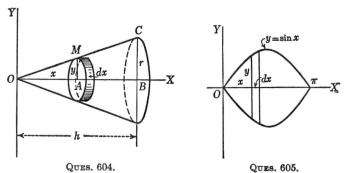

QUES. 604.   QUES. 605.

605. What is the volume generated by revolving the sine curve from 0 to π about the x axis?

$$V = \pi \int_0^\pi y^2 \cdot dx = \pi \int_0^\pi \sin^2 x \cdot dx$$

$$= \pi \int_0^\pi \left(\frac{1}{2} - \frac{1}{2} \cos 2x \right) \cdot dx = \pi \left[\frac{x}{2} - \frac{1}{4} \sin 2x \right]_0^\pi$$

$$= \pi \left(\frac{\pi}{2} - 0 - 0 + 0 \right) = \frac{\pi^2}{2} = \text{the number of cubic units}$$

in the volume provided that the unit on the x axis to represent 1 radian is the length of the unit cube

606. What is the volume formed by revolving the hyperbola $xy = 6$ from $x = 2$ to $x = 4$?

$dV = \pi y^2 \cdot dx =$ the volume differential of the circular
strip of the generated figure

Now $\qquad xy = 6 \qquad$ or $\qquad y = \dfrac{6}{x}$

Then $\qquad V = \int_2^4 \pi \cdot \left(\dfrac{6}{x} \right)^2 \cdot dx = 36\pi \int_2^4 x^{-2} \cdot dx$

$$= 36\pi \left[\frac{x^{-2+1}}{-2+1} \right]_2^4 = -36\pi \left[\frac{1}{x} \right]_2^4$$

$$= -36\pi \left(\frac{1}{4} - \frac{1}{2} \right) = -36\pi \cdot -\frac{1}{4} = 9\pi$$

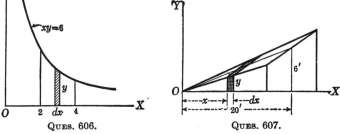

QUES. 606.   QUES. 607.

607. What is the volume of a right pyramid having a square base 12 feet on a side and an altitude of 20 feet?

The figure shows one-half the pyramid above the x axis. The area of the face of an elementary slice $= (2y)^2$, and $dV = (2y)^2 \cdot dx =$ the volume of the elementary slice.

Now, by similar triangles

$$\frac{y}{x} = \frac{6}{20} \quad \text{or} \quad y = \frac{6x}{20} = \frac{3}{10}x$$

Then $V = \int_0^{20} (2y)^2 \cdot dx = \int_0^{20} \left(2 \cdot \frac{3}{10}x\right)^2 \cdot dx$

$$= \int_0^{20} (0.6x)^2 \cdot dx = 0.36 \int_0^{20} x^2 \cdot dx = 0.36\left[\frac{x^3}{3}\right]_0^{20}$$

$$= 0.36 \cdot \frac{8,000}{3} = 960 \text{ cu. ft.}$$

which can readily be checked by geometry.

608. How much metal is removed from a circular shaft 24 inches in diameter by a notch formed by two planes intersecting at the center if one is perpendicular to the axis and the other is at 45 degrees with the first?

Assume the volume of the notch is generated by the area of a triangle always perpendicular to the edge XP,

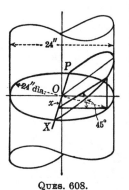

and suppose the triangle, which is always a 45-deg. right triangle, starts at center O and moves toward X. The base of the triangle in any position = $\sqrt{12^2 - x^2}$. This is also equal to the altitude of the triangle. Then the area of the triangle

$$= \tfrac{1}{2} \cdot \sqrt{12^2 - x^2} \cdot \sqrt{12^2 - x^2}$$
$$= \tfrac{1}{2}(12^2 - x^2)$$
$$\therefore dV = \tfrac{1}{2}(12^2 - x^2) \cdot dx$$
$$= \text{the differential of the volume}$$

QUES. 608.

The total volume is twice this volume where x varies from 0 to 12.

$$\therefore V = 2\int_0^{12} \tfrac{1}{2}(12^2 - x^2)\,dx = \int_0^{12} 144 \cdot dx - \int_0^{12} x^2 \cdot dx$$
$$= \left[144 \cdot x - \frac{x^3}{3}\right]_0^{12} = 1,728 - 576 = 1,152 \text{ cu. in.}$$

609. What is the volume of a solid ring (a torus) generated by revolving the area of the circle $x^2 + (y - b)^2 = a^2$, where $b > a$, about the x axis?

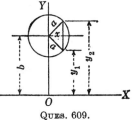

QUES. 609.

The volume to be found is equal to the volume generated by the area under the upper half of the circle minus the volume generated by the area under the lower half of the circle.

$y_2 = b + \sqrt{a^2 - x^2}$ = for the upper half of the circle

$y_1 = b - \sqrt{a^2 - x^2}$ = for the lower half of the circle

$$\therefore V = \pi \int (y_2{}^2 - y_1{}^2)\, dx = \pi \int_{-a}^{a} (b + \sqrt{a^2 - x^2})^2 \cdot dx$$
$$- \pi \int_{-a}^{a} (b - \sqrt{a^2 - x^2})^2 \cdot dx$$

This reduces to

$$V = 4\pi b \int_{-a}^{a} \sqrt{a^2 - x^2} \cdot dx = 4\pi b \left[\frac{x}{2} \sqrt{a^2 - x^2} \right.$$
$$\left. + \frac{a^2}{2} \sin^{-1} \frac{x}{a} \right]_{-a}^{a} \quad \text{(by formula VII)}$$

$$= 4\pi b \left(\frac{a}{2} \sqrt{a^2 - a^2} + \frac{a^2}{2} \sin^{-1} \frac{a}{a} + \frac{a}{2} \sqrt{a^2 - a^2} \right.$$
$$\left. - \frac{a^2}{2} \sin^{-1} \frac{-a}{a} \right)$$

$$= 4\pi b \left(0 + \frac{a^2}{2} \cdot \frac{\pi}{2} + 0 - \frac{a^2}{2} \cdot \frac{3\pi}{2} \right) = 4\pi b \left(\frac{\pi a^2}{4} - \frac{3\pi a^2}{4} \right)$$

$$= 4\pi b \cdot - \frac{\pi a^2}{2}$$

$$\therefore V = 2\pi^2 a^2 b$$

PROBLEMS

1. What is the area of the surface generated by revolving one loop of $8y^2 = x^2 - x^4$ about the x axis?

2. What is the area of the surface generated by revolving about the x axis the arc of the parabola $y = x^2$ from (0, 0) to (2, 4)?

3. What is the area of the surface generated by revolving about the x axis the loop of $9y^2 = x(3 - x)^2$?

4. What is the area of the surface generated by revolving about the y axis the curve $x = y^3$ from $y = 0$ to $y = 2$?

5. What is the area of the surface of a cone generated by revolving about the x axis the line joining the origin to the point $(3, 4)$?

6. What is the area of the surface of revolution generated by revolving the hypocycloid $x^{\frac{2}{3}} + y^{\frac{2}{3}} = a^{\frac{2}{3}}$ about the x axis?

7. What is the area of the surface generated by revolving about the y axis the arc of the parabola $y = x^2$ from $y = 0$ to $y = 2$?

8. What is the area of the surface generated by revolving about the y axis the curve $6xy = x^4 + 3$ from $x = 1$ to $x = 3$?

9. What is the area of the surface generated by revolving about the x axis the arc of the parabola $y^2 = 2x$ from $x = 0$ to $x = 4$?

10. What is the area of the surface generated by revolving about the x axis the cycloid $x = a(\theta - \sin \theta)$, $y = a(1 - \cos \theta)$?

11. Find the volume of a sphere generated by revolving the area of the upper half of the circle $x^2 + y^2 = 4$ about the x axis.

12. Find the volume generated by revolving the area under one arch of $y = \cos x$ about the x axis.

13. Find the volume generated by revolving the area under the upper half of the parabola $y^2 = 2x$ from $x = 1$ to $x = 4$ about the x axis.

14. Find the volume when the area between the curve $y^2 = 2x$, the y axis, and the lines from $y = 1$ to $y = 4$ is revolved about the y axis.

15. Find the volume of a right circular cone with radius of base $= 6$ in. and altitude $= 10$ in.

16. Find the volume of the segment of a sphere, 20 in. in diameter, formed by a plane cutting the sphere 6 in. from the center.

17. Find the volume of the solid of revolution generated when the area bounded by the semicubical parabola $2y^2 = x^3$, the y axis, and the line $BC(y = 2)$ is revolved about the line BC.

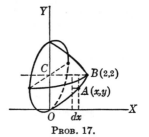

PROB. 17.

18. Find the volume of the oblate spheroid generated by revolving the area bounded by the ellipse $b^2x^2 + a^2y^2 = a^2b^2$ about the y axis.

19. If a tree 30 in. in diameter has a notch cut to the center, what is the volume of the wood removed if the lower plane of the notch is horizontal and the upper plane makes an angle of 45 deg. with the lower plane?

20. What is the volume generated by revolving about the x axis the area bounded by one arch of $y = \cos 2\theta$?

21. What is the volume generated by revolving about the x axis the area bounded by $y^2 = (3 - x)^2$, $y = 0$, $x = 0$?

22. Find the volume generated by revolving about the y axis the area bounded by $y = \epsilon^x$, $y = 0$, $x = 0$.

23. What is the volume generated by revolving one arch of the cycloid $x = a(\theta - \sin \theta)$, $y = a(1 - \cos \theta)$, about its base OX?

NOTE.—The limits are $x = 0$, $x = 2\pi$.

CHAPTER **XXXII**

SUCCESSIVE AND PARTIAL INTEGRATION

610. What is successive integration?

Integration of the derived function successively a number of times as indicated.

The reverse of successive differentiation.

611. If $\dfrac{d^3y}{dx^3} = 10x$, what is y?

$$\frac{d^3y}{dx^3} = \frac{d\left(\dfrac{d^2y}{dx^2}\right)}{dx} = 10x$$

and $\quad d\left(\dfrac{d^2y}{dx^2}\right) = 10x \cdot dx = $ differential-equation form

Integrating, $\dfrac{d^2y}{dx^2} = \displaystyle\int 10x \cdot dx = \dfrac{10x^2}{2} = 5x^2 + C_1$

Now $\quad \dfrac{d^2y}{dx^2} = \dfrac{d\left(\dfrac{dy}{dx}\right)}{dx} = 5x^2 + C_1$

and $\quad d\left(\dfrac{dy}{dx}\right) = (5x^2 + C_1)\, dx = $ differential-equation

$$\text{form}$$

Integrating, $\dfrac{dy}{dx} = \displaystyle\int (5x^2 + C_1)\, dx = \dfrac{5x^3}{3} + C_1x + C_2$

and $\quad dy = \left(\dfrac{5x^3}{3} + C_1x + C_2\right) dx$

Integrating, $\quad y = \displaystyle\int \left(\dfrac{5x^3}{3} + C_1x + C_2\right) dx = \dfrac{5}{3}\cdot\dfrac{x^4}{4}$

$$+ \frac{C_1x^2}{2} + C_2x + C_3$$

612. What are the symbols for successive integration?

$y = \iiint 10x \cdot dx \cdot dx \cdot dx$ means "Integrate $10x$ three times with respect to x."

$y = \iint f(x) \cdot dx \cdot dx$ means "Integrate $f(x)$ twice with respect to x."

613. If the acceleration of a moving point $= \alpha =$ constant, what is the expression for $s =$ distance traversed?

Given the acceleration $\alpha = \dfrac{d^2s}{dt^2} =$ constant

or $\qquad \dfrac{d\left(\dfrac{ds}{dt}\right)}{dt} = \alpha$

Then $d\left(\dfrac{ds}{dt}\right) = \alpha \cdot dt =$ differential-equation form

$\therefore \dfrac{ds}{dt} = \int \alpha \cdot dt = \alpha t + C_1$

Now $\qquad ds = (\alpha t + C_1)\, dt =$ differential-equation form

$\therefore s = \int (\alpha t + C_1)\, dt = \dfrac{\alpha t^2}{2} + C_1 t + C_2$

or, in symbols for successive integration,

$$s = \iint \alpha \cdot dt \cdot dt$$

which means "Integrate twice with respect to t."

Integrating once, $\qquad s = \int (\alpha t + C_1)\, dt$

Integrating again,

$s = \dfrac{\alpha t^2}{2} + C_1 t + C_2 =$ the distance traversed for constant acceleration

614. What is y if $y = \iiint x^3 \cdot dx \cdot dx \cdot dx$?

$y = \iiint x^3 \cdot dx \cdot dx \cdot dx$. Integrate once to get

$$y = \int \int \left(\frac{x^{3+1}}{3+1} + C_1 \right) dx \cdot dx = \int \int \left(\frac{x^4}{4} + C_1 \right) dx \cdot dx$$

Now integrate again to get

$$y = \int \left(\frac{1}{4} \cdot \frac{x^{4+1}}{4+1} + C_1 x + C_2 \right) dx = \int \left(\frac{x^5}{20} + C_1 x + C_2 \right) dx$$

Integrate once more, and get

$$y = \frac{x^6}{120} + C_1 \frac{x^2}{2} + C_2 x + C_3$$

615. When do the constants of integration disappear in successive integration?

When the successive integrations are performed between limits.

Example

$$\int_0^3 \int_1^4 \int_1^3 3x \cdot dx \cdot dx \cdot dx = \int_0^3 \int_1^4 \left[\frac{3x^2}{2} \right]_1^3 \cdot dx \cdot dx$$

$$= \int_0^3 \int_1^4 \left(\frac{3 \cdot 3^2}{2} + C - \frac{3 \cdot 1}{2} - C \right) \cdot dx \cdot dx$$

$$= \int_0^3 \int_1^4 \left(\frac{27}{2} - \frac{3}{2} \right) dx \cdot dx$$

$$= \int_0^3 \int_1^4 12 \cdot dx \cdot dx = \int_0^3 \left[12x \right]_1^4 \cdot dx$$

$$= \int_0^3 \left(12 \cdot 4 + C - 12 \cdot 1 - C \right) dx$$

$$= \int_0^3 36 \, dx = \left[36x \right]_0^3 = 36 \cdot 3 = 108$$

616. What is the equation of the elastic curve and what is the maximum deflection of a simply supported beam uniformly loaded with w pounds per unit length when the differential equation of its elastic curve is $EI \dfrac{d^2y}{dx^2} = \dfrac{w}{2} \left(\dfrac{l^2}{4} - x^2 \right)$?

$$EI \frac{d^2y}{dx^2} = \frac{w}{2} \left(\frac{l^2}{4} - x^2 \right) \qquad (1)$$

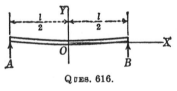

Ques. 616.

The constants are

E = modulus of elasticity

I = moment of inertia of the cross section of the beam

l = length of beam

Divide Eq. (1) by EI to get

$$\frac{d^2y}{dx^2} = \frac{wl^2}{8EI} - \frac{wx^2}{2EI} \qquad \text{or} \qquad \frac{d}{dx}\left(\frac{dy}{dx}\right) = \frac{wl^2}{8EI} - \frac{wx^2}{2EI}$$

and $d\left(\dfrac{dy}{dx}\right) = \dfrac{wl^2}{8EI}\cdot dx - \dfrac{wx^2}{2EI}\cdot dx = $ differential-equation

form

Integrating, you get

$$\frac{dy}{dx} = \int \frac{wl^2}{8EI}\cdot dx - \int \frac{wx^2}{2EI}\cdot dx$$

or $\qquad \dfrac{dy}{dx} = \dfrac{wl^2}{8EI}x - \dfrac{wx^3}{6EI} + C_1 \hfill (2)$

Again, $dy = \dfrac{wl^2}{8EI}x\cdot dx - \dfrac{wx^3}{6EI}\cdot dx + C_1 dx = $ differential-

equation form

and integrating, $y = \displaystyle\int \frac{wl^2}{8EI}x\cdot dx - \int \frac{wx^3}{6EI}\cdot dx + \int C_1 dx$

or $\qquad y = \dfrac{wl^2}{16EI}\cdot x^2 - \dfrac{wx^4}{24EI} + C_1 x + C_2 \hfill (3)$

To find the constants C_1 and C_2 you will note that the slope $\dfrac{dy}{dx}$ is zero when $x = 0$ and that $y = 0$ when $x = 0$ as the curve passes through the origin.

Substituting $\dfrac{dy}{dx} = 0$ and $x = 0$ in Eq. (2),

$$0 = 0 - 0 + C_1$$
$$\therefore C_1 = 0$$

and $y = 0$, $x = 0$ in Eq. (3),

$$0 = 0 - 0 + 0 + C_2$$
$$\therefore C_2 = 0$$

$\therefore y = \dfrac{wl^2x^2}{16EI} - \dfrac{wx^4}{24EI} = \dfrac{w}{48}(3l^2x^2 - 2x^4) = $ the equation of

the elastic curve

Now, for a maximum deflection, $x = \dfrac{l}{2}$. Then

$$y = \frac{wl^4}{64EI} - \frac{wl^4}{384} = \frac{5wl^4}{384} = \text{maximum deflection}$$

617. What is partial integration?

It is the reverse operation from partial differentiation. This means integrating first with only one variable considered as varying, all the others remaining constant. Then the result of this integration is again integrated, and another variable is considered as varying, all the others remaining constant, etc., until all the variables have been accounted for.

618. What is the general illustration of partial integration?

$u = \iint f(x, y) \, dy \cdot dx$ means that you are to find a function u in y and x so that its second partial differential $\dfrac{\partial^2 u}{\partial x \partial y} = f(x, y) = $ the function within the integral signs.

If you integrate first with respect to y, considering x as constant, the constant of integration may involve x or it may be a constant, C.

Then $\displaystyle\int \frac{\partial^2 u}{\partial x \, \partial y} = \frac{\partial u}{\partial x} = \int f(x, y) + \phi(x)$ or

$$\frac{\partial u}{\partial x} = \int f(x, y) + C$$

As the differentiation of either resulting expression with respect to y gives the same result, $f(x, y)$, we can then assume a general case.

$$\therefore \frac{\partial u}{\partial x} = \int f(x, y) + \phi(x) = \text{general expression}$$

where $\phi(x)$ is an arbitrary function of x.

619. What is u, given $\dfrac{\partial u}{\partial x} = 3x + 2y + 4$?

Integrating with respect to x, considering y as constant, you get $u = \dfrac{3x^2}{2} + 2yx + 4x + \phi$ where ϕ means the constant of integration.

Now as y was regarded constant during the integration, ϕ may involve y and should be replaced by the symbol ϕy to indicate the dependency of ϕ on y.

$\therefore u = \frac{3}{2}x^2 + 2yx + 4x + \phi(y) =$ the general form for u where $\phi(y)$ means an arbitrary function of y.

620. What is the solution of u in $u = \iint (x^3 + y^3) \, dy \cdot dx$?

This means "What is u if $\dfrac{\partial^2 u}{\partial x \cdot \partial y} = x^3 + y^3$?"

Integrate first with respect to y, regarding x as constant, getting $\dfrac{\partial u}{\partial x} = x^3 y + \dfrac{y^4}{4} + \psi(x)$ where $\psi(x)$ is an arbitrary function x.

Now integrate this result with respect to x, regarding y as a constant, getting

$$u = \frac{x^4 y}{4} + \frac{xy^4}{4} + \psi_1(x) + \phi(y)$$

where $\phi(y)$ is an arbitrary function of y and

$$\psi_1(x) = \int \psi(x) \, dx$$

621. If $u = \iint \epsilon^{4x} \cdot y^3 \cdot dy \cdot dx$, what is u?

Now $\dfrac{\partial^2 u}{\partial x \cdot \partial y} = \epsilon^{4x} \cdot y^3 =$ the expression within the integral signs

Integrating with respect to y, while treating x as constant,

$$\frac{\partial u}{\partial x} = \epsilon^{4x} \cdot \frac{y^4}{4} + \phi(x)$$

Then integrating with respect to x, while treating y as constant,

$$u = \frac{\epsilon^{4x}}{4} \cdot \frac{y^4}{4} + \int \phi(x) \cdot dx + F(y)$$

And since $\phi(x)$ was arbitrary, therefore, $\int \phi(x) \cdot dx$ is arbitrary, so that

$$u = \frac{\epsilon^{4x}y^4}{16} + \psi_1(x) + \psi_2(y)$$

where both $\psi_1(x)$ and $\psi_2(y)$ are arbitrary.

622. How would you illustrate partial integration geometrically?

$z = f(x, y) =$ the equation of the surface AB. Now assume one elementary column with a curvilinear surface at the top and a base area of $\Delta x \cdot \Delta y$. The height of the column $= z = f(x, y)$. Therefore

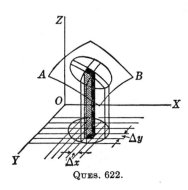

Ques. 622.

$f(x, y) \cdot \Delta x \cdot \Delta y =$ the volume of this elementary column, and $V = \iint f(x, y) \cdot dx \cdot dy =$ the volume of all such elementary columns for the entire figure, by the principle of summation.

Now you integrate or sum up first with respect to x, assuming y as a constant, and the result you integrate with respect to y, keeping x as constant, or vice versa.

623. What is a surface integral in multiple integration?

A surface integral is one that deals with areas of surfaces and solids. You integrate once for length and again for breadth to obtain the total area.

$\iint u \cdot dx \cdot dy =$ area = a surface integral ($u =$ a property that depends at each point on both x and y).

The value of all elements, as $u \cdot \Delta x \cdot \Delta y$, or the value of u over a little rectangle Δx long and Δy broad, must be summed up over the entire length and entire breadth of the area.

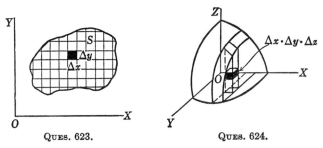

QUES. 623.   QUES. 624.

624. What is a volume integral in multiple integration?

A volume integral is one that deals with solids, or three dimensions. Elementary cubes $\Delta x \cdot \Delta y \cdot \Delta z$ are added up in the three directions x, y, z.

$$\iiint f(x, y, z) \cdot dx \cdot dy \cdot dz = \text{a volume integral}$$

625. What are the requirements for integration of areas and volumes?

a. Appropriate limits in each dimension must be given or determined.

b. The manner in which the boundaries of the surface depend on x, y, and z must be given or determined

or $\qquad \int_{z_1}^{z_2} \int_{y_1}^{y_2} \int_{x_1}^{x_2} f(x, y, z) \cdot dx \cdot dy \cdot dz = \text{volume}$

The innermost integral is usually placed to apply to the innermost differential.

PROBLEMS

1. What is y if $\dfrac{d^2y}{dx^2} = 5x$?

2. What is y if $\dfrac{d^4y}{dx^4} = 0$?

3. Find y in $y = \iint (2x^4 + x^2)\, dx \cdot dx$.

4. Find y in $y = \int_1^3 \int_2^4 \int_0^3 x^3 \cdot dx \cdot dx \cdot dx$.

5. Find y in $y = \int\int\int \epsilon^{2x} \cdot dx \cdot dx \cdot dx$.

6. What is the equation of a curve if the rate of change of its slope at every point is twice the abscissa of the point and if it passes through points $(1, 3)$ and $(-3, -1)$?

7. Find u when $\dfrac{\partial u}{\partial x} = 6x - 5y - 3$.

8. What is u in $u = \int\int (2x^4 - 3y^2)\, dy \cdot dx$?

Solve

9. $\displaystyle\int_0^4 \int_0^x y \cdot dy \cdot dx$ **10.** $\displaystyle\int_{-1}^1 \int_0^{x^2} (x + y)\, dy \cdot dx$

11. $\displaystyle\int_0^1 \int_0^x \int_0^{x+y} \epsilon^{x+y+s} \cdot dz \cdot dy \cdot dx$

12. $\displaystyle\int_1^2 \int_0^z \int_0^{x\sqrt{3}} \left(\dfrac{x}{x^2 + y^2}\right) \cdot dy \cdot dx \cdot dz$

13. $\displaystyle\int_0^2 \int_1^y y \cdot dx \cdot dy$

Chapter XXXIII

PLANE AREAS BY DOUBLE INTEGRATION

626. How do we find a plane area by double integration?

Divide the distance between $x = a$ and $x = b$ into a number of narrow strips Δx distance apart. Cut up a strip into small rectangles Δy high.

$\Delta x \cdot \Delta y$ = area of one
               such rectangle

Sum up the small rectangles for one strip, getting

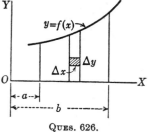

Ques. 626.

$$\Delta x \sum_{0}^{f(x)} \Delta y = \text{area of one strip}$$

Now sum up all strips between $x = a$ and $x = b$.

$$\sum_{a}^{b} \left(\sum_{0}^{f(x)} \Delta y \right) \cdot \Delta x = \text{area of all the strips}$$

Pass to the limit, first letting Δy approach zero, and then letting Δx approach zero.

Then $\quad A = \int_{a}^{b} \int_{0}^{f(x)} dy \cdot dx = \text{the required area}$

The inside integral sign belongs to dy and the outside to dx.

627. How do we find a plane area by double integration in polar coordinates?

Sector $AOD = \frac{1}{2}\rho \cdot \rho \, \Delta\theta = \frac{1}{2}\rho^2 \, \Delta\theta$
Sector $BOC = \frac{1}{2}(\rho + \Delta\rho)^2 \cdot \Delta\theta$
∴ Area $ABCD = \frac{1}{2}(\rho + \Delta\rho)^2 \cdot \Delta\theta - \frac{1}{2}\rho^2 \, \Delta\theta$

$$= \frac{\rho^2}{2} \Delta\theta + \rho \cdot \Delta\rho \cdot \Delta\theta + \frac{(\Delta\rho)^2}{2} \cdot \Delta\theta - \frac{\rho^2 \Delta\theta}{2}$$

$$= \left(\rho + \frac{\Delta\rho}{2} \right) \Delta\rho \cdot \Delta\theta$$

Now keep $\Delta\theta$ constant, and sum up all such elements of area as $ABCD$ to get the area of sector NOM, or

$$\Delta\theta \cdot \lim_{\Delta\rho \to 0} \sum_{\rho=0}^{\rho=f(\theta)} \left(\rho + \tfrac{1}{2} \cdot \Delta\rho\right) \Delta\rho = \Delta\theta \int_0^{f(\theta)} \rho \cdot d\rho$$

Now sum up all the sectors to get the whole area.

$$A = \lim_{\Delta\theta \to 0} \sum_{\theta_1}^{\theta_2} \Delta\theta \int_0^{f(\theta)} \rho \cdot d\rho = \int_{\theta_1}^{\theta_2} \int_0^{f(\theta)} \rho \cdot d\rho \cdot d\theta$$

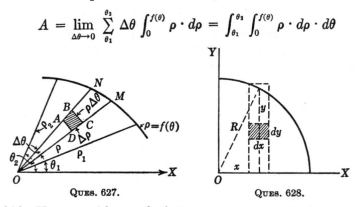

QUES. 627.   QUES. 628.

628. How would you find the area of a circle by double integration?

$$\frac{A}{4} = \int_0^R \int_0^{f(x)} dy \cdot dx = \text{area of a quarter circle}$$

Now $x^2 + y^2 = R^2$ and $y = (R^2 - x^2)^{\frac{1}{2}} = f(x)$

$$\therefore \frac{A}{4} = \int_0^R \int_0^{(R^2-x^2)^{\frac{1}{2}}} dy \cdot dx = \int_0^R \left[y\right]_0^{(R^2-x^2)^{\frac{1}{2}}} \cdot dx$$

Integrate by parts, letting $u = (R^2 - x^2)^{\frac{1}{2}}$ and $dv = dx$;

then $du = -\dfrac{x \cdot dx}{(R^2 - x^2)^{\frac{1}{2}}}$ and $v = x$

$$\therefore \frac{A}{4} = \int_0^R (R^2 - x^2)^{\frac{1}{2}} \cdot dx = \int u \cdot dv = uv - \int v \cdot du$$

$$= (R^2 - x^2)^{\frac{1}{2}} \cdot x + \int \frac{x \cdot x \cdot dx}{(R^2 - x^2)^{\frac{1}{2}}}$$

or $\dfrac{A}{4} = \displaystyle\int_0^R (R^2 - x^2)^{\frac{1}{2}} \cdot dx = x(R^2 - x^2)^{\frac{1}{2}}$

$$+ \int \frac{x^2 \cdot dx}{(R^2 - x^2)^{\frac{1}{2}}} \quad (1)$$

Also, $\dfrac{A}{4} = \displaystyle\int_0^R (R^2 - x^2)^{\frac{1}{2}} \cdot dx = \int \dfrac{(R^2 - x^2)\,dx}{(R^2 - x^2)^{\frac{1}{2}}}$

$$= \int \dfrac{R^2 \cdot dx}{(R^2 - x^2)^{\frac{1}{2}}} - \int \dfrac{x^2 \cdot dx}{(R^2 - x^2)^{\frac{1}{2}}} \quad (2)$$

Now add Eqs. (1) and (2), getting

$$\dfrac{A}{2} = 2 \int_0^R (R^2 - x^2)^{\frac{1}{2}} \cdot dx = x(R^2 - x^2)^{\frac{1}{2}} + \int \dfrac{R^2 \cdot dx}{(R^2 - x^2)^{\frac{1}{2}}}$$

But $\displaystyle\int \dfrac{dx}{(R^2 - x^2)^{\frac{1}{2}}} = \text{arc sin } \dfrac{x}{R} = \sin^{-1} \dfrac{x}{R}$

$$\therefore \dfrac{A}{4} = \int_0^R (R^2 - x^2)^{\frac{1}{2}} \cdot dx = \left[\dfrac{x}{2} (R^2 - x^2)^{\frac{1}{2}} \right.$$

$$\left. + \dfrac{R^2}{2} \sin^{-1} \dfrac{x}{R} \right]_0^R$$

$$= \left(0 + \dfrac{R^2}{2} \cdot \dfrac{\pi}{2} - 0 - 0 \right) = \dfrac{\pi R^2}{4}$$

$\therefore A = $ area of the entire circle
$= \pi R^2$

629. What is the area between the parabolas $y^2 = x$ and $y = x^2$?

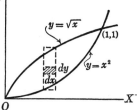

QUES. 629.

$$A = \int_a^b \int_{f(x)}^{F(x)} dy \cdot dx$$

Solving as simultaneous equations to get the limits for x,

$$y = \sqrt{x}, \qquad y = x^2 \qquad \text{(given)}$$

$$\therefore \sqrt{x} = x^2, \qquad x = x^4, \qquad x(1 - x^3) = 0 \qquad \text{or}$$

$$x = 0, \qquad x = 1$$

$$\therefore A = \int_0^1 \int_{x^2}^{\sqrt{x}} dy \cdot dx = \int_0^1 \left[y \right]_{x^2}^{\sqrt{x}} = \int_0^1 (\sqrt{x} - x^2)\,dx$$

$$= \int_0^1 x^{\frac{1}{2}} \cdot dx - \int_0^1 x^2 \cdot dx = \left[\dfrac{2}{3} x^{\frac{3}{2}} - \dfrac{x^3}{3} \right]_0^1 = \dfrac{2}{3} - \dfrac{1}{3}$$

$$= \dfrac{1}{3} \text{ square units}$$

630. How do we reduce the expression of an area for double integration to that for single integration?

If $A = \iint dy \cdot dx$ is the expression for an area by double integration, then, by integrating with respect to y, we get the form for an area by single integration, or $A = \int y \cdot dx$.

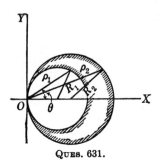

QUES. 631.

631. What is the area, by polar coordinates, between two circles tangent internally and having radii R_1 and R_2, respectively?

From trigonometry,

$$\rho_1 = 2R_1 \cos \theta$$
$$\rho_2 = 2R_2 \cos \theta$$

For $\dfrac{\text{area}}{2}$, integrate between the limits of $\theta = \dfrac{\pi}{2}$ and $\theta = 0$.

Now $\dfrac{A}{2} = \displaystyle\int_0^{\frac{\pi}{2}} \int_{\rho_1 = 2R_1 \cos \theta}^{\rho_2 = 2R_2 \cos \theta} \rho \cdot d\rho \cdot d\theta$

or $\quad = \displaystyle\int_0^{\frac{\pi}{2}} \left[\frac{\rho^2}{2} \right]_{2R_1 \cos \theta}^{2R_2 \cos \theta} \cdot d\theta$

$\quad = \displaystyle\int_0^{\frac{\pi}{2}} \left[\frac{4R_2{}^2 \cos^2 \theta - 4R_1{}^2 \cos^2 \theta}{2} \right] d\theta$

$\quad = \displaystyle\int_0^{\frac{\pi}{2}} 2(R_2{}^2 - R_1{}^2) \cos^2 \theta \cdot d\theta$

$\quad = 2(R_2{}^2 - R_1{}^2) \left[\dfrac{\theta}{2} + \dfrac{\sin 2\theta}{4} \right]_0^{\frac{\pi}{2}}$

$\quad = 2(R_2{}^2 - R_1{}^2) \left(\dfrac{\pi}{4} - 0 \right)$

$\quad = \dfrac{\pi}{2} (R_2{}^2 - R_1{}^2)$

$\therefore A = \pi(R_2{}^2 - R_1{}^2)$

632. What is the area included between the parabola $y^2 = 5x$ and the circle $x^2 + y^2 = 36$ in the first quadrant?

Here it is best to make the first summation of an elementary horizontal strip, for then each strip begins at the parabola and ends at the circle while some of the vertical strips end on the parabola and some on the circle.

Now $\quad x^2 + y^2 = 36 \quad$ or $\quad x = \sqrt{36 - y^2}$

$$y^2 = 5x \quad \text{or} \quad x = \frac{y^2}{5}$$

These are the x limits for the horizontal strips. The y limit is found by solving the two equations simultaneously.

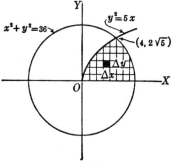

QUES. 632.

Substituting $y^2 = 5x$ in the equation of the circle, you get

$$x^2 + 5x = 36 \quad \text{or} \quad x^2 + 5x - 36 = 0$$

Then $x = 4$ for the first and fourth quadrant.

Substituting $x = 4$ in $y^2 = 5x$ gives

$$y^2 = 5 \cdot 4 = 20 \quad \text{or} \quad y = \pm 2\sqrt{5}$$

and this is the upper limit for y in the first quadrant.

$$\therefore A = \text{area} = \int_0^{2\sqrt{5}} \int_{\frac{y^2}{5}}^{\sqrt{36-y^2}} dx \cdot dy = \int_0^{2\sqrt{5}} \left[x \right]_{\frac{y^2}{5}}^{\sqrt{36-y^2}} \cdot dy$$

$$= \int_0^{2\sqrt{5}} \left(\sqrt{36 - y^2} - \frac{y^2}{5} \right) dy = \int_0^{2\sqrt{5}} \sqrt{36 - y^2} \cdot dy$$

$$- \frac{1}{5} \int_0^{2\sqrt{5}} y^2 \cdot dy$$

This is of the form $\int \sqrt{a^2 - u^2} \cdot du = \frac{u}{2}\sqrt{a^2 - u^2}$
$$+ \frac{a^2}{2}\sin^{-1}\frac{u}{a}.$$

$$\therefore A = \left[\frac{y}{2}\sqrt{36 - y^2} + \frac{36}{2}\sin^{-1}\frac{y}{6} - \frac{y^3}{15}\right]_0^{2\sqrt{5}}$$

$$= \frac{2\sqrt{5}}{2}\sqrt{36 - 20} + 18\sin^{-1}\frac{2\sqrt{5}}{6} - \frac{(2\sqrt{5})^3}{15}$$

$$= \sqrt{5} \cdot 4 + 18\sin^{-1}\frac{\sqrt{5}}{3} - \frac{40\sqrt{5}}{15}$$

$$= 2.236 \cdot 4 + 18 \cdot \sin^{-1}\frac{2.236}{3} - \frac{8 \cdot 2.236}{3}$$

$$= 8.944 + 18\sin^{-1} 0.7453 - 5.963$$

Now $\sin^{-1} 0.7453 = 48.183° = \frac{48.183}{57.29} = 0.841$ radian

$$\therefore A = 8.944 + 18 \cdot 0.841 - 5.963 = 18.119 \text{ square units}$$

633. How would you determine the area in the first quadrant bounded by the parabolas $x^2 = 5y$, $y^2 = 5x$ and the circle $x^2 + y^2 = 36$?

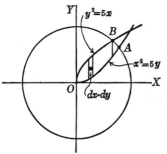

Solve $x^2 = 5y$ and $x^2 + y^2 = 36$ for the point of intersection. $5y + y^2 = 36$ gives $y = 4$ in the first quadrant.

Then $x^2 = 5y = 5 \cdot 4 = 20$
and $x = 2\sqrt{5}$

Therefore the coordinates of the point of intersection A

QUES. 633.

are $(2\sqrt{5}, 4)$. Solving $y^2 = 5x$ and $x^2 + y^2 = 36$ gives the coordinates of the point of intersection B as $(4, 2\sqrt{5})$.

Note that the vertical strips extend from parabola OA to parabola OB while x varies from 0 to 4 and from parabola OA to the arc of the circle AB while x varies from 4 to $2\sqrt{5}$, which means that the area is to be expressed in two double integrals

$$A = \int_0^4 \int_{\frac{x^2}{5}}^{\sqrt{5x}} dy \cdot dx + \int_4^{2\sqrt{5}} \int_{\frac{x^2}{5}}^{\sqrt{36-x^2}} dy \cdot dx$$

634. What is the area of the upper half of the cardioid $\rho = 2a(1 - \cos \theta)$ by double integration?

$$A = \int_{\theta_1}^{\theta_2} \int_0^{f(\theta)} \rho \cdot d\rho \cdot d\theta$$

$$= \int_0^\pi \int_0^{2a(1-\cos\theta)} \rho \cdot d\rho \cdot d\theta$$

$$= \int_0^\pi \left[\frac{\rho^2}{2}\right]_0^{2a(1-\cos\theta)} \cdot d\theta$$

$$= 2a^2 \int_0^\pi (1 - \cos\theta)^2 \cdot d\theta$$

$$= 2a^2 \int_0^\pi (1 - 2\cos\theta + \cos^2\theta) \, d\theta$$

$$= 2a^2 \left[\theta - 2\sin\theta + \frac{\theta}{2} + \frac{\sin 2\theta}{4}\right]_0^\pi$$

$$= 2a^2 \left(\pi - 0 + \frac{\pi}{2} + 0\right) = 2a^2 \cdot \frac{3\pi}{2} = 3\pi a^2$$

QUES. 634.

PROBLEMS

Find the following areas by double integration:

1. The circle $x^2 + y^2 = 9$.

2. The area between the parabolas $y^2 = 2x$ and $y = 4x^2$.

3. The parabola $y^2 = 6x$ and the circle $x^2 + y^2 = 25$ in the first quadrant.

4. The area in the first quadrant bounded by the parabolas $x^2 = 6y$ and $y^2 = 6x$ and the circle $x^2 + y^2 = 25$.

5. The area of the upper half of $\rho = a(1 - \cos\theta)$.

6. The area bounded by the circles $\rho = 4\sin\theta$ and $\rho = 8\sin\theta$.

7. The area of the ellipse $\frac{x^2}{9} + \frac{y^2}{4} = 1$.

8. The area bounded by the two coordinate axes and the line $y = 3x + 5$.

9. The segment of the circle $x^2 + y^2 = 25$, cut off by the line $x + y = 5$.

10. The area of the circle $\rho = 4\cos\theta$.

VOLUMES BY TRIPLE INTEGRATION

635. What is the procedure for finding volumes by triple integration?

Divide the solid into elementary rectangular parallelepipeds each having volume $\Delta x \cdot \Delta y \cdot \Delta z$.

First sum up all the elementary volumes of a column. Then sum up all the columns to form a slice. Then sum up all slices.

$$V = \int_0^a \int_0^{y=F(x)} \int_0^{z=f(x,y)} \cdot dz \cdot dy \cdot dx$$
$$\text{(slices) (slice) (column)}$$

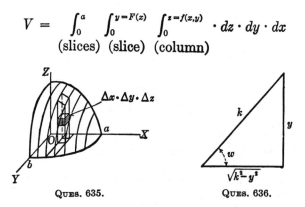

QUES. 635. QUES. 636.

636. What is the volume of an ellipsoid by triple integration?

$\dfrac{x^2}{a^2} + \dfrac{y^2}{b^2} + \dfrac{z^2}{c^2} = 1$ is the equation of an ellipsoid

Now consider one quadrant, or one-eighth of the volume.

The z limits are from $z = 0$ to $z = c\sqrt{1 - \dfrac{x^2}{a^2} - \dfrac{y^2}{b^2}}$; this sums up the elementary volumes into a column.

288

The y limits are from $y = 0$ to $y = b\sqrt{1 - \dfrac{x^2}{a^2}}$; this sums up the elementary columns into a slice $(z = 0)$ because the boundary of the curve is in the xy plane.

The x limits are from $x = 0$ to $x = a = Oa$; this sums up all the slices (see figure, Question 635).

$$\therefore \frac{V}{8} = \int_{x=0}^{x=a} \int_{y=0}^{y=b\sqrt{1-\frac{x^2}{a^2}}} \int_{z=0}^{z=c\sqrt{1-\frac{x^2}{a^2}-\frac{y^2}{b^2}}} dz \cdot dy \cdot dx$$

Solving,

$$\frac{V}{8} = c \int_0^a \int_0^{b\sqrt{1-\frac{x^2}{a^2}}} \left(1 - \frac{x^2}{a^2} - \frac{y^2}{b^2}\right)^{\frac{1}{2}} \cdot dy \cdot dx$$

Now, find the integral of

$$\int_0^{b\sqrt{1-\frac{x^2}{a^2}}} \left(1 - \frac{x^2}{a^2} - \frac{y^2}{b^2}\right)^{\frac{1}{2}} \cdot dy$$

or

$$\frac{1}{b} \int_0^{b\sqrt{1-\frac{x^2}{a^2}}} \left(b^2 - \frac{x^2 b^2}{a^2} - y^2\right)^{\frac{1}{2}} \cdot dy$$

Let

$$\left(b^2 - \frac{x^2 b^2}{a^2}\right) = k^2$$

$$\therefore \frac{1}{b} \int_0^{b\sqrt{1-\frac{x^2}{a^2}}} (k^2 - y^2)^{\frac{1}{2}} \cdot dy \qquad (1)$$

Now let $\qquad\qquad y = k \sin w$

Then $\qquad\qquad dy = k \cos w \cdot dw$

and

$(k^2 - y^2)^{\frac{1}{2}} = (k^2 - k^2 \sin^2 w)^{\frac{1}{2}} = k(1 - \sin^2 w)^{\frac{1}{2}} = k \cos w$

Substitute this in Eq. (1).

$$\frac{1}{b} \int_0^{b\sqrt{1-\frac{x^2}{a^2}}} (k^2 - y^2)^{\frac{1}{2}} \, dy =$$

$$\frac{1}{b} \int_0^{b\sqrt{1-\frac{x^2}{a^2}}} k \cos w \cdot k \cos w \cdot dw$$

$$= \frac{k^2}{b} \int_0^{b\sqrt{1-\frac{x^2}{a^2}}} \cos^2 w \cdot dw,$$

but $\qquad\qquad \cos^2 w = \frac{1}{2}(1 + \cos 2w)$

$$\frac{1}{b} \int_0^{b\sqrt{1-\frac{x^2}{a^2}}} (k^2 - y^2)^{\frac{1}{2}} \cdot dy = \frac{k^2}{2b} \int_0^{b\sqrt{1-\frac{x^2}{a^2}}} (\cos 2w + 1) \, dw$$

$$= \frac{k^2}{2b} \left[\frac{1}{2} \sin 2w + w \right]_0^{b\sqrt{1-\frac{x^2}{a^2}}}$$

$$= \frac{b^2 - \frac{x^2 b^2}{a^2}}{2b} \left[\sin w \cos w + w \right]_0^{b\sqrt{1-\frac{x^2}{a^2}}}$$

$$= \frac{b^2 - \frac{x^2 b^2}{a^2}}{2b} \left[\frac{y}{k} \cdot \frac{(k^2 - y^2)^{\frac{1}{2}}}{k} + \sin^{-1} \frac{y}{k} \right]_0^{b\sqrt{1-\frac{x^2}{a^2}}}$$

(See above triangle.)

$$= \frac{b^2 - \frac{x^2 b^2}{a^2}}{2b} \left[\frac{y(k^2 - y^2)^{\frac{1}{2}}}{k^2} + \sin^{-1} \frac{y}{k} \right]_0^{b\sqrt{1-\frac{x^2}{a^2}}}$$

$$= \left(\frac{b^2 - \frac{x^2 b^2}{a^2}}{2b} \right) \left\{ \frac{b\sqrt{1 - \frac{x^2}{a^2}} \left[b^2 - \frac{x^2 b^2}{a^2} - b^2 \left(1 - \frac{x^2}{a^2} \right) \right]^{\frac{1}{2}}}{\left(b^2 - \frac{x^2 b^2}{a^2} \right)} \right.$$

$$\left. + \sin^{-1} \frac{b\sqrt{1 - \frac{x^2}{a^2}}}{b\sqrt{1 - \frac{x^2}{a^2}}} \right\}$$

The   term $\dfrac{b\sqrt{1-\dfrac{x^2}{a^2}}\left[b^2 - \dfrac{x^2b^2}{a^2} - b^2\left(1 - \dfrac{x^2}{a^2}\right)\right]^{\frac{1}{2}}}{\left(b^2 - \dfrac{x^2b^2}{a^2}\right)} = 0.$

The term $\sin^{-1}\dfrac{b\sqrt{1-\dfrac{x^2}{a^2}}}{b\sqrt{1-\dfrac{x^2}{a^2}}} = \dfrac{\pi}{2}$ because an angle whose

sine is $1 = \dfrac{\pi}{2}.$

$$\therefore \frac{1}{b}\int_0^{b\sqrt{1-\frac{x^2}{a^2}}}(k^2 - y^2)^{\frac{1}{2}}\cdot dy = \frac{b^2 - \dfrac{x^2b^2}{a^2}}{2b}\cdot\frac{\pi}{2}$$

or $\dfrac{1}{b}\displaystyle\int_0^{b\sqrt{1-\frac{x^2}{a^2}}}(k^2 - y^2)^{\frac{1}{2}}\cdot dy = \dfrac{b^2a^2 - x^2b^2}{2ba^2}\cdot\dfrac{\pi}{2}$

$$= \frac{\pi b(a^2 - x^2)}{4a^2}$$

Substitute this, and continue.

$$\frac{V}{8} = \frac{\pi cb}{4a^2}\int_0^a (a^2 - x^2)\cdot dx = \frac{\pi cb}{4a^2}\left[a^2x - \frac{x^3}{3}\right]_0^a$$

or $\dfrac{V}{8} = \dfrac{\pi cb}{4a^2}\left(a^3 - \dfrac{a^3}{3}\right) = \dfrac{\pi cb}{4a^2}\cdot\dfrac{2a^3}{3} = \dfrac{\pi abc}{6}$

$$\therefore V = 8\cdot\frac{\pi abc}{6} = \frac{4}{3}\pi abc = \text{the volume of the entire}$$

ellipsoid

PROBLEMS

Find the volumes by triple integration:

1. Find the volume above the xy plane and bounded by it; the plane $z = 5x$, and the cylinder $x^2 + y^2 = 64$; $V = 2\displaystyle\int_0^8\int_0^{\sqrt{64-x^2}}\int_0^{5x} dz\cdot dy\cdot dx.$

2. Find the volume of the solid bounded by the elliptic paraboloids $z = 4 - x^2 - \dfrac{y^2}{4}$ and $z = 3x^2 + \dfrac{y^2}{4}.$

3. Find the volume bounded by the coordinate planes and the plane $\dfrac{x}{2} + \dfrac{y}{3} + \dfrac{z}{4} = 1.$

4. Find the volume of the smaller segment of the sphere

$$x^2 + y^2 + z^2 = 25,$$

cut off by the plane $x = 3$.

5. Find the volume of the solid below the cylindrical surface $z = 4 - x^2$ above the plane $z = 2 - x$ and included between the planes $y = 0$ and $y = 3$; $V = \int_0^3 \int_{-1}^2 \int_{2-x}^{4-x^2} \cdot dz \cdot dx \cdot dy$.

6. Find the volume below the plane $z = x$ and above the elliptic paraboloid $z = x^2 + y^2$.

CHAPTER XXXV

CENTER OF GRAVITY

637. What is the moment of a force?

The product of the force and the perpendicular distance from its line of action to the place of reference about which the moment is desired.

$Ma = Pb$ = moments of the forces M and P about the fulcrum

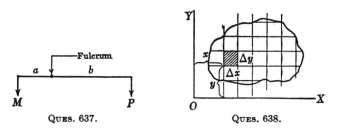

QUES. 637.   QUES. 638.

638. What is the moment of an area?

The summation of all the moments of the elementary rectangles $\Delta x \cdot \Delta y$.

$M_x = \iint y \cdot dy \cdot dx$ = the moment of the area with respect to the x axis, or OX

$M_y = \iint x \cdot dy \cdot dx$ = the moment of the area with respect to the y axis, or OY

The proper limits must be supplied.

639. What is the center of gravity (c.g.) of an area?

The *moment* of an area with respect to an axis divided by the *area*.

293

Then the x coordinate $= \bar{x} = \dfrac{\iint x \cdot dy \cdot dx}{\iint dy \cdot dx}$

and the y coordinate $= \bar{y} = \dfrac{\iint y \cdot dy \cdot dx}{\iint dy \cdot dx}$

The point $(\bar{x}, \bar{y}) =$ the c.g. of the area; it determines the average distance at which the entire area can be concentrated to give the same moment. If the area, length, or volume can be more readily found by a single integration, then the coordinates of the c.g. are found by dividing the moment of the area, length, or volume with respect to the desired axis by the area, length, or volume, as

$$\bar{y} = \frac{\int y \cdot dA}{\int dA} \quad \text{and} \quad \bar{x} = \frac{\int x \cdot dA}{\int dA}$$

640. What is the center of mass?

The area can be considered as a thin material plate in which case $(\bar{x}, \bar{y})$ is the center of mass, or *centroid*, of the plate.

641. Why is it that there is no tendency for a body to rotate if it is suspended from its center of gravity?

The body is in perfect balance because the entire mass can be considered as concentrated at this point, and therefore the moment arm of the force is zero.

642. Why does a line or plane of symmetry pass through the center of gravity?

Because there is no tendency to rotate about this line or plane, and the moment on one side of the line or plane counterbalances the moment on the other side.

643. Where is the point of symmetry?

At the c.g. Therefore, any figure that has a center of symmetry has its c.g. at that point.

644. What are the expressions for the determination of the center of gravity in polar coordinates?

The area of an element of area $= \rho \cdot \Delta\theta \cdot \Delta\rho$.

Now $\qquad\qquad\qquad x = \rho \cos \theta$

and $\qquad\qquad\qquad y = \rho \sin \theta$

Having passed to the limit,

then the x coordinate $\bar{x} = \dfrac{\iint \rho \cos \theta \cdot \rho \cdot d\rho \cdot d\theta}{\iint \rho \cdot d\rho \cdot d\theta}$

or $\qquad\qquad \bar{x} = \dfrac{\iint \rho^2 \cos \theta \cdot d\rho \cdot d\theta}{\iint \rho \cdot d\rho \cdot d\theta}$

The y coordinate $= \bar{y} = \dfrac{\iint \rho \cdot \sin \theta \cdot \rho \cdot d\rho \cdot d\theta}{\iint \rho \cdot d\rho \cdot d\theta}$

$\qquad\qquad\qquad\qquad = \dfrac{\iint \rho^2 \sin \theta \cdot d\rho \cdot d\theta}{\iint \rho \cdot d\rho \cdot d\theta}$

Here also, by single integration,

$$\bar{x} = \frac{\int \rho \cos \theta \cdot dA}{\int dA} \qquad \text{and} \qquad \bar{y} = \frac{\int \rho \sin \theta \cdot dA}{\int dA}$$

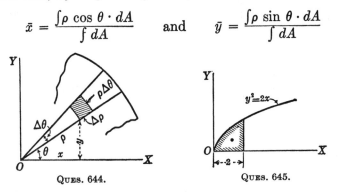

QUES. 644. $\qquad\qquad\qquad$ QUES. 645.

645. What is the center of gravity of the area bounded by $y^2 = 2x$ and $x = 2$, $y = 0$?

Now $M_y = \displaystyle\int_0^2 \int_0^{\sqrt{2x}} x \cdot dy \cdot dx =$ the moment about the y axis

or $\qquad M_y = \displaystyle\int_0^2 \Big[x \cdot y \Big]_0^{\sqrt{2x}} \cdot dx = \int_0^2 x \sqrt{2x} \cdot dx$

$\qquad\qquad = \sqrt{2} \displaystyle\int_0^2 x^{\frac{3}{2}} \cdot dx$

Finally, $M_y = \left[\sqrt{2} \cdot x^{\frac{5}{2}} \cdot \dfrac{2}{5} \right]_0^2 = \dfrac{2\sqrt{2}}{5} \left[x^{\frac{5}{2}} \right]_0^2$

$$= \dfrac{2\sqrt{2}}{5} \cdot \sqrt{2^5} = \dfrac{16}{5}$$

Also, $\quad M_x = \displaystyle\int_0^2 \int_0^{\sqrt{2x}} y \cdot dy \cdot dx = $ the moment about the

$$x \text{ axis}$$

or $\quad M_x = \displaystyle\int_0^2 \left[\dfrac{y^2}{2} \right]_0^{\sqrt{2x}} \cdot dx = \int_0^2 \dfrac{2x}{2} \cdot dx = \int_0^2 x \cdot dx$

$$= \left[\dfrac{x^2}{2} \right]_0^2$$

Finally, $\qquad\qquad M_x = \tfrac{4}{2} = 2$

Now $\ A = \text{area} = \displaystyle\int_0^2 \int_0^{\sqrt{2x}} dy \cdot dx = \int_0^2 \left[y \right]_0^{\sqrt{2x}} \cdot dx$

or $\qquad A = \displaystyle\int_0^2 \sqrt{2x} \cdot dx = \sqrt{2} \int_0^2 x^{\frac{1}{2}} \cdot dx$

Finally, $A = \sqrt{2} \left[x^{\frac{3}{2}} \cdot \dfrac{2}{3} \right]_0^2 = \dfrac{2\sqrt{2}}{3} \left[x^{\frac{3}{2}} \right]_0^2 = \dfrac{2\sqrt{2}}{3} \cdot \sqrt{2^3}$

$$= \dfrac{2}{3} \cdot 4 = \dfrac{8}{3}$$

$\therefore\ \bar{x} = \dfrac{M_y}{A} = \dfrac{16}{5} \cdot \dfrac{3}{8} = \dfrac{6}{5} = $ the x coordinate

$\quad \bar{y} = \dfrac{M_x}{A} = 2 \cdot \dfrac{3}{8} = \dfrac{3}{4} = $ the y coordinate

Therefore, the c.g. is at $\bar{x} = \tfrac{6}{5}$, $\bar{y} = \tfrac{3}{4}$.

646. Where is the center of gravity of the first quadrant of a circle?

The x coordinate $= \bar{x} = \dfrac{\displaystyle\int_0^R \int_0^{\sqrt{R^2 - x^2}} x \cdot dy \cdot dx}{\displaystyle\int_0^R \int_0^{\sqrt{R^2 - x^2}} dy \cdot dx} = \dfrac{M_y}{A}$

The y coordinate $= \bar{y} = \dfrac{\displaystyle\int_0^R \int_0^{\sqrt{R^2 - x^2}} y \cdot dy \cdot dx}{\displaystyle\int_0^R \int_0^{\sqrt{R^2 - x^2}} dy \cdot dx} = \dfrac{M_x}{A}$

The denominator in each must equal $\left(\dfrac{\pi R^2}{4}\right)$, which is one-fourth the area of the circle $= A$.

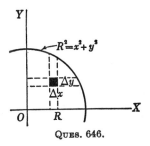

QUES. 646.

Also, because of symmetry, the value of $\bar{x}$ is the same as that of $\bar{y}$. The numerators are therefore equal, and it is necessary to evaluate only one of them.

Now $M_y = \displaystyle\int_0^R \int_0^{\sqrt{R^2 - x^2}} x \cdot dy \cdot dx = \int_0^R \sqrt{R^2 - x^2} \cdot x \cdot dx$

or $\quad M_y = -\frac{1}{2} \displaystyle\int_0^R (R^2 - x^2)^{\frac{1}{2}}(-2x) \cdot dx$

$$= -\frac{1}{2} \cdot \frac{2}{3} \left[(R^2 - x^2)^{\frac{3}{2}} \right]_0^R$$

Finally, $M_y = -\dfrac{1}{3} [(R^2 - R^2)^{\frac{3}{2}} - (R^2 - 0)^{\frac{3}{2}}]$

$$= -\frac{1}{3} \cdot -R^3 = \frac{R^3}{3}$$

and $\quad \bar{x} = \dfrac{M_y}{A} = \dfrac{\dfrac{R^3}{3}}{\frac{1}{4}\pi R^2} = \dfrac{R^3}{3} \cdot \dfrac{4}{\pi R^2} = \dfrac{4R}{3\pi}$

Therefore, the c.g. of the quarter circle is at $\left(\dfrac{4R}{3\pi}, \dfrac{4R}{3\pi}\right)$.

647. Where is the center of gravity of the arc of a semicircle?

OX is a line of symmetry, and therefore the c.g. is on this line at some point $\bar{x}$.

Now, $\bar{x} = \dfrac{\text{sum of all the elementary moments } x \cdot \Delta s}{\text{length of the semicircle}}$

$$= \frac{\displaystyle\int_{-\frac{\pi}{2}}^{\frac{\pi}{2}} x \cdot ds}{\pi R} \quad \text{for the limit of the summation}$$

As $\qquad x = R \cos \theta \qquad$ and $\qquad ds = R \cdot d\theta$

then $\bar{x} = \dfrac{\displaystyle\int_{-\frac{\pi}{2}}^{\frac{\pi}{2}} R \cdot \cos \theta \cdot R \cdot d\theta}{\pi R} = \dfrac{R^2 \displaystyle\int_{-\frac{\pi}{2}}^{\frac{\pi}{2}} \cos \theta \cdot d\theta}{\pi R}$

or $\qquad \bar{x} = \dfrac{R^2 \Big[\sin \theta\Big]_{-\frac{\pi}{2}}^{\frac{\pi}{2}}}{\pi R} = \dfrac{R^2(1 + 1)}{\pi R} = \dfrac{2R^2}{\pi R} = \dfrac{2R}{\pi}$

Therefore, the c.g. of the arc of the semicircle is at $\left(\dfrac{2R}{\pi}, 0\right)$

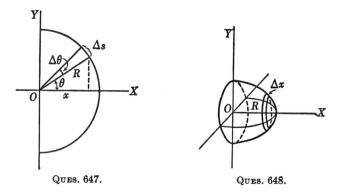

QUES. 647.   QUES. 648.

648. Where is the center of gravity of a hemisphere?

Place the hemisphere so that the y and z axes are axes of symmetry.

$$\therefore \quad \bar{y} = 0 \qquad \text{and} \qquad \bar{z} = 0 \quad \text{from symmetry}$$

Take the elementary slice, which has an area πy^2 and is Δx thick.

$$\therefore \Delta V = \pi y^2 \cdot \Delta x = \text{the element of the volume}$$

Now $x^2 + y^2 = R^2$ or $y^2 = R^2 - x^2$ and $V = \frac{2}{3}\pi R^3$, the volume of the hemisphere.

Passing to the limit,

$$\bar{x} = \frac{\int x \cdot dV}{V} = \frac{\int x \cdot \pi y^2 \cdot dx}{\frac{2}{3}\pi R^3} = \frac{\int_0^R \pi x (R^2 - x^2) \cdot dx}{\frac{2}{3}\pi R^3}$$

$$\text{or } \bar{x} = \frac{\int_0^R R^2 x \cdot dx}{\frac{2}{3}R^3} - \frac{\int_0^R x^3 \cdot dx}{\frac{2}{3}R^3} = \frac{\left[R^2 \cdot \frac{x^2}{2} \right]_0^R}{\frac{2}{3}R^3} - \frac{\left[\frac{x^4}{4} \right]_0^R}{\frac{2}{3}R^3}$$

$$\text{Finally, } \bar{x} = \frac{R^4}{2} \cdot \frac{3}{2R^3} - \frac{R^4}{4} \cdot \frac{3}{2R^3} = \frac{3}{4}R - \frac{3}{8}R = \frac{3}{8}R$$

Therefore, the c.g. of a hemisphere is at $\frac{3}{8}R$ from the origin O.

649. What are the coordinates of the center of gravity of a circular arc of radius ρ subtending an angle 2β at the center, using polar coordinates?

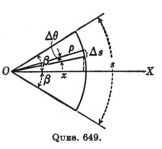

QUES. 649.

$\bar{y} = 0$ because there is a line of symmetry and the c.g. will be on OX.

Here we can find the arc and moment by a single integration.

$$\text{Moment arm } x = \rho \cos \theta$$

$$\text{while} \quad \Delta s = \rho \cdot \Delta \theta \quad \text{and} \quad s = 2\rho\beta$$

Passing to the limit,

$$\bar{x} = \frac{\int x \, ds}{s} = 2\frac{\int_0^\beta \rho \cos \theta \cdot \rho \cdot d\theta}{2\rho\beta} = 2\frac{\int_0^\beta \rho^2 \cos \theta \cdot d\theta}{2\rho\beta}$$

$$= \frac{2\rho^2 \sin \beta}{2\rho\beta} = \frac{\rho \sin \beta}{\beta}$$

650. Where is the center of gravity of a solid cone of revolution?

The equation of the element OA is

$$\frac{y}{x} = \frac{r}{h} \quad \text{or} \quad y = \frac{xr}{h}$$

Since the x axis is a line of symmetry $\bar{y} = 0$, then find $\bar{x}$ only.

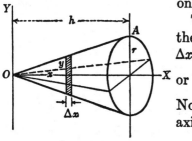

QUES. 650.

The element of volume = the area of the circle πy^2 times Δx

or $\quad \Delta V = \pi y^2 \cdot \Delta x$

Now the moment about the y axis is

$$\Delta M_y = x \cdot \pi y^2 \cdot \Delta x$$

Passing to the limit,

$$M_y = \pi \int_0^h x \cdot y^2 \cdot dx$$

But $y^2 = \dfrac{x^2 r^2}{h^2}$ from above. Then

$$M_y = \frac{\pi r^2}{h^2} \int_0^h x^3 \cdot dx = \frac{\pi r^2}{h^2} \left[\frac{x^4}{4} \right]_0^h = \frac{\pi r^2 h^4}{4h^2} = \frac{\pi r^2 h^2}{4}$$

Now the volume of the entire cone is $\frac{1}{3}\pi r^2 h$.

$$\therefore \bar{x} = \frac{M_y}{V} = \frac{\pi r^2 h^2}{4} \cdot \frac{3}{\pi r^2 h} = \frac{3}{4} h$$

651. What are the coordinates of the center of gravity of the area under one arch of the sine curve $y = \sin x$?

$$\Delta A = y \, \Delta x = \sin x \cdot \Delta x$$

The moment of ΔA about its x axis is

$$\Delta M_x = \frac{y}{2} \cdot y \, \Delta x \quad \text{or} \quad \Delta M_x = \frac{y^2}{2} \cdot \Delta x = \frac{1}{2} \sin^2 x \cdot \Delta x$$

The moment of ΔA about the y axis is

$$\Delta M_y = x \cdot y \, \Delta x = x \sin x \cdot \Delta x$$

The limits are $x = 0$ and $x = \pi$. Passing to the limit,

$$M_y = \int_0^\pi x \cdot \sin x \cdot dx$$

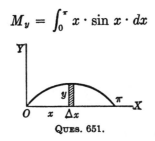

QUES. 651.

Let $u = x$ and $dv = \sin x \cdot dx$; then $du = dx$ and $v = \int \sin x \cdot dx = -\cos x$.

Then $\int x \cdot \sin x \cdot dx = uv - \int v \cdot du = -x \cos x$
$$- \int -\cos x \cdot dx$$

or $M_y = \left[-x \cos x + \sin x \right]_0^\pi = \left[-\pi \cdot (-1) + 0 + 0 \cdot 1 \right.$
$$\left. - 0 \right] = \pi$$

$$M_x = \tfrac{1}{2} \int_0^\pi \sin^2 x \cdot dx$$

But $\sin^2 x = \tfrac{1}{2} - \tfrac{1}{2} \cos 2x$.

Then $M_x = \dfrac{1}{4} \int dx - \dfrac{1}{4} \int \cos 2x \cdot dx = \left[\dfrac{x}{4} - \dfrac{1}{8} \sin 2x \right]_0^\pi$
$$= \left(\dfrac{\pi}{4} - \dfrac{1}{8} \sin 2\pi - 0 + \dfrac{1}{8} \sin 0 \right) = \dfrac{\pi}{4}$$

Now $A = \text{area} = \int_0^\pi \sin x \cdot dx = \left[-\cos x \right]_0^\pi = -\cos \pi$
$$+ \cos 0 = 1 + 1 = 2.$$

$$\therefore \ \bar{x} = \frac{M_y}{A} = \frac{\pi}{2} \ \left.\begin{array}{c} \\ \\ \\ \\ \end{array}\right\}$$

$$\bar{y} = \frac{M_x}{A} = \frac{\dfrac{\pi}{4}}{2} = \frac{\pi}{8}$$

652. What is the center of gravity of the area between the parabola $y^2 = 2px$ and the lines $y = b$ and $x = a$?

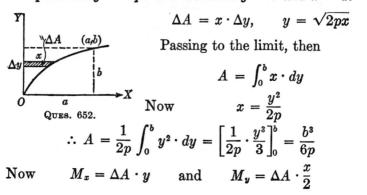

QUES. 652.

$$\Delta A = x \cdot \Delta y, \qquad y = \sqrt{2px}$$

Passing to the limit, then

$$A = \int_0^b x \cdot dy$$

Now

$$x = \frac{y^2}{2p}$$

$$\therefore A = \frac{1}{2p} \int_0^b y^2 \cdot dy = \left[\frac{1}{2p} \cdot \frac{y^3}{3}\right]_0^b = \frac{b^3}{6p}$$

Now $\qquad M_x = \Delta A \cdot y \qquad$ and $\qquad M_y = \Delta A \cdot \dfrac{x}{2}$

Passing to the limit,

$$M_x = \int_0^b y \cdot x \cdot dy = \int_0^b \frac{y^2}{2p} \cdot y \cdot dy = \frac{1}{2p} \int_0^b y^3 \cdot dy$$

$$= \frac{1}{2p}\left[\frac{y^4}{4}\right]_0^b$$

$$= \frac{b^4}{8p}$$

$$M_y = \int_0^b \frac{x}{2} \cdot x \cdot dy = \frac{1}{2}\int_0^b \frac{y^4}{4p^2} \cdot dy = \frac{1}{8p^2}\int_0^b y^4 \cdot dy$$

$$= \frac{1}{8p^2}\left[\frac{y^5}{5}\right]_0^b$$

$$= \frac{1}{8p^2} \cdot \frac{b^5}{5} = \frac{b^5}{40p^2}$$

Then $\qquad \bar{x} = \dfrac{M_y}{A} = \dfrac{b^5}{40p^2} \cdot \dfrac{6p}{b^3} = \dfrac{3b^2}{20p}$

$$\bar{y} = \frac{M_x}{A} = \frac{b^4}{8p} \cdot \frac{6p}{b^3} = \frac{3b}{4}$$

But $x = a$, $y = b$, and $y^2 = 2px$; therefore, $b^2 = 2pa$.

$$\therefore \bar{x} = \frac{3b^2}{20p} = \frac{3 \cdot 2pa}{20p} = \frac{3a}{10}$$

$$\bar{y} = \frac{3b}{4}$$

PROBLEMS

1. Where is the c.g. of the area of a semicircle?

2. Where is the c.g. of the arc of one-fourth of a circle of radius r?

3. Where is the c.g. of the segment of the parabola $y^2 = 4x$, cut off by the line $x = 3$?

4. Where is the c.g. of the area bounded by the cycloid

$$x = a(\theta - \sin \theta), \ y = a(1 - \cos \theta)$$

and the x axis?

5. Where is the c.g. of the area of a quadrant of the ellipse

$$\frac{x^2}{a^2} + \frac{y^2}{b^2} = 1?$$

6. What is the distance from the vertex of a right circular cone, of altitude h and radius of base r, to its c.g.?

7. Where is the c.g. of the area included by the parabolas $y^2 = ax$ and $x^2 = by$?

8. Where is the c.g. of the area bounded by the cardioid

$$\rho = a(1 + \cos \theta)?$$

9. Where is the c.g. of the area bounded by one loop of the curve $\rho = a \cos 3\theta$?

10. Where is the c.g. of the area bounded by the loop of the curve $y^2 = 4x^2 - x^3$?

11. Where is the c.g. of the area bounded by $y = 2x + 3$ and $y = x^2$?

12. Where is the c.g. of the solid formed by revolving about the x axis the part of the area of the ellipse $\frac{x^2}{a^2} + \frac{y^2}{b^2} = 1$ that is in the first quadrant?

13. Where is the c.g. of the solid formed by revolving about the x axis the area bounded by lines $y = 0$, $x = \frac{\pi}{4}$ and the curve $y = \sin x$?

14. Where is the c.g. of the solid formed by revolving about the x axis the area in the first quadrant bounded by lines $y = 0$, $x = 2a$ and the hyperbola $\frac{x^2}{a^2} - \frac{y^2}{b^2} = 1$?

15. Where is the c.g. of the solid formed by revolving about the x axis the curve $x^2 = ay$ and $x = a$?

CHAPTER XXXVI

MOMENT OF INERTIA

653. What is moment of inertia?

It is a moment of the second order where the second power of the distance is used.

654. How is mass taken care of in moments of inertia of material bodies?

By introducing the factor μ = mass per unit of length, area, or volume when the mass is uniformly distributed throughout the body.

655. What are the formulas for moment of inertia of a body considered as a plane area with mass uniformly distributed?

$I_x = \int\int \mu y^2 \cdot dy \cdot dx$ = the moment of inertia with respect to the x axis

$I_y = \int\int \mu x^2 \cdot dy \cdot dx$ = the moment of inertia with respect to the y axis

$I_0 = \int\int \mu (x^2 + y^2)\, dy \cdot dx$ = the moment of inertia with respect to the origin

Here also the area or length may sometimes be found by a single integration.

656. What are rectangular moments of inertia?

The moments of inertia with respect to each of two axes at right angles to each other and lying in the plane of the area considered.

657. What is polar moment of inertia?

The moment of inertia of an area with respect to an axis perpendicular to its plane.

658. How do you know that the polar moment of inertia is equal to the sum of the rectangular moments of inertia?

Let I_o = the polar moment of inertia with respect to an axis perpendicular to the plane of the area and through point O.

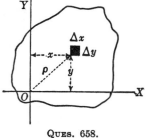

Then $I_o = \int\int \mu\rho^2 \cdot dy \cdot dx$

Now $\rho^2 = x^2 + y^2$

Then $I_o = \int\int \mu(x^2 + y^2)\, dy \cdot dx$

$\quad = \int\int \mu x^2\, dy \cdot dx + \int\int \mu y^2\, dy \cdot dx$

$\therefore I_o = I_x + I_y$ = the sum of the rectangular moments of inertia

QUES. 658.

659. What is the moment of inertia of a circular plate of radius R with respect to a diameter as an axis?

The moment of inertia of the entire plate is four times the moment of inertia of one quadrant

or $\quad I_x = 4 \int_0^R \int_0^{\sqrt{R^2-x^2}} \mu y^2\, dy \cdot dx$

Then $I_x = 4\mu \int_0^R \left[\tfrac{1}{3} \cdot y^3 \right]_0^{\sqrt{R^2-x^2}} \cdot dx = \tfrac{4}{3}\mu \int_0^R (R^2 - x^2)^{\frac{3}{2}} \cdot dx$

$\quad = \dfrac{4}{3}\mu \left[\dfrac{1}{8} x(5R^2 - 2x^2)\sqrt{R^2 - x^2} + \dfrac{3}{8} R^4 \sin^{-1}\dfrac{x}{R} \right]_0^R$

$\quad = \tfrac{4}{3}\mu(\tfrac{3}{8}R^4 \cdot \tfrac{1}{2}\pi) = \tfrac{1}{4}\mu\pi R^4$

But $\quad\quad\quad\quad \mu\pi R^2 = M = $ mass

$\quad\quad\quad\quad\quad \therefore I_x = \tfrac{1}{4}MR^2$

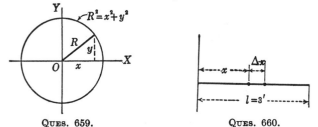

QUES. 659.   QUES. 660.

660. What is the moment of inertia of a thin wire 3 feet long weighing 0.01 pound per foot with respect to an axis through one end?

Now, $\qquad I = \int_0^l \mu x^2 \cdot dx = \mu \int_0^l x^2 \cdot dx$

or $I = \mu \dfrac{x^3}{3}\Big]_0^l = \mu \dfrac{l^3}{3} = $ moment of inertia about one end

But $\mu l = $ total mass or weight $= M = 0.01 \cdot 3 = 0.03$ lb.

$$\therefore I = \frac{Ml^2}{3} = \frac{0.03 \cdot (3)^2}{3} = 0.09 \text{ lb-ft.}^3$$

661. What is the moment of inertia of a 14- by 6-inch rectangular plate weighing 0.5 pound per square inch, with respect to its base as an axis?

Now, $I = \int_0^a \int_0^b \mu y^2 \cdot dy \cdot dx = $ the moment of inertia

about its base as an axis

Then $\qquad I = \mu \int_0^a \left[\dfrac{y^3}{3} \right]_0^b \cdot dx = \mu \dfrac{b^3}{3} \int_0^a dx$

$$= \mu \frac{b^3}{3} \Big[x \Big]_0^a = \mu \frac{ab^3}{3}$$

But $\qquad\qquad M = $ mass $= \mu ab$

$\therefore I = \dfrac{Mb^2}{3} = $ the expression for the moment of inertia

about its base

Finally, $I = 14 \cdot 6 \cdot 0.5 \cdot \dfrac{(6)^2}{3} = 42 \cdot 12 = 504$ lb.-in.^{3}

$= $ moment of inertia

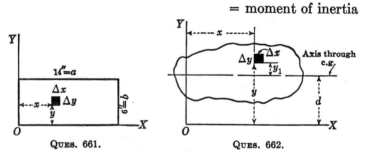

QUES. 661.   QUES. 662.

662. Can you show that the moment of inertia of a body with respect to any axis equals moment of inertia with respect to a parallel axis through its center of gravity plus its mass times the square of the distance between the axes?

Let I = moment of inertia with respect to any axis
and I_g = moment of inertia with respect to a parallel axis through the c.g.
and d = the distance between the axes
Then $I = \iint \mu y^2 \cdot dy \cdot dx$ = moment of inertia about the OX axis

Now $y = y_1 + d$
$\therefore I = \iint \mu (y_1 + d)^2 \cdot dy \cdot dx$
$= \iint \mu y_1{}^2 \cdot dy \cdot dx + 2\mu d \iint y_1 \cdot dy \cdot dx + d^2 \iint \mu \cdot dy \cdot dx$
$\therefore I = I_g + 0 + Md^2$
$(2\mu d \iint y_1 \cdot dy \cdot dx$ = moment about c.g. = 0)

663. **What is the moment of inertia of a right circular cone with respect to an axis OY through its vertex and perpendicular to its own axis?**

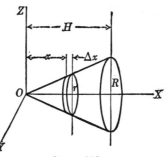

QUES. 663.

Cut thin circular plates of width $dx = \Delta x$.

Now, $I_{c.g.} = \frac{1}{4}\mu \pi r^4 \cdot dx$ = moment of inertia of a circular plate about its diameter (from Question 659)
Then $I_{OY} = \frac{1}{4}\mu \pi r^4 \cdot dx + x^2 \cdot \mu \pi r^2 \cdot dx$ = moment of inertia about a parallel axis through O, at x distance away

Summing up,

$$I_{OY} = \int_0^H \mu (\tfrac{1}{4}\pi r^4 + \pi x^2 r^2) \, dx$$

Now, $\dfrac{r}{R} = \dfrac{x}{H}$ or $r = \dfrac{Rx}{H}$ from similar triangles

$$\therefore I_{or} = \pi\mu \int_0^H \left(\frac{R^4 x^4}{4H^4} + \frac{R^2 x^4}{H^2} \right) dx$$

$$= \frac{\mu\pi R^2}{H^2} \left(\frac{R^2}{4H^2} + 1 \right) \int_0^H x^4 \cdot dx$$

$$\text{or } I_{or} = \frac{\mu\pi R^2}{H^2} \left(\frac{R^2}{4H^2} + 1 \right) \cdot \left[\frac{x^5}{5} \right]_0^H$$

$$= \frac{\mu\pi R^2 H^5}{H^2 \cdot 4H^2 \cdot 5} (R^2 + 4H^2) = \frac{\mu\pi R^2 H}{20} (R^2 + 4H^2)$$

But the mass $= M = \frac{1}{3}\mu\pi R^2 H =$ the volume of a right circular cone.

$\therefore I_{or} = \frac{3}{20}M(R^2 + 4H^2) =$ moment of inertia about an OY axis through O and perpendicular to the axis of the cone

664. What is the polar moment of inertia of a circular plate with respect to its center?

From Question 659, the moment of inertia of a circular plate with respect to a diameter as an axis is $\frac{1}{4}MR^2$

or $\qquad I_x = \frac{1}{4}MR^2 \qquad$ and $\qquad I_y = \frac{1}{4}MR^2$
$\therefore I_o = I_x + I_y = \frac{1}{2}MR^2 =$ the polar moment of inertia

665. What are the formulas for the moments and moments of inertia, using polar coordinates?

$$M_x = \iint \rho^2 \sin\theta \cdot d\rho \cdot d\theta$$
$$M_y = \iint \rho^2 \cos\theta \cdot d\rho \cdot d\theta$$
$$I_x = \iint \rho^3 \sin^2\theta \cdot d\rho \cdot d\theta$$
$$I_y = \iint \rho^3 \cos^2\theta \cdot d\rho \cdot d\theta$$
$$I_o = \iint \rho^3 d\rho \cdot d\theta$$

666. What is I_o over the area bounded by the circle $\rho = 2r \cos\theta$?

Summing up the elementary rectangles in the triangular-shaped strip OS, the ρ limits are 0 and $2r \cos\theta$ from the equation of the circle.

Now sum up all the triangular strips between $\theta = -\frac{\pi}{2}$ and $\frac{\pi}{2}$.

Then

$$I_o = \int_{-\frac{\pi}{2}}^{\frac{\pi}{2}} \int_0^{2r \cos \theta} \rho^3 \cdot d\rho \cdot d\theta = \int_{-\frac{\pi}{2}}^{\frac{\pi}{2}} \left[\frac{\rho^4}{4} \right]_0^{2r \cos \theta} \cdot d\theta$$

$$= \int_{-\frac{\pi}{2}}^{\frac{\pi}{2}} \left[\frac{16r^4 \cos^4 \theta}{4} \right] d\theta = 4r^4 \int_{-\frac{\pi}{2}}^{\frac{\pi}{2}} \cos^4 \theta \cdot d\theta$$

$$= 4r^4 \int_{-\frac{\pi}{2}}^{\frac{\pi}{2}} \cos^2 \theta \cdot \cos^2 \theta \cdot d\theta \qquad (1)$$

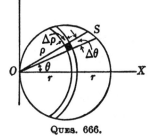

QUES. 666.

To integrate, substitute $\cos^2 \theta = \frac{1}{2} + \frac{1}{2} \cos 2\theta$.

$$\therefore \int \cos^4 \theta \cdot d\theta = \int (\tfrac{1}{2} + \tfrac{1}{2} \cos 2\theta)^2 \cdot d\theta$$
$$= \int (\tfrac{1}{4} + \tfrac{1}{2} \cdot \cos 2\theta + \tfrac{1}{4} \cos^2 2\theta) \, d\theta$$
$$\int \cos^4 \theta \cdot d\theta = \frac{\theta}{4} + \frac{1}{4} \sin 2\theta + \frac{1}{4} \int \cos^2 2\theta \cdot d\theta \qquad (2)$$

Now, to integrate, $\cos^2 2\theta = \frac{1}{2} + \frac{1}{2} \cos 4\theta$.

$$\therefore \tfrac{1}{4} \int \cos^2 2\theta d\theta = \tfrac{1}{4} \int (\tfrac{1}{2} + \tfrac{1}{2} \cos 4\theta) \, d\theta = \int (\tfrac{1}{8} + \tfrac{1}{8} \cos 4\theta) \, d\theta$$
$$\frac{1}{4} \int \cos^2 2\theta d\theta = \frac{\theta}{8} + \frac{1}{32} \sin 4\theta \qquad (3)$$

Now combine Eqs. (2) and (3) and substitute in Eq. (1).

$$\therefore \ 4r^4 \int_{-\frac{\pi}{2}}^{\frac{\pi}{2}} \cos^4 \theta \cdot d\theta = 4r^4 \left[\frac{\theta}{4} + \frac{1}{4} \sin 2\theta + \frac{\theta}{8} \right.$$

$$\left. + \frac{1}{32} \sin 4\theta \right]_{-\frac{\pi}{2}}^{\frac{\pi}{2}}$$

$$= 4r^4 \left[\tfrac{3}{8}\theta + \tfrac{1}{4} \sin 2\theta + \tfrac{1}{32} \sin 4\theta \right]_{-\frac{\pi}{2}}^{\frac{\pi}{2}}$$

$$= 4r^4 \left[\frac{3}{8} \cdot \frac{\pi}{2} + \frac{1}{4} \sin \pi + \frac{1}{32} \sin 2\pi \right.$$

$$- \left(-\frac{3}{8} \cdot \frac{\pi}{2} + \frac{1}{4} \sin - \pi \right.$$

$$\left. \left. + \frac{1}{32} \cdot \sin - 2\pi \right) \right]$$

$$= 4r^4 (\tfrac{3}{16}\pi + 0 + 0 + \tfrac{3}{16}\pi - 0 - 0)$$

$$\therefore \ I_o = 4r^4 \cdot \tfrac{3}{8}\pi = \tfrac{3}{2}\pi r^4$$

667. What is meant by radius of gyration?

It is the distance from the axis at which the area or volume would be considered as concentrated, the moment of inertia remaining the same.

Distance k = radius of gyration = $\sqrt{\dfrac{I}{A}}$

$$= \sqrt{\frac{\text{moment of inertia}}{\text{area}}}$$

668. What are I_x, I_y, and the corresponding radii of gyration for the area bounded by the semicubical parabola $y^2 = x^3$ and the straight line $y = x$?

The y limits are found by solving simultaneously the two curves $y = x$ and $y = \sqrt{x^3}$ to get the point of intersection.

$$x = \sqrt{x^3} \quad \text{or} \quad x^2 = x^3, \quad x^3 - x^2 = 0$$

$$\therefore \ x = 0 \quad \text{and} \quad x = 1$$

Substitute in $y = \sqrt{x^3}$ to get $y = 1$.

Then $I_x = \int_0^1 \int_y^{y^{\frac{2}{3}}} y^2 \cdot dx \cdot dy = \int_0^1 y^2 \Big[x \Big]_y^{y^{\frac{2}{3}}} \cdot dy$

$\qquad = \int_0^1 y^2(y^{\frac{2}{3}} - y) \, dy = \int_0^1 (y^{\frac{8}{3}} - y^3) \, dy$

$\qquad = \left[\dfrac{3y^{\frac{11}{3}}}{11} - \dfrac{y^4}{4} \right]_0^1 = \left(\dfrac{3}{11} - \dfrac{1}{4} \right) = \dfrac{1}{44}$

and $\quad I_y = \int_0^1 \int_y^{y^{\frac{2}{3}}} x^2 \cdot dx \cdot dy = \int_0^1 \left[\dfrac{x^3}{3} \right]_y^{y^{\frac{2}{3}}} \cdot dy$

$\qquad\qquad\qquad\qquad\qquad\qquad = \int_0^1 \left(\dfrac{y^2}{3} - \dfrac{y^3}{3} \right) dy$

$\qquad = \left[\dfrac{y^3}{9} - \dfrac{y^4}{12} \right]_0^1 = \left(\dfrac{1}{9} - \dfrac{1}{12} \right) = \dfrac{1}{36}$

Now $A = \int_0^1 \int_y^{y^{\frac{2}{3}}} dx \cdot dy = \int_0^1 \Big[x \Big]_y^{y^{\frac{2}{3}}} \cdot dy = \int_0^1 (y^{\frac{2}{3}} - y) \, dy$

$\qquad = \left[\dfrac{3}{5} y^{\frac{5}{3}} - \dfrac{y^2}{2} \right]_0^1 = \dfrac{3}{5} - \dfrac{1}{2} = \dfrac{1}{10}$

$\therefore k_x = \sqrt{\dfrac{I_x}{A}} = \sqrt{\dfrac{\frac{1}{44}}{\frac{1}{10}}} = \sqrt{\dfrac{5}{22}} = 0.477$

$k_y = \sqrt{\dfrac{I_y}{A}} = \sqrt{\dfrac{\frac{1}{36}}{\frac{1}{10}}} = \sqrt{\dfrac{5}{18}} = 0.527$

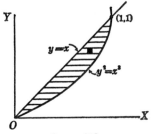

QUES. 668.

PROBLEMS

1. What is the moment of inertia of a triangle of base b and altitude h with respect to its base?

$$\frac{b'}{b} = \frac{h - y}{h} \qquad \text{or} \qquad b' = b - \frac{b}{h} y$$

Then $\qquad I_z = \int_0^h y^2 \left(b - \frac{by}{h}\right) dy = \frac{1}{12} bh^3$

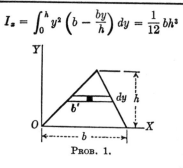

PROB. 1.

2. Find I_z, I_y, and I_0 for the area of the ellipse $\dfrac{x^2}{a^2} + \dfrac{y^2}{b^2} = 1$.

3. Find I_z, I_y, and I_0 for the area included between the circle $x^2 + y^2 = 36$ and the circle $x^2 + (y+3)^2 = 4$.

4. Find I_z, I_y, and I_0 for the semicircle to the right of the y axis and bounded by $x^2 + y^2 = 4$.

5. Find I_z, I_y, and I_0 for the entire area bounded by $x^{\frac{2}{3}} + y^{\frac{2}{3}} = a^{\frac{2}{3}}$.

6. Find I_z and I_y for the area included between the circle

$$x^2 + y^2 = 3y$$

and the ellipse $\dfrac{x^2}{16} + \dfrac{y^2}{25} = 1$.

7. Find I_z and I_y in the first quadrant bounded by $y^2 = 4x$ and $x = 4$, $y = 0$.

8. Find I_z, I_y, and the radii of gyration k_y and k_z for the parabolic segment above and below the x axis of $y^2 = 2px$ and $x = a$, $y = b$.

9. Find I_y for the lemniscate $\rho^2 = a^2 \cos 2\theta$.

10. Find I_z and I_y for one loop of $\rho = a \cos 2\theta$.

CHAPTER XXXVII

INTRODUCTION TO DIFFERENTIAL EQUATIONS

669. What is a differential equation?

It is an equation involving derivatives or differentials.

Examples

(a) $$\frac{dy}{dx} = 8x$$

(b) $$y'' = \frac{3a}{y^2}$$

(c) $$2x \cdot dx + 3y^3 \cdot dy - ax \cdot dy + ay \cdot dx = 0$$
(d) $$y^{(n)} = a^n \epsilon^{bx}$$

These are differential equations.

670. What are partial differential equations?

Equations containing partial derivatives are called partial differential equations.

Examples

(a) $$\frac{\partial y}{\partial x} = 2y + x^2$$

(b) $$\frac{\partial^2 y}{\partial x^2} + \frac{\partial^2 y}{\partial z^2} = 0$$

671. What is meant by the order of a differential equation?

It is the order of the highest derivative that occurs in the equation.

Examples

(a) $\dfrac{dy}{dx} = 8x$, (b) $\dfrac{dy}{dx} = \dfrac{2x^2}{3y}$ are equations of the first order.

(c) $\dfrac{d^2y}{dx^2} + 8y = 0$, (d) $2\dfrac{dy}{dx} + \dfrac{d^2y}{dx^2} = 0$ are equations of the second order.

(e) $y^{(n)} = a^n\epsilon^{bx}$ is an equation of the nth order.

672. What is meant by the degree of a differential equation?

The largest exponent with which the derivative of the highest order is affected gives the degree of the equation.

Example

$$\left(\frac{d^2y}{dx^2}\right)^2 = \left[3 + \left(\frac{dy}{dx}\right)^2\right]^4$$

The highest order of the derivative is 2, and its exponent is 2; therefore, the equation is of the second order and of the second degree.

673. What is meant by a solution of a differential equation?

A solution is an integral of the differential equation. Generally, it is any relation between the variables by which the equation is satisfied, although sometimes the actual integration cannot be effected.

Example

Integrating the differential equation $\dfrac{dy}{dx} = 8x$, you get

$y = \dfrac{8x^2}{2} = 4x^2 + C =$ the solution. C is an arbitrary constant of integration.

674. What is meant by (a) a complete or general solution; (b) a particular solution of a differential equation?

a. A general solution is one that contains arbitrary essential constants of integration the number of which is equal to the order of the equation.

b. A particular solution is obtained by giving particular values to the constants. Given conditions of a problem are satisfied by the particular solution.

Example

If $x \cdot dy + y \cdot dx = 0$, divide by xy to get

$$\frac{dy}{y} + \frac{dx}{x} = 0$$

Integrate to get

$$\log_\epsilon y + \log_\epsilon x = \log_\epsilon C$$

or $$\log_\epsilon xy = \log_\epsilon C$$

Therefore, $xy = C$ is a general solution.

Now $xy = 0$ and $xy = 3$ are particular solutions with particular values for the constant of integration C.

675. How many solutions can a differential equation have?

It can have an infinity of solutions since the solution may involve one or more arbitrary constants of integration and you can assign different values to the arbitrary constants, obtaining an infinity of solutions.

Example

If $\frac{d^2y}{dx^2} = 3\epsilon^{-x}$, then $y = 3\epsilon^{-x} + C_1 x + C_2$ is a solution that satisfies the equation and the constants may have any values given to them.

676. In the solution of a differential equation does the constant of integration always appear as an additive constant?

Not always. The constant often enters in other ways.

Example

If
$$y = C\epsilon^{3x} \tag{1}$$

is a general solution of a differential equation, then, differentiating, you get

$$dy = 3C\epsilon^{3x} \cdot dx$$

Now eliminate C by dividing by $C\epsilon^{3x}$.

$$\frac{dy}{C\epsilon^{3x}} = 3 \cdot dx \quad \text{or} \quad \frac{dy}{y} = 3 \cdot dx = \text{the differential equation whose solution is Eq. (1)}$$

Equation (1) has a multiplying constant.

677. In what general form can every differential equation of the first order and of the first degree be written?

In the form $M \cdot dx + N \cdot dy = 0$, where, in general, M and N are functions of x and y.

678. How can you often transform the above equation so that it can be solved by a simple integration?

By separation of the variables it is often possible to make M a function of x alone and N a function of y alone, thus permitting a simple integration.

679. What is the solution of $2xy \cdot dx + (x^2 + 1) \, dy = 0$?

First divide by $y(x^2 + 1)$, obtaining

$$\frac{2x \cdot dx}{(x^2 + 1)} + \frac{dy}{y} = 0$$

Now integrate, getting

$$\log_\epsilon (x^2 + 1) + \log_\epsilon y = C$$

or $\qquad \log_\epsilon y(x^2 + 1) = C$

$\qquad\qquad y(x^2 + 1) = \epsilon^c$ by definition of a logarithm

or $\qquad\qquad y(x^2 + 1) = C' \qquad$ where $\qquad C' = \epsilon^c$

680. What is meant by an exact differential equation?

It is an equation of the type $M \cdot dx + N \cdot dy = 0$ where the entire left-hand member is exactly the derivative of some function of x and y, as $\phi(x, y)$, in which case the solution can be at once given as $\phi(x, y) = C$.

681. What is the solution of
$$4x^3y \cdot dx + x^4 \cdot dy = 2x \cdot dy + 2y \cdot dx?$$

It can be seen at once that this equation is the derivative of the equation $x^4y = 2xy + C$, which is a solution.

This can be proved by working backward. The derivative of $x^4y = 4x^3y \cdot dx + x^4 \cdot dy$, and the derivative of $2xy + C = 2x \cdot dy + 2y \cdot dx$.

682. What is meant by an integration factor?

The quantity or factor, in general, involving x and y, that is multiplied by both members of the equation to make it exact. The equation becomes exact after separation of the variables, and the factor that separates the variables is an integrating factor.

Example

In the solution of $2xy \cdot dx + (x^2 + 1) \, dy = 0$, $\dfrac{1}{y(x^2 + 1)}$ is an integrating factor that when multiplied by the equation makes the equation exact (see Question 679).

683. What is the solution of $\dfrac{dy}{dx} = \dfrac{2 + 3y^2}{(2 + x^2)xy}$?

First clear fractions.

$$(2 + x^2)xy \cdot dy = (2 + 3y^2) \, dx$$
or
$$(2 + 3y^2) \, dx - x(2 + x^2)y \cdot dy = 0$$

Separate the variables by means of the integrating factor $\dfrac{1}{(2 + 3y^2)(2 + x^2)x}$ to bring the equation to the form

$$f(x) \cdot dx + f(y) \cdot dy = 0$$

getting
$$\frac{dx}{x(2 + x^2)} - \frac{y \, dy}{(2 + 3y^2)} = 0$$

Now break up $\dfrac{1}{x(2 + x^2)}$ into two proper fractions

or
$$\frac{1}{x(2 + x^2)} = \frac{A}{x} + \frac{Bx + C}{(2 + x^2)}$$

and clear fractions, getting

$$1 = x^2(A + B) + Cx + 2A$$

Equate coefficients of like powers.

$$0 = (A + B)$$
$$\therefore A = -B$$
$$C = 0$$
$$2A = 1 \quad \text{or} \quad A = \tfrac{1}{2}$$
$$\therefore B = -\tfrac{1}{2}$$
$$\therefore \frac{1}{x(2 + x^2)} = \frac{1}{2x} + \frac{-\tfrac{1}{2}x + 0}{(2 + x^2)} = \frac{1}{2x} - \frac{x}{2(2 + x^2)}$$

Then $\dfrac{dx}{x(2 + x^2)} - \dfrac{y \cdot dy}{(2 + 3y^2)} = \dfrac{dx}{2x} - \dfrac{x \cdot dx}{2(2 + x^2)}$

$$- \frac{y \, dy}{(2 + 3y^2)} = 0$$

Now integrate, getting

$$\tfrac{1}{2} \log_\epsilon x - \tfrac{1}{4} \log_\epsilon (2 + x^2) - \tfrac{1}{6} \log_\epsilon (2 + 3y^2) = C$$
or $6 \log_\epsilon x - 12C = 3 \log_\epsilon (2 + x^2) + 2 \log_\epsilon (2 + 3y^2)$

The arbitrary constant may be assumed as $\log_\epsilon C$.

$\therefore 6 \log_\epsilon x + \log_\epsilon C = 3 \log_\epsilon (2 + x^2) + 2 \log_\epsilon (2 + 3y^2)$
or $\log_\epsilon (2 + x^2)^3(2 + 3y^2)^2 = \log_\epsilon Cx^6 = $ the solution

684. Can the integrating factor work in cases where the variables cannot be separated?

Yes, in many cases.

Example

What is the solution of $4y \cdot dx + x \cdot dy = dx$? The sum in the left member suggests the derivative of a product, and the coefficient 4 indicates that one of the factors of this product is a cube. Then x^3 is seen to be an integrating factor, and the original equation is multiplied by x^3 and is transformed to

$$4x^3y \cdot dx + x^4 \cdot dy = x^3 \cdot dx$$

Therefore, the solution is

$$x^4y = \tfrac{1}{4}x^4 + C$$

685. What is the solution of $2x \cdot dy - 2y \cdot dx = 4x^2 \cdot dx$?

The difference of the two terms of the left member suggests the derivative of a quotient of $\dfrac{y}{2x}$, and it would seem that $\dfrac{1}{4x^2}$ is an integrating factor and

$$\frac{2x \cdot dy - 2y \cdot dx}{4x^2} = \frac{4x^2 \cdot dx}{4x^2} = dx$$

$$\therefore \frac{y}{2x} = x + C \text{ is a direct solution}$$

Proved by differentiating,

$$\frac{y}{2x} = x + C \quad \text{or} \quad \frac{2x \cdot dy - 2y \cdot dx}{4x^2} = dx$$

which was the expression to be integrated.

686. What is the solution of $6x \cdot dy - 6y \cdot dx = 4x^2y^2 \cdot dy$?

The left member indicates a quotient; the right member suggests an integrating factor of $\dfrac{1}{4x^2}$;

then
$$\frac{6x \cdot dy - 6y \cdot dx}{4x^2} = y^2 \cdot dy$$

The denominator of the solution could be the square root of $4x^2$ (or $2x$), and the numerator would have to be $3y$ to give $6x \cdot dy$. The right member now suggests a y^3 term.

$$\therefore \frac{3y}{2x} = \frac{y^3}{3} + C = \text{the solution}$$

Proved by differentiating the above,

$$\frac{2x \cdot 3\, dy - 3y \cdot 2\, dx}{4x^2} = \frac{3y^2}{3} \cdot dy$$

or $\dfrac{6x \cdot dy - 6y \cdot dx}{4x^2} = y^2 \cdot dy = $ the original expression that

was to be integrated

687. What would be the distance s that a falling airplane would drop in 3 seconds if it started downward with an initial velocity v_o of 1,000 feet per second and if the velocity v at any time t seconds after it starts to fall is given by $v = gt + v_o = 32t + v_o$?

Now $\qquad\qquad v = \dfrac{ds}{dt} = 32t + v_o$

or $\quad ds = (32t + v_o)\, dt = $ differential-equation form
Integrating, $s = \int 32t \cdot dt + v_o \int dt$

$$= 32 \cdot \frac{t^2}{2} + v_o t = 16t^2 + v_o t + C$$

Now $s = 0$ when $t = 0$, which we substitute.

$$\therefore\ 0 = 0 + 0 + C \qquad \text{and} \qquad C = 0$$

Then $\qquad\qquad s = 16t^2 + v_o t$

Now when $t = 3$ and $v_o = 1,000$, as given,

then $s = 16 \cdot (3)^2 + 1,000 \cdot 3 = 3,144$-ft. drop in 3 sec.

688. If the acceleration of a moving body is proportiona to the time and if $v = v_o$ and $s = s_o$ when $t = 0$ (where $v_o =$ initial velocity and $s_o =$ initial distance).

what would be the total distance s the body will have moved at the end of t seconds?

$$\frac{dv}{dt} = \frac{d^2s}{dt^2} = \text{acceleration}$$

Now $\quad \dfrac{dv}{dt} = kt \quad$ given as proportional to the time

or $\quad dv = kt \cdot dt = $ differential-equation form

Integrating, $\quad v = k \displaystyle\int t \cdot dt = \dfrac{kt^2}{2} + C_1$

Given $v = v_o$ when $t = 0$; substitute this to get

$$v_o = 0 + C_1 \quad \text{or} \quad C_1 = v_o$$
$$\therefore v = \frac{ds}{dt} = \frac{kt^2}{2} + v_o$$

or $\quad ds = \dfrac{kt^2}{2}\,dt + v_o\,dt = $ differential-equation form

Integrating, $\quad s = \dfrac{k}{2}\displaystyle\int t^2 \cdot dt + v_o \int dt = \dfrac{k}{2}\cdot\dfrac{t^3}{3} + v_o t$

$$= \frac{kt^3}{6} + v_o t + C_2$$

Given $s = s_o$ when $t = 0$.

$$\therefore s_o = 0 + 0 + C_2 \quad \text{or} \quad C_2 = s_o$$

Then $\quad\quad s = \dfrac{kt^3}{6} + v_o t + s_o$

689. How long will it take a flywheel to make one revolution if it starts and continues at an angular speed in radians per second of $\omega = 0.1t$?

Let $\theta = $ the number of radians the flywheel turned.

Then $\quad\quad \omega = \dfrac{d\theta}{dt} = 0.1t$

or $\quad\quad d\theta = 0.1t \cdot dt$

and $\quad\quad \theta = 0.1\displaystyle\int t \cdot dt = 0.1 \cdot \dfrac{t^2}{2} + C$

Now $\theta = 0$ when $t = 0$, or $0 = 0 + C$.   Therefore, $C = 0$.

Then $\theta = 0.1\dfrac{t^2}{2}$ or $t = \sqrt{20 \cdot \theta}$

Now when $\theta = 2\pi =$ one revolution,

then $t = \sqrt{20 \cdot 2\pi} = 11.21$ sec. = the time for the first
revolution

**690. If it is given that the velocity of a falling body is
32t feet per second for any time t seconds after it
starts to fall, what is the distance from the surface
of the earth of a falling airplane 13 seconds after it
starts to fall from a height of 3,000 feet above the
earth?**

Let $s =$ the number of feet the damaged plane is above
the earth's surface at any moment.   Then, as s is decreasing
when t is increasing,

$v =$ velocity $= \dfrac{ds}{dt} = -32t =$ the rate at which the plane
is falling

or $ds = -32t \cdot dt =$ differential-equation form

Integrating, $s = -32 \displaystyle\int t \cdot dt = -32\dfrac{t^2}{2} = -16t^2 + C$

Given $s = 3,000$ when $t = 0$.   Then

$$3,000 = -0 + C   \text{or}   C = 3,000$$

Therefore, at any time t sec. after the body starts to fall,

$$s = -16t^2 + 3,000$$

Now, when $t = 13$ sec.,

$s = -16 \cdot (13)^2 + 3,000 = -2,704 + 3,000 = 296$ ft.
above the earth's surface

**691. What is the velocity of a bullet after it has risen
40,000 feet when it is fired directly upward with a
muzzle velocity of 1,600 feet per second?**

INTRODUCTION TO DIFFERENTIAL EQUATIONS 323

The bullet will decelerate at the rate of $g = 32$ ft./sec.^{2}

or
$$\frac{dv}{dt} = -32, \qquad dv = -32 \cdot dt$$
$$v = -32\int dt = -32t + C_1$$

Now $v = 1{,}600$ when $t = 0$, or
$$1{,}600 = -0 + C_1, \qquad \text{and} \qquad C_1 = 1{,}600$$
$$\therefore v = -32t + 1{,}600 \tag{1}$$

Also, $v = \dfrac{ds}{dt} = -32t + 1{,}600 \qquad$ or $\qquad ds = -32t \cdot dt$
$$+ 1{,}600\, dt$$

and $s = -32 \displaystyle\int t \cdot dt + 1{,}600 \int dt = -32\dfrac{t^2}{2} + 1{,}600t$
$$= -16t^2 + 1{,}600t + C_2$$

Now $s = 0$ when $t = 0$, or
$$0 = -0 + 0 + C_2$$
$$\therefore C_2 = 0$$
and
$$s = -16t^2 + 1{,}600t$$

Now when $s = 40{,}000$,
$$40{,}000 = -16t^2 + 1{,}600t$$
and
$$(4t - 200)^2 = 0$$
whence
$$4t = 200 \qquad \text{or} \qquad t = 50$$

Substitute this in Eq. (1) to get
$$v = -32 \cdot (50) + 1{,}600 = 0 \text{ ft./sec.}$$

692. Can you show how the differential equation $\dfrac{dy}{dx} = ky$ is connected with the term "compound-interest law"?

$$\frac{dy}{dx} = ky \tag{1}$$

means that the rate of change of the function y with respect to the variable x is proportional to the corresponding value of the function y.

Now solve Eq. (1) by first separating the variables and then integrating.

$$\frac{dy}{y} = k \cdot dx$$

$$\log_\epsilon y = kx + \log_\epsilon C$$

or $y = C\epsilon^{kx} =$ a solution of Eq. (1) where C is an arbitrary constant

Now if $y =$ a sum of money in dollars accumulating at compound interest

and $\quad\quad i =$ the interest in dollars on \$1 for a year

and $\quad\quad dt =$ an interval of time measured in years

and $\quad\quad dy =$ the interest or growth of y dollars for the interval of time dt

then $\quad\quad\quad dy = i \cdot y \cdot dt \quad\quad$ or $\quad\quad \frac{dy}{dt} = iy$

which means that the average rate of growth of y for the period of time dt is proportional to y itself. Since interest is added to the principal at certain times, as yearly, quarterly, semiannually, etc., the function y changes or grows discontinuously with t. In nature, however, growth is continuous so that $\frac{dy}{dt}$ becomes a true value of the growth at any time, as dy and dt are infinitesimal.

Therefore, the rate of change of y is proportional to y and agrees with Eq. (1) if $k = i$. This is then the compound-interest type of growth.

PROBLEMS

Solve the following:

1. $\frac{dy}{dx} = 4x$

2. $2x \cdot dy + 2y \cdot dx = 0$

3. $\frac{d^2y}{dx^2} = 4\epsilon^{-2x}$

4. $3xy \cdot dx + 2(x^2 + 2)\, dy = 0$

5. $2x^5 \cdot dy + 10x^4y \cdot dx = 3x \cdot dy + 3y \cdot dx$

6. $\frac{dy}{dx} = \frac{3 + 4y^2}{4xy(3 + x^2)}$

7. $12x \cdot dy - 12y \cdot dx = 32x^2 \cdot dx$

8. What is the differential equation whose general solution is $y = \sin^2 2\theta + C$?

9. What is the differential equation whose solution is $y = C\epsilon^x$?

10. Find the differential equation whose solution is $x = Ct + C^2$.

11. Find the differential equation whose solution is $x^2 = 2Cy + C^2$.

12. Find the differential equation whose solution is $C(x + y) = xy$.

Solve the following:

13. $dy = 2xy \cdot dx$

14. $\dfrac{dx}{d\theta} = 1 - \sin 2\theta$

15. $\dfrac{dx}{dt} = tx^2$

16. $x^2 \cdot dy - xy^2 \cdot dx = 2ay^2 \cdot dx$

17. $\sin \theta \cdot \cos \beta \cdot d\theta = \cos \theta \sin \beta \cdot d\beta$

18. $(1 + y^2) \, dx + (1 + x^2) \, dy = 0$

SECTION III

APPLICATIONS OF CALCULUS

CHAPTER XXXVIII

APPLICATIONS OF CALCULUS

693. What is the distance BC that a paratrooper must travel to reach objective C after he is picked up at B by a landplane traveling at 200 miles per hour, assuming that he has been dropped at B by a seaplane traveling at 120 miles per hour from a carrier at A, which is 600 miles from shore line $OC = 1,500$ miles, also assuming that the entire route from A to B to C is to be covered in a minimum of time?

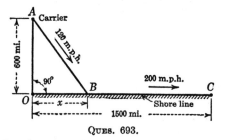

QUES. 693.

Let $t =$ the time it takes to cover the entire route
$$x = OB$$

Then $t = \dfrac{AB}{120} + \dfrac{BC}{200}$

or $\quad t = \dfrac{(\overline{600^2 + x^2})^{\frac{1}{2}}}{120} + \dfrac{1,500 - x}{200}$

and $\dfrac{dt}{dx} = \dfrac{x}{120}(\overline{600^2 + x^2})^{-\frac{1}{2}} - \dfrac{1}{200} = 0$ for a minimum

Solving, $x(\overline{600^2 + x^2})^{-\frac{1}{2}} = \frac{120}{200} = \frac{3}{5}$

$$5x = 3(\overline{600^2 + x^2})^{\frac{1}{2}}$$

$$25x^2 = 9(\overline{600^2 + x^2}) = 9x^2 + 9 \cdot \overline{600^2}$$

$$16x^2 = 9 \cdot \overline{600^2}$$

$$4x = 3 \cdot 600 = 1,800$$

$$x = 450 \text{ miles}$$

$$\therefore BC = 1,500 - 450 = 1,050 \text{ miles}$$

694. At what rate does the paved area of a landing field increase 4 hours after work has started on paving around a rectangular building 100 feet long, 50 feet wide, if the increased length of paving is 50 feet per hour and increased width is 30 feet per hour?

QUES. 694.

Let x = the width of the field
     y = the length of the field
Then z = area = xy

$$\frac{dz}{dt} = \frac{\partial z}{\partial x} \cdot \frac{dx}{dt} + \frac{\partial z}{\partial y} \cdot \frac{dy}{dt} \qquad (1)$$

Now $\dfrac{\partial z}{\partial x} = y_o$ $(y = \text{constant})$

and $\dfrac{\partial z}{\partial y} = x_o$ $(x = \text{constant})$

But $\dfrac{dx}{dt} = 30$ ft./hr. = rate of increase of width (given)

and $\dfrac{dy}{dt} = 50$ ft./hr. = rate of increase of length (given)

Substitute in Eq. (1), getting

$$\frac{dz}{dt} = y_o \cdot 30 + x_o \cdot 50$$

Now $y_o = 100 + 50 \cdot 4 = 300 =$ the length 4 hr. after
work has started

and $x_o = 50 + 30 \cdot 4 = 170 =$ the width 4 hr. after work
has started

$$\therefore \frac{dz}{dt} = 300 \cdot 30 + 170 \cdot 50 = 9{,}000 + 8{,}500 = 17{,}500$$

sq. ft./hr.

The area of paving increases 17,500 sq. ft./hr.

695. If a ship *A*, sailing eastward from dock *O* at 10 miles per hour, travels 6 hours before another ship *B* from the north, traveling at 14 miles per hour, reaches dock *O*, how fast is the distance between the ships changing 3 hours after ship *A* leaves dock *O*?

Let *x* = the distance of ship *A* from dock *O*, after 3 hr. of sailing

   y = the distance of ship *B* from dock *O* at the same time

   z = the distance between ships *A* and *B*

Then $z^2 = x^2 + y^2$

which we differentiate with respect to time.

$$2z \cdot \frac{dz}{dt} = 2x \cdot \frac{dx}{dt} + 2y \cdot \frac{dy}{dt}$$

$$z \frac{dz}{dt} = x \frac{dx}{dt} + y \frac{dy}{dt} \tag{1}$$

Now $x = 10t = 10 \cdot 3 = 30$ miles after 3 hr. of sailing
and $\frac{dx}{dt} = 10$

Also, ship $B = 6 \cdot 14 = 84$ miles from dock *O* when ship *A* is at *O* since ship *A* travels 6 hr. before *B* reaches *O*.

Now $y = 84 - 14t = 84 - 14 \cdot 3 = 42$ miles after 3 hr. of travel

and $\frac{dy}{dt} = -14$

But $z = \sqrt{x^2 + y^2} = \sqrt{30^2 + 42^2} = \sqrt{2{,}664} = 51.61$

miles

Substituting in Eq. (1), we get

$$51.61 \frac{dz}{dt} = 30 \cdot 10 + 42(-14) = -288$$

or $$\frac{dz}{dt} = -\frac{288}{51.61} = -5.58 \text{ m.p.h.}$$

Therefore, the distance between ships A and B is decreasing at the rate of 5.58 m.p.h. at the instant 3 hr. after ship A has left the dock.

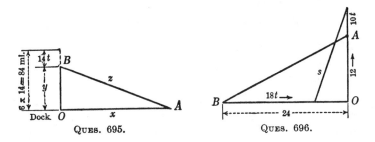

QUES. 695.   QUES. 696.

696. At what rate is the distance between ships A and B decreasing and how far does ship B travel before the distance between them begins to increase if A sails north at 10 miles per hour and B, which is 12 miles south and 24 miles west of A, sails east at 18 miles per hour?

Let s = the distance between ships in the position after they have traveled for t hr. from their original locations. Then

$$s = \sqrt{(24 - 18t)^2 + (12 + 10t)^2}$$

or $$s = \sqrt{720 - 624t + 424t^2}$$

Now $$\frac{ds}{dt} = \frac{1 \cdot (-624 + 848t)}{2\sqrt{720 - 624t + 424t^2}}$$

$$= \frac{-312 + 424t}{\sqrt{720 - 624t + 424t^2}}$$

With the ships at their original position, $t = 0$.

Then $\dfrac{ds}{dt} = \dfrac{-312 + 424 \cdot 0}{\sqrt{720 - 624 \cdot 0 + 424 \cdot 0}}$

$$= -\frac{312}{\sqrt{720}} = -\frac{312}{26.83}$$

or $\dfrac{ds}{dt} = -11.63$

The distance between the ships is decreasing, and they are approaching each other at 11.63 m.p.h. when they are at their original positions.

The distance between the ships will stop decreasing when $\dfrac{ds}{dt} = 0$, which means that

$$-312 + 424t = 0 \quad \text{or} \quad t = \tfrac{312}{424} = 0.736 \text{ hr.}$$

In other words, 0.736 hr. after they have left their original positions, the distance between the ships begins to increase and ship B has traveled $0.736 \cdot 18 = 13.25$ miles before the distance begins to increase.

697. What is the most economical speed in still water of a steamer if the cost per hour for fuel is proportional to the cube of the speed and is $30 per hour for a speed of 12 knots and the other expenses are $200 per hour?

$$c = kv^3$$

where $v =$ knots.

$c =$ cost of fuel.

Now $\$30 = k \cdot \overline{12}^3 \quad \text{or} \quad k = 0.0174$

Then $c = 0.0174v^3$

Cost of operation per hour $= 0.0174v^3 + 200$

Total cost per knot $= C = \dfrac{0.0174v^3 + 200}{v}$

$$= 0.0174v^2 + 200v^{-1}$$

Now $\dfrac{dC}{dv} = 0.0348v + (-200v^{-2}) = 0.0348v - 200v^{-2}$

For a minimum, set the above equal to zero.

Then $\qquad 0.0348v - 200v^{-2} = 0$

or $\qquad\qquad 0.0348v = \dfrac{200}{v^2}$

$$v^3 = \dfrac{200}{0.0348} = 5{,}747.1$$

∴ $v = 17.91$ knots = the most economical speed in still water

698. At what peripheral speed is the power output of an impulse turbine a maximum if it is driven by a jet of water or steam flowing at 30 feet per second?

Let v = peripheral velocity of the turbine and 30 ft./sec. = the peripheral velocity of the jet (given).

Now the impulse force = $k(30 - v)$ since the direction of the jet and the peripheral motion are in the same direction. But the power = P = impulse × velocity, or

$$P = k(30 - v)v = k(30v - v^2)$$

Now $\qquad\qquad \dfrac{dP}{dv} = k(30 - 2v)$

and $\qquad\qquad k(30 - 2v) = 0 \quad$ for a maximum

or $\qquad 30 = 2v \quad$ and $\quad v = \tfrac{30}{2} = 15$ ft./sec.

Therefore, the output is a maximum when the peripheral velocity of the wheel equals one-half the velocity of the jet.

699. What is the maximum height to which a projectile will rise and what is the range (neglecting air resistance) if it is fired with a velocity of 1,000 feet per second at an angle of 30 degrees when its path is described by $y = \tan \alpha \cdot x - \dfrac{g}{2v^2 \cos^2 \alpha} \cdot x^2$?

If $\qquad\qquad y = \tan \alpha \cdot x - \dfrac{g}{2v^2 \cdot \cos^2 \alpha} \cdot x^2$

then $\qquad\qquad \dfrac{dy}{dx} = \tan \alpha - \dfrac{2g}{2v^2 \cdot \cos^2 \alpha} \cdot x$

and $$\tan \alpha - \frac{g}{v^2 \cos^2 \alpha} \cdot x = 0 \text{ for a maximum}$$

or $$x = \frac{\tan \alpha \cdot v^2 \cdot \cos^2 \alpha}{g} = \frac{\sin \alpha}{\cos \alpha} \cdot \frac{v^2 \cos^2 \alpha}{g}$$

Finally, $x = \dfrac{v^2}{g} \cdot \sin \alpha \cdot \cos \alpha = $ the point where the maximum height occurs

and $$2x = \frac{2v^2}{g} \cdot \sin \alpha \cdot \cos \alpha = \frac{v^2}{g} \cdot \sin 2\alpha = \text{the range}$$

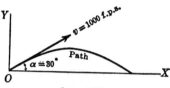

Ques. 699.

Substitute this value of x in the expression for y above getting

$$y = \frac{\sin \alpha}{\cos \alpha} \cdot \sin \alpha \cdot \cos \alpha \frac{v^2}{g} - \frac{g}{2v^2 \cos^2 \alpha} \cdot \frac{v^4}{g^2} \sin^2 \alpha \cos^2 \alpha$$

Finally, $y = \dfrac{v^2}{g} \sin^2 \alpha - \dfrac{v^2}{2g} \sin^2 \alpha = \dfrac{v^2}{2g} \sin^2 \alpha = $ maximum height

For $v = 1{,}000$ ft. per sec., $\alpha = 30°$, $g = 32.2$ ft./sec.^{2},

$$\sin \alpha = 0.5, \qquad \sin 2\alpha = \sin 60° = 0.866$$

Then $y = \dfrac{1{,}000^2}{2 \cdot 32.2} \cdot \left(\dfrac{1}{2}\right)^2 = \dfrac{1{,}000{,}000}{8 \cdot 32.2} = 3{,}882$ ft. $=$ the maximum height

and $2x = \dfrac{1{,}000{,}000}{32.2} \cdot 0.866 = 26{,}894$ ft. $=$ the range

700. What are the equations of motion of a projectile fired with a speed of 1,800 feet per second at an inclination of 27 degrees?

Ignore air resistance.

Now $\dfrac{d^2y}{dt^2} = -32.2$ ft./sec.^{2} = the vertical acceleration due to gravity

and $\dfrac{d^2x}{dt^2} = 0 =$ the horizontal acceleration

Integrating the above equations twice,

$$\frac{dy}{dt} = -32.2t + C_1$$

or $\qquad\qquad y = -16.1t^2 + C_1t + K_1 \qquad\qquad (1)$

$$\frac{dx}{dt} = 0 + C$$

or $\qquad\qquad x = Ct + K \qquad\qquad\qquad\qquad (2)$

Now, when $t = 0$, $x = 0$ and $y = 0$.

$\therefore K = 0 \qquad$ and $\qquad K_1 = 0$

Also, when $t = 0$,

$\dfrac{dx}{dt} = C \qquad$ and $\qquad \dfrac{dy}{dt} = C_1$

QUES. 700.

But $\dfrac{dy}{dt} = V_y =$ the vertical component of velocity

$$= 1{,}800 \sin 27°$$

$$V_y = 1{,}800 \cdot 0.454 = 817.2 = C_1$$

and $\dfrac{dx}{dt} = V_x =$ the horizontal component of velocity =

$$1{,}800 \cos 27°$$

$$V_x = 1{,}800 \cdot 0.891 = 1{,}603.8 = C$$

Substituting in Eqs. (1) and (2),

$\left.\begin{array}{l} x = 1{,}603.8t \\ y = 817.2t - 16.1t^2 \end{array}\right\}$ equations of motion of projectile

701. How do you find by integration the distance traveled by a projectile during any interval of time when the velocity at every instant is known?

$s = \int v \cdot dt$ = distance traveled = sum of all the products of the elements of time and velocity

QUES. 701.

Now, if we are given $x = \dfrac{t^2}{2} - t$ and $y = \dfrac{t^2}{2} + t$, then $\dfrac{dx}{dt} = v_x = t - 1$ = velocity in x direction and $\dfrac{dy}{dt} = v_y = t + 1$ = velocity in y direction

Then $v = \sqrt{v_x{}^2 + v_y{}^2}$ = resultant velocity

or $\quad v = \sqrt{(t - 1)^2 + (t + 1)^2}$

$\quad\quad = \sqrt{t^2 - 2t + 1 + t^2 + 2t + 1} = \sqrt{2t^2 + 2}$

$\therefore s = \int v \cdot dt = \sqrt{2} \int \sqrt{t^2 + 1} \cdot dt$

$\quad\quad = \sqrt{2} \left[\dfrac{t}{2} \sqrt{t^2 + 1} + \dfrac{1}{2} \log_e (t + \sqrt{t^2 + 1}) \right] + C$

$$(1)$$

When $t = 0$, then $s = 0$ and $C = 0$ in Eq. (1)

Finally $s = \sqrt{2} \left[\dfrac{t}{2} \sqrt{t^2 + 1} + \dfrac{1}{2} \log_e (t + \sqrt{t^2 + 1}) \right]$

702. What is the velocity of a bullet when it has risen 8,000 feet after it was fired directly upward with a muzzle velocity of 1,000 feet per second?

Neglecting air resistance, the acceleration = -32 ft./sec.^{2} = α.

Then $\quad\quad\quad\quad\quad \alpha = \dfrac{dv}{dt} = -32$

and
$$dv = -32\, dt$$
$$\int dv = -32 \int dt$$
or
$$v = -32t + C_1$$

When $t = 0$, $v = 1{,}000$.

$$\therefore\ C_1 = 1{,}000$$

And
$$v = \frac{ds}{dt} = -32t + 1{,}000$$

or,
$$ds = -32t \cdot dt + 1{,}000\, dt$$
$$\int ds = -32 \int t \cdot dt + 1{,}000 \int dt$$

Then
$$s = -32\frac{t^2}{2} + 1{,}000t + C_2$$

Now $s = 0$ when $t = 0$.

$$\therefore\ C_2 = 0$$

And
$$s = -16t^2 + 1{,}000t$$
$$\therefore\ 8{,}000 = -16t^2 + 1{,}000t$$
$$16t^2 - 1{,}000t + 8{,}000 = 0$$
$$4t^2 - 250t + 2{,}000 = 0$$

Now $t = \dfrac{-b \pm \sqrt{b^2 - 4ac}}{2a} =$ the root of a quadratic equation

Where $a = 4$,
   $b = -250$,
   $c = 2{,}000$,

then
$$t = \frac{+250 \pm \sqrt{62{,}500 - 32{,}000}}{8}$$

or
$$t = \frac{250 \pm \sqrt{30{,}500}}{8} = \frac{250 \pm 174.6}{8}$$

Finally, $t = 53.1$ or 9.4 sec. to reach 8,000 ft.

Substituting in $v = -32t + 1{,}000$,

$$v = -32 \cdot 9.4 + 1{,}000 = -300.8 + 1{,}000 = 700^- \text{ ft./sec.}$$

The value of $t = 53.1$ gives a negative velocity and is not used.

703. For maximum power, how must a storage battery of n cells be connected in a series-parallel arrangement with a constant external resistance of R ohms, each cell having e volts with an internal resistance of r ohms?

If $\qquad x$ = number of cells in parallel

Then $\qquad \dfrac{n}{x}$ = number of cells in series

and $\quad \dfrac{\dfrac{n}{x} \cdot r}{x} = \dfrac{nr}{x^2}$ ohms = the total internal resistance

Then $\quad \dfrac{nr}{x^2} + R$ = the total resistance of circuit

and $\qquad \dfrac{n}{x} \cdot e$ = the total voltage of the circuit

$\therefore I = \dfrac{\dfrac{ne}{x}}{\dfrac{nr}{x^2} + R} = \dfrac{nex}{nr + Rx^2}$ = the total current delivered

Now $P = RI^2 = \dfrac{Rn^2e^2x^2}{(nr + Rx^2)^2}$ = the power produced

P will be a maximum when $\dfrac{x^2}{(nr + Rx^2)^2}$ is a maximum.

Then $\dfrac{d}{dx}\left(\dfrac{x^2}{(nr + Rx^2)^2} \right)$

$\qquad = \dfrac{(nr + Rx^2)^2 \cdot 2x - x^2 \cdot 2(nr + Rx^2) \cdot 2Rx}{(nr + Rx^2)^4} = 0$

$\qquad = \dfrac{(2x)(nr + Rx^2)[(nr + Rx^2) - 2Rx^2]}{(nr + Rx^2)^4} = 0$

$\qquad = \dfrac{2x(nr - Rx^2)}{(nr + Rx^2)^3} = 0$

Then $2x(nr - Rx^2) = 0$ $\qquad$ or $\qquad x = \sqrt{\dfrac{nr}{R}}$ for maximum

$\qquad\qquad\qquad\qquad\qquad\qquad\qquad\qquad\qquad\qquad$ power

704. Show that the maximum current is obtained from a battery of n cells in series parallel when the internal resistance equals the external resistance R?

From the previous question,
$$x = \text{number of cells in parallel}$$

and
$$I = \frac{nex}{nr + Rx^2} = \text{total current delivered}$$

Then
$$\frac{dI}{dx} = \frac{(nr + Rx^2)ne - nex \cdot 2Rx}{(nr + Rx^2)^2}$$

Set
$$\frac{dI}{dx} = \frac{n^2er - neRx^2}{(nr + Rx^2)^2} = 0 \qquad \text{for a maximum}$$

or
$$n^2er - neRx^2 = 0$$

$\therefore$ External $R = \dfrac{n^2er}{nex^2} = \dfrac{nr}{x^2}$ = which is also the internal resistance

705. At what current I is the efficiency of a transformer a maximum if the constant impressed electromotive force = 3,000 volts, the constant loss (independent of output) = 300 watts, and the internal resistance $r = 30$ ohms?

$$E = \text{efficiency} = \frac{\text{output}}{\text{input}} = \frac{3{,}000I - 300 - 30I^2}{3{,}000I}$$
$$= \frac{\text{input} - \text{losses}}{\text{input}}$$

$$E = \frac{100I - 10 - I^2}{100I}$$

Then
$$\frac{dE}{dI} = \frac{100I(100 - 2I) - (100I - 10 - I^2)100}{10{,}000I^2}$$
$$= \frac{10{,}000I - 200I^2 - 10{,}000I + 1{,}000 + 100I^2}{10{,}000I^2}$$
$$= \frac{1{,}000 - 100I^2}{10{,}000I^2} = \frac{10 - I^2}{100I^2}$$

or
$$\frac{dE}{dI} = \frac{10 - I^2}{100I^2} = 0 \qquad \text{for a maximum}$$

Then $10 - I^2 = 0$ or $I^2 = 10$ and $I = \sqrt{10}$

Finally, $I = 3.16$ amp.

Substitute this in the original equation.

$$\therefore E = \frac{100 \cdot 3.16 - 10 - 10}{100 \cdot 3.16} = \frac{296}{316} = 0.937 \text{ or } 93.7 \text{ per}$$

cent

706. What formula would give the current I at any time after an electromotive force is removed if given the equation $\frac{dI}{dt} = -40I$ and $I = 20$ amperes when $t = 0$?

Given $\frac{dI}{dt} = -40I$ = rate of change of current with respect to time.   Then

$$\frac{dI}{I} = -40 \, dt = \text{differential-equation form}$$

and $$\int \frac{dI}{I} = -40 \int dt$$

or $\log_\epsilon I = -40t + C$

Then $I = \epsilon^{-40t+c}$ from the definition of a log

or $I = \epsilon^{-40t} \cdot \epsilon^c = C_1 \epsilon^{-40t}$

But $I = 20$ when $t = 0$ (given).

Then $20 = C_1 \epsilon^0$ or $C_1 = 20$

$$\therefore I = 20\epsilon^{-40t} = \text{the formula for the current}$$

707. In the above what is the current $\frac{1}{40}$ second after the electromotive force is shut off?

$$I = 20\epsilon^{-40t} = 20\epsilon^{-40 \cdot \frac{1}{40}} = 20\epsilon^{-1}$$

$$= \frac{20}{2.718} = 7.36 \text{ amp.}$$

708. What is the work done in moving a positive charge of electricity of strength P_1 from a distance a to a greater distance b away from another positive charge of strength P_2?

Like charges of electricity repel each other with a force that is directly proportional to the product of the charges and inversely proportional to the square of the distance between them, or

$$F = \text{force} = k \cdot \frac{P_1 \cdot P_2}{s^2}$$

Now $\qquad dW = F \cdot ds$

or differential of work = force × differential of distance moved

or $\qquad dW = k \cdot \dfrac{P_1 \cdot P_2}{s^2} \cdot ds$

Then $\quad W = kP_1 \cdot P_2 \displaystyle\int_a^b \frac{ds}{s^2} = kP_1 \cdot P_2 \displaystyle\int_a^b s^{-2} \cdot ds$

or $\qquad W = kP_1 \cdot P_2 \left[-\dfrac{1}{s} \right]_a^b$

Finally, $W = kP_1 \cdot P_2 \left(\dfrac{1}{a} - \dfrac{1}{b} \right)$ = work = dyne centimeters

if s = centimeters and F = dynes

The unit strength of a charge = the repulsion force of 1 dyne at 1 cm. distance. Then two charges are equal and of unit strength.

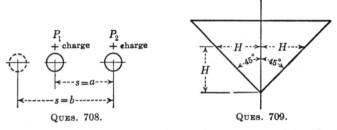

P₁ + charge   P₂ + charge
|← s = a →|
|← s = b →|

QUES. 708.   QUES. 709.

709. **At what rate is the surface of the water rising in the vessel whose form is that of an inverted right circular cone when the water is 20 feet deep and is flowing in at a uniform rate of 40 cubic feet per minute?**

$$\text{Volume} = \frac{1}{3} H \cdot \pi H^2 = \frac{\pi H^3}{3}$$

$$\frac{dV}{dH} = \frac{3\pi H^2}{3} = \pi H^2 = \text{rate of change of volume with}$$

respect to height

Given $\dfrac{dV}{dt} = 40$ cu. ft./min. Now

$$\frac{dV}{dt} = \frac{dV}{dH} \cdot \frac{dH}{dt}$$

or $\qquad 40 = \pi H^2 \cdot \dfrac{dH}{dt}$

$$\frac{dH}{dt} = \frac{40}{\pi H^2}$$

When $H = 20$ ft.,

$\dfrac{dH}{dt} = \dfrac{40}{\pi \cdot 400} = 0.03183$ ft./min. $=$ rate of rising of

water level when $H = 20$ ft.

710. In the above question, at what rate is the area of the surface of the water increasing?

Area $= \pi H^2 = A$

$\dfrac{dA}{dH} = 2\pi H =$ rate of change of area with respect to height

Now $\qquad \dfrac{dA}{dt} = \dfrac{dA}{dH} \cdot \dfrac{dH}{dt}$

From the above question,

$$\frac{dH}{dt} = \frac{40}{\pi H^2}$$

$$\therefore \frac{dA}{dt} = 2\pi H \cdot \frac{40}{\pi H^2} = \frac{80}{H}$$

When $H = 20$ ft.,

$\dfrac{dA}{dt} = \dfrac{80}{20} = 4$ sq. ft./min. $=$ rate of increase of surface

area of the water at $H = 20$ ft.

711. How long will it take the surface of the water in a cylindrical tank having an opening in the side to lower from 60 feet above the opening to 40 feet above the opening if the tank cross section is 100 square feet and the jet cross section is 0.10 square foot?

$v = \sqrt{2gh}$ = theoretical velocity of jet, feet per second.
Now $dV = -100\,dh$ = a differential of volume of water flowing out in a differential of a drop of level dh
Also, $dV = 0.10\sqrt{2gh} \cdot dt$ = the same differential of volume of water running out through the opening in a differential of time dt

$$\therefore\ -100\,dh = 0.10\sqrt{2gh} \cdot dt$$

or
$$dt = -\frac{1,000}{\sqrt{2g}}\,h^{-\frac{1}{2}} \cdot dh$$

Now sum up all the differentials of time during which the water is flowing between 60 and 40 ft. above the opening,

or
$$\int dt = -\int_{40}^{60} \frac{1,000}{\sqrt{2g}}\,h^{-\frac{1}{2}} \cdot dh$$

$$t = -\frac{1,000}{\sqrt{2g}}\int_{40}^{60} h^{-\frac{1}{2}} \cdot dh = \left[-\frac{1,000}{\sqrt{2g}} \cdot h^{\frac{1}{2}} \cdot 2\right]_{40}^{60}$$

$$t = \left[-\frac{2,000}{\sqrt{2g}} \cdot h^{\frac{1}{2}}\right]_{40}^{60} = -\frac{2,000}{\sqrt{2g}}\left[h^{\frac{1}{2}}\right]_{40}^{60}$$

Finally, $t = -249.22\,[\sqrt{60} - \sqrt{40}]$
$$\therefore\ t = 354.2 \text{ sec.} = 5.9^{+} \text{ min.}$$

As dV was assumed negative, $-t$ merely means the time required for outflow.

712. **What is the total pressure on a triangular dam when the water level is at the top, with dimensions of dam as shown in the accompanying illustration?**

A differential of area $dA = mn \cdot dH$.

Now $\dfrac{mn}{60} = \dfrac{20 - H}{20}$ or $mn = \dfrac{60(20 - H)}{20}$

Then $dA = \dfrac{60(20 - H)}{20} \cdot dH$

and $dP = \dfrac{62.5 \cdot 60H(20 - H)}{20} \cdot dH$ = the pressure on the differential of area

$\therefore P = 187.5 \displaystyle\int_0^{20} (20H - H^2) \cdot dH$ = the total pressure on the dam

Total pressure $P = 187.5 \left[\dfrac{20H^2}{2} - \dfrac{H^3}{3} \right]_0^{20}$

Finally, $\qquad P = 187.5 \left(10 \cdot \overline{20}^2 - \dfrac{\overline{20}^3}{3} \right) = 250{,}000$ lb.

QUES. 712.

QUES. 713.

713. What is the pressure on a gate closing a circular main 4 feet in diameter when the main is half full?

$y \cdot dx$ = a differential of area

xw = fluid pressure per square foot

$\quad$ = depth $\times$ weight of 1 cu. ft. of fluid, w

$\therefore xw \cdot y \cdot dx$ = pressure on the differential of area

Summing up the differential fluid pressures, $w\!\int y \cdot x \cdot dx$. For water, $w = 62.5$ lb./cu. ft.

Now $\qquad\qquad\qquad y = \sqrt{4 - x^2}$

$\therefore P = 62.5 \displaystyle\int_0^2 \sqrt{4 - x^2} \cdot x \cdot dx$ = pressure on one-half of the water area

To integrate, let $u = 4 - x^2$.

Then $\qquad\qquad\qquad du = -2x \cdot dx$

and $P = \frac{2}{2} \cdot 62.5 \int_0^2 \sqrt{4 - x^2} \cdot x \cdot dx$

$$= -\tfrac{1}{2} \cdot 62.5 \int \sqrt{4 - x^2} \cdot - 2x \cdot dx$$

or　$P = -\dfrac{62.5}{2} \int u^{\frac{1}{2}} \cdot du$

$$= -\frac{62.5}{2} \cdot \frac{u^{\frac{3}{2}}}{\frac{3}{2}} = \left[-\frac{62.5}{3} (4 - x^2)^{\frac{3}{2}} \right]_0^2$$

$$= -\frac{62.5}{3} (4 - 4)^{\frac{3}{2}} + \frac{62.5}{3} (4 - 0)^{\frac{3}{2}}$$

$$= \frac{62.5}{3} \sqrt{4^3} = \frac{62.5}{3} \cdot 8 = \frac{500}{3} = 167 \quad \text{lb.} \quad \text{for one-}$$

half of the water area

$\therefore P = 2 \cdot 167 = 334$ lb. pressure on the gate

714. **What is the work done in pumping out a cylindrical tank that is 10 feet in diameter and 15 feet high when the water is 10 feet deep in the tank and is to be lifted 15 feet above the top of the tank?**

A differential of work $= dW = F \cdot ds$ where $F =$ the weight of water in the differential of height $ds = 62.5\pi r^2 \cdot ds$ and $s =$ the total height this element of the weight is raised. Then

$$dW = 62.5\pi r^2 \cdot ds \cdot s$$

QUES. 714.

The total work $= W = 62.5\pi r^2 \displaystyle\int_{20 \text{ ft.}}^{30 \text{ ft.}} s \cdot ds$

$$= 62.5\pi r^2 \left[\frac{s^2}{2} \right]_{20 \text{ ft.}}^{30 \text{ ft.}}$$

The limits are 20 ft. and 30 ft. down from the line to which the water is to be raised.

$W = 31.25\pi r^2 \left[s^2 \right]_{20}^{30} = 31.25\pi r^2 (900 - 400)$

$\quad = 31.25 \cdot 3.1416 \cdot 25 \cdot 500 = 1,227,187.5$ ft.-lb. $=$ work

715. What distances in earth and rock will give a minimum cost for excavation of a pipe line to run between points *A* and *B*, 400 feet apart horizontally and 100 feet vertically, with earth excavation at $8 per running foot and rock $20 per foot?

If x = the horizontal distance in rock

then $400 - x$ = distance in earth

and $\sqrt{x^2 + 10,000}$ = actual distance in rock

$\therefore C$ = cost = $20\sqrt{x^2 + 10,000} + 8(400 - x)$

or $C = 20(x^2 + 10,000)^{\frac{1}{2}} + 3,200 - 8x$

and $\dfrac{dC}{dx} = \dfrac{20x}{(x^2 + 10,000)^{\frac{1}{2}}} - 8 = 0$ for a minimum

$$20x = 8\sqrt{x^2 + 10,000}$$
$$5x = 2\sqrt{x^2 + 10,000}$$
$$25x^2 = 4(x^2 + 10,000) = 4x^2 + 40,000$$
$$21x^2 = 40,000$$
$$x^2 = \frac{40,000}{21} = 1,904.76$$
$$x = 43.64 \text{ ft.}$$

$\therefore$ Distance in rock = $\sqrt{x^2 + 10,000}$

$$= \sqrt{1,904.76 + 10,000}$$
$$= \sqrt{11,904.76} = 109.11 \text{ ft.}$$

Distance in earth = $400 - x = 400 - 43.64 = 356.36$ ft.

Cost = $20 \cdot 109.11 + 8 \cdot 356.36 = \$2,182.20 + \$2,850.88$
$$= \$5,033.08$$

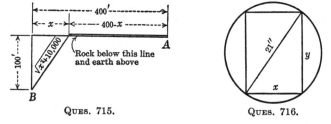

Ques. 715.   Ques. 716.

716. What are the dimensions of the strongest rectangular beam that can be cut from a round log 21 inches in.

diameter if the strength is proportional to the width and the square of the depth?

Given s = strength = kxy^2 where x = width and y = depth. But

$$x^2 + y^2 = \overline{21}^2 = 441$$

or
$$y^2 = 441 - x^2$$
$$\therefore s = kxy^2 = kx(441 - x^2) = 441kx - kx^3$$

and $\dfrac{ds}{dx} = 441k - 3kx^2 = 0$ for a maximum or minimum

$$3kx^2 = 441k$$
$$x^2 = 147$$
$$x = \textbf{12.12 in.}$$

which is a maximum because $\dfrac{d^2s}{dx^2} = -6kx$ is $(-)$ negative for $x = 12.12$.

Finally, $x^2 + y^2 = 441,$ $\overline{12.12}^2 + y^2 = 441$
or
$$y^2 = 441 - 147 = 294$$
$$\therefore y = \textbf{17.15 in.}$$

The beam should be 12.12 in. wide and 17.15 in. deep.

717. Can you show that the derivative of the equation for the moment of a beam is equal to the equation for the shear?

Assume a beam simply supported with uniform load = w lb. per ft.

$$P_1, P_2, P_3, = \text{concentrated loads}$$
$$R_L, R_R = \text{left and right reactions}$$

For shear at a point beyond the P_3 load,

$$V = R_L - P_1 - P_2 - P_3 - wx = \text{shear equation}$$

Now the bending moment at this point is

$$M = R_L x - P_1(x - a) - P_2(x - b) - P_3(x - c) - \frac{wx^2}{2}$$

Now differentiate M, getting

$$\frac{dM}{dx} = R_L - P_1 - P_2 - P_3 - wx = \text{shear equation}$$

Reversing the operation, given the shear curve, the moment curve is found by integration.

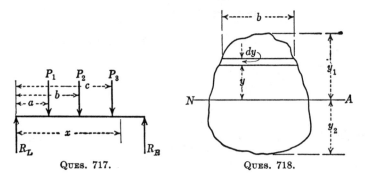

QUES. 717.   QUES. 718.

718. How is the expression for the resisting moment of a beam arrived at?

The sum of the moments of the differentials of areas $y \cdot b \cdot dy$ above the neutral axis (N.A.) will balance the moments below the N.A.   The assumptions are: (a) stress varies as strain; (b) modulus of elasticity is the same for compression and tension; (c) original radius of curvature of beam is very large as compared with the beam section.

Let $\qquad f$ = stress at unit distance from N.A.

$\therefore f \cdot y$ = stress at y

$f \cdot y \cdot b \cdot dy$ = stress $\times$ differential of area = the resisting force of a small strip

$f \cdot y \cdot b \cdot dy \cdot y = fb \, dy \cdot y^2$ = the moment of a small strip about N.A.

Summing up, $f\int b \cdot dy \cdot y^2$ = the moment of resistance

But $\int b \cdot dy \cdot y^2 = \int$ area $\times \overline{\text{distance}}^2 = I$ = moment of inertia

$\therefore$ Resisting moment of beam = fI

Now the external bending moment $= M =$ the resisting moment of the beam, or

$$M = fI$$

Now $S_1 = fy_1 =$ the maximum stress at the extreme tension fiber

and $S_2 = fy_2 =$ the maximum stress at the extreme compression fiber

$$\therefore f = \frac{S_1}{y_1} = \frac{S_2}{y_2}$$

Finally, $M = \dfrac{S}{C} \cdot I$ or $\dfrac{M}{I} = \dfrac{S}{C}$

where $C =$ the distance of the fiber carrying the greatest fiber stress, from the N.A. C can be either y_1 or y_2.

719. How do you arrive at the equation of the elastic curve and the expression for the maximum deflection of a simply supported beam having a load P at the middle when given the differential equation for the beam $EI\dfrac{d^2y}{dx^2} = \dfrac{P}{2}\left(\dfrac{l}{2} - x\right)$?

Divide by EI.

$$\frac{d^2y}{dx^2} = \frac{Pl}{4EI} - \frac{Px}{2EI}$$

$$\frac{d}{dx}\left(\frac{dy}{dx}\right) = \frac{Pl}{4EI} - \frac{Px}{2EI}$$

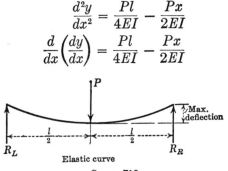

Elastic curve

Ques. 719.

Multiply by dx.

$$d\left(\frac{dy}{dx}\right) = \frac{Pl}{4EI} \cdot dx - \frac{Px}{2EI} \cdot dx$$

Integrating,

$$\frac{dy}{dx} = \frac{Pl}{4EI} \cdot x - \frac{Px^2}{4EI} + C \tag{1}$$

Multiplying by dx,

$$dy = \frac{Pl}{4EI} \cdot x \cdot dx - \frac{Px^2}{4EI} \cdot dx + C \cdot dx$$

Integrating,

$$y = \frac{Pl}{8EI} \cdot x^2 - \frac{Px^3}{12EI} + Cx + C_1 \tag{2}$$

When slope $\frac{dy}{dx} = 0$, $x = 0$, $y = 0$; and, substituting in Eqs. (1) and (2),

$$0 = 0 - 0 + C \qquad \therefore C = 0$$
$$0 = 0 - 0 + 0 + C_1 \qquad \therefore C_1 = 0$$
$$\therefore y = \frac{Plx^2}{8EI} - \frac{Px^3}{12EI} = \text{equation of the elastic curve}$$

For maximum deflection at the ends, substitute $x = \frac{l}{2}$ in the preceding equation.

Then $y = \dfrac{Pl}{8EI}\left(\dfrac{l}{2}\right)^2 - \dfrac{P}{12EI}\left(\dfrac{l}{2}\right)^3 = \dfrac{Pl^3}{32EI} - \dfrac{Pl^3}{96EI}$

or $\quad y = \dfrac{Pl^3}{48EI} = \text{maximum deflection at the ends}$

720. **What is the relative rate of increase and the percentage rate of increase of the number of bacteria in a culture described by formula $N = 2,000\epsilon^{0.6t}$ where $N =$ number and $t =$ time in hours?**

Given $N = 2,000\epsilon^{0.6t}$.   Then

$$\log_\epsilon N = \log_\epsilon 2,000 + 0.6t \log_\epsilon \epsilon$$

and, differentiating,

$$\frac{1}{N} \cdot \frac{dN}{dt} = 0.6 = \text{relative rate of increase}$$

and $100 \cdot 0.6 = 60$ per cent $=$ percentage rate of increase

721. Can you show that the quantity of salt in a solution at any time is reduced, by water running into a tank, according to the compound-interest law, when the volume v of the mixture is kept constant?

Let x = amount of water that has run through

s = quantity of salt in the mixture of total volume v

Then $\dfrac{s}{v}u$ = quantity of salt in any other volume u of the mixture

If a differential of volume dx of the mixture is removed from the tank, then

$$ds = -\frac{s}{v} \cdot dx = \text{the differential of salt that is removed} \quad (1)$$

Now if a differential volume of water dx is added to fill the tank to its original volume v, then, from Eq. (1),

$$\frac{ds}{dx} = -\frac{s}{v} = \text{the ratio of the amount of salt removed to the}$$

volume of water added

$\therefore s$ ($=$ amount of salt removed) changes according to the compound-interest law, as $\dfrac{ds}{dx} = -ks$

In this case, k = constant = $\dfrac{1}{v}\left(= \dfrac{1}{\text{volume}}\right)$.

722. Can you give the relation of the building up of a salt solution by adding salt while maintaining the volume constant?

Let v = the constant volume
y = the amount of salt in the tank at any moment
x = the salt added from the beginning

Then $v - y$ = quantity of water or other vehicle in v

$$\therefore \frac{v - y}{v} \cdot u = \text{quantity of water in any other volume } u$$

If a differential of mixture dx is removed from the tank, then

$$dy = \left(\frac{v - y}{v}\right) dx = \text{a differential of water removed} \quad (1)$$

Now if a differential volume of salt is to replace the water removed, then, from Eq. (1),

$$\frac{dy}{dx} = \frac{v - y}{v} = \frac{1}{v}(v - y)$$

which shows the relation or ratio of the amount of salt in the tank at any time to the amount of salt added at the beginning. This is also according to the compound-interest law.

723. What is the work done in stretching a spring from a length of 8 to 11 inches if it is 6 inches long with no force applied and a 5-pound weight stretches it 1 inch?

$F = ks$, or the force varies with the stretch or elongation.
5 lb. = $k \cdot 1$ in.   (5 lb. stretches spring 1 in.)

$$\therefore k = 5 \quad \text{and} \quad F = 5s$$

$dW = F \cdot ds$.   A differential of work = force $\times$ differential of elongation.

$$\therefore dW = 5s \cdot ds$$

and
$$W = 5 \int_{2 \text{ in.}}^{5 \text{ in.}} s \cdot ds$$

When the spring is 8 in. long, the stretch is 2 in.; when 11 in. long, the stretch is 5 in. Therefore, add up the differentials of work between 2 and 5.

$$W = \left[5 \cdot \frac{s^2}{2}\right]_2^5 = \left(5 \cdot \frac{25}{2} - 5 \cdot \frac{4}{2}\right) = 52.5 \text{ in.-lb.}$$
$$= 4.38 \text{ ft.-lb. of work}$$

724. What is the relation between the maximum power and the centrifugal tension in a belt drive?

Let C = centrifugal tension

T_2 = greatest tension on tight side of belt

w = weight of 1 ft. of belt of 1 sq. in. in section

v = velocity, feet per second

Now $C = \dfrac{wv^2}{g}$ = centrifugal tension (from mechanics)

Then $E = T_2 - \dfrac{wv^2}{g}$ = effective tension for transmission of power

and $P = E \cdot v = T_2 v - \dfrac{wv^3}{g}$ = power transmitted per square inch of belt section

Then $\dfrac{dP}{dv} = T_2 - \dfrac{3wv^2}{g} = 0$ for a maximum

or $T_2 = \dfrac{3wv^2}{g}$

But $\dfrac{wv^2}{g} = C$

$\therefore T_2 = 3C$

$C = \dfrac{T_2}{3}$ = centrifugal tension for maximum power

Therefore, maximum power results when the centrifugal tension is one-third of the greatest tension (working tension) on the tight side of the belt.

725. How do you arrive at the expression for the kinetic energy of a body at any instant?

Kinetic energy is the work a body can do in giving up all its velocity. And work = force × distance

$\therefore dw$ = differential of work = $F \cdot ds$ = force × differential of distance = $F(v \cdot dt)$

Also Force = F = mass × acceleration = $\dfrac{W}{g} \cdot \dfrac{dv}{dt}$

where W = weight.

$$\therefore \ dw = \frac{W}{g} \cdot \frac{dv}{dt} \cdot v \cdot dt$$

or

$$\frac{dw}{dv} = \frac{Wv}{g}$$

Then, integrating, we get

$$w = \frac{Wv^2}{2g} = \text{work or kinetic energy, foot-pounds}$$

The limits of integration may be from one velocity to another as v_1 to v or from rest when $v = 0$ to v.

726. How do you obtain the expression for the relation of the tight and slack tensions of a pulley drive?

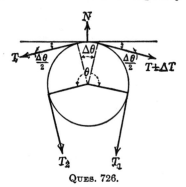

QUES. 726.

Assume an element of belt as a free body held in equilibrium by the applied forces.

T = tension on the slack end of the element of belt
considered

$T + \Delta T$ = tension on the tight end
N = normal force
μ = coefficient of friction

$$\mu N = (T + \Delta T) \cos \frac{\Delta \theta}{2} - T \cos \frac{\Delta \theta}{2} \qquad \text{or}$$

$$N = \frac{\Delta T}{\mu} \cos \frac{\Delta \theta}{2} \qquad (1)$$

The friction force = the difference of the horizontal components of the tensions.

Now equate the vertical forces.

$$N = (T + \Delta T) \sin \frac{\Delta\theta}{2} + T \sin \frac{\Delta\theta}{2}$$

$$= 2T \sin \frac{\Delta\theta}{2} + \Delta T \sin \frac{\Delta\theta}{2} \qquad (2)$$

Combine Eqs. (1) and (2). Then

$$\frac{\Delta T}{\mu} \cos \frac{\Delta\theta}{2} = 2T \sin \frac{\Delta\theta}{2} + \Delta T \sin \frac{\Delta\theta}{2}$$

or

$$\Delta T = \frac{\mu(2T + \Delta T) \sin \dfrac{\Delta\theta}{2}}{\cos \dfrac{\Delta\theta}{2}}$$

Divide both sides by $\dfrac{\Delta\theta}{2}$ to get

$$2 \frac{\Delta T}{\Delta\theta} = \frac{\mu(2T + \Delta T)}{\cos \dfrac{\Delta\theta}{2}} \cdot \frac{\sin \dfrac{\Delta\theta}{2}}{\dfrac{\Delta\theta}{2}}$$

As $\Delta\theta$ approaches zero, $\cos \dfrac{\Delta\theta}{2}$ approaches 1,

$\dfrac{\sin \dfrac{\Delta\theta}{2}}{\dfrac{\Delta\theta}{2}}$ approaches 1, and $\dfrac{\Delta T}{\Delta\theta}$ approaches $\dfrac{dT}{d\theta}$.

Now, taking the limit of both sides,

$$2 \frac{dT}{d\theta} = \mu \cdot 2T$$

and

$$\frac{dT}{d\theta} = \mu T$$

or

$$\frac{dT}{T} = \mu \, d\theta$$

Integrating between the limits T_1 and T_2 and 0 and θ,

$$\int_{T_2}^{T_1} \frac{dT}{T} = \mu \int_0^\theta d\theta$$

Then
$$\log_\epsilon T_1 - \log_\epsilon T_2 = \mu\theta$$

or
$$\log_\epsilon \frac{T_1}{T_2} = \mu\theta$$

$$\therefore \frac{T_1}{T_2} = \epsilon^{\mu\theta} = \frac{\text{tension of tight belt}}{\text{tension of slack belt}}$$

727. What is the relation of the belt tensions when two drives A and B are geared?

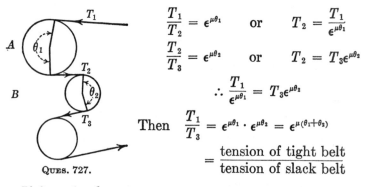

$$\frac{T_1}{T_2} = \epsilon^{\mu\theta_1} \quad \text{or} \quad T_2 = \frac{T_1}{\epsilon^{\mu\theta_1}}$$

$$\frac{T_2}{T_3} = \epsilon^{\mu\theta_2} \quad \text{or} \quad T_2 = T_3\epsilon^{\mu\theta_2}$$

$$\therefore \frac{T_1}{\epsilon^{\mu\theta_1}} = T_3\epsilon^{\mu\theta_2}$$

Then
$$\frac{T_1}{T_3} = \epsilon^{\mu\theta_1} \cdot \epsilon^{\mu\theta_2} = \epsilon^{\mu(\theta_1+\theta_2)}$$

$$= \frac{\text{tension of tight belt}}{\text{tension of slack belt}}$$

QUES. 727.

If $\theta_1 = \theta_2$, then

$$\frac{T_1}{T_3} = \epsilon^{\mu 2\theta}$$

728. What is the relation of belt tensions when three drive pulleys are used?

By analogy from the preceding question,

$$\frac{\text{Tension of tight belt}}{\text{Tension of slack belt}} = \epsilon^{\mu(\theta_1+\theta_2+\theta_3)}$$

729. What is the pressure of a gas and at what rate is it decreasing if, at a certain instant that the volume is increasing at 0.3 cubic foot per minute, the temperature is increasing at 0.4 degree per minute, the volume is

20 cubic feet, the temperature is 280 degrees absolute, and k = constant = 30 in the formula $Pv = k\phi$?

P = pressure, v = volume, and ϕ = temperature, absolute.   Now

$$P = \frac{k\phi}{v} = \frac{30 \cdot 280}{20} = 420 \text{ lb./sq. ft.} = \text{pressure of the gas}$$

Given $\dfrac{dv}{dt} = 0.3$ = rate of increase of volume

and $\dfrac{d\phi}{dt} = 0.4$ = rate of increase of temperature

Now, from $P = \dfrac{30\phi}{v}$,

$\dfrac{\partial P}{\partial \phi} = \dfrac{30}{v}$ = partial derivative (ϕ varies, while v = constant.)

$\dfrac{\partial P}{\partial v} = -\dfrac{30\phi}{v^2}$ = partial derivative (v varies, while ϕ = constant.)

Then $\dfrac{dP}{dt} = \dfrac{\partial P}{\partial \phi} \cdot \dfrac{d\phi}{dt} + \dfrac{\partial P}{\partial v} \cdot \dfrac{dv}{dt}$ = the total derivative

or $\dfrac{dP}{dt} = \dfrac{30}{20} \cdot 0.4 + \left(-\dfrac{30 \cdot 280}{20^2} \cdot 0.3\right)$

Finally, $\dfrac{dP}{dt} = 0.6 - 6.3 = -5.7$ lb./sq. ft. = rate of decrease of pressure

730. **What is the work done by air in expanding against a movable piston in a closed cylinder from a volume of 6 cubic feet to a volume of 10 cubic feet if the expansion is according to the law $pV^{1.41} = k$ and the pressure is 60 pounds per square inch when the volume is 6 cubic feet?**

Let A = area of piston.   Then

Force $F = pA$ = unit pressure × area

But $pV^{1.41} = k$

$\therefore p = \dfrac{k}{V^{1.41}}$ = unit pressure

$\therefore F = \dfrac{kA}{V^{1.41}}$

Now $dw = F \cdot ds$ (differential of work = force $\times$ differential of expansion)

or $\quad dw = \dfrac{kA}{V^{1.41}} \cdot ds$ $\hfill$ (1)

Now express s (= expansion) in terms of V (= volume).

$$V(= \text{volume of cylinder}) = \text{area } A \cdot s$$

$$\therefore s = \frac{V}{A}$$

and, differentiating this, we get

$$ds = \frac{dV}{A}$$

Substitute in Eq. (1), and get

$$dw = \frac{kA}{V^{1.41}} \cdot \frac{dV}{A} = \frac{k}{V^{1.41}} \cdot dV$$

which is to be integrated. Then

$$w = k \int_{6}^{10} V^{-1.41} \cdot dV = \left[-\frac{k}{0.41} \cdot V^{-0.41} \right]_{6}^{10}$$

or $\quad w = -\dfrac{k}{0.41}(10^{-0.41} - 6^{-0.41})$ $\hfill$ (2)

Now find k from $pV^{1.41} = k$, when given $p = 60$ lb./sq. in. = 8,640 lb./sq. ft.

Then $\hspace{4em} k = 8{,}640 \cdot 6^{1.41}$

Substitute this in Eq. (2) to get

$$w = -\frac{8{,}640 \cdot 6^{1.41}}{0.41}(10^{-0.41} - 6^{-0.41})$$

$$= -\frac{8{,}640 \cdot 6^{1.41}}{0.41}(0.3890 - 0.4797)$$

$$= \frac{8{,}640 \cdot 6^{1.41} \cdot 0.0907}{0.41} = 23{,}907 \text{ ft.-lb. work}$$

731. What is the velocity of a falling body at the end of 10 seconds?

A body falling from rest in a vacuum near the earth's surface follows the law $s = 16.1t^2$, where s = distance fallen in feet and t = time in seconds.

Now $\dfrac{ds}{dt}$ = rate of change of distance with respect to time

= velocity = v

Differentiating s, we get

$$\frac{ds}{dt} = 2 \cdot 16.1t = 32.2t = v$$

$$\therefore v = 32.2 \cdot 10 = 322 \text{ ft./sec. at end of 10 sec.}$$

732. To what height will a ball rise when thrown into the air if it follows the law $s = 60t - 16.1t^2$?

Let s = height

$$\alpha = \text{rate of change of velocity} = \frac{dv}{dt} = \frac{d\left(\dfrac{ds}{dt}\right)}{dt} = \frac{d^2s}{dt^2}$$

Now $\qquad v = \dfrac{ds}{dt} = 60 - 32.2t$ ft./sec.

and $\qquad \alpha = \dfrac{d^2s}{dt^2} = -32.2$ ft./sec.^{2}

Therefore, velocity decreases as the ball rises and the acceleration is negative. The ball rises until the velocity = 0.

$$\therefore \frac{ds}{dt} = 0 = 60 - 32.2t \qquad \text{or} \qquad t = \frac{60}{32.2} = 1.86 \text{ sec. for}$$

the ball to rise

Finally, $s = 60t - 16.1t^2 = 60 \cdot 1.86 - 16.1 \cdot \overline{1.86}^2$

$= 111.6 - 55.7 = 55.9$ ft. = height to

which the ball will rise

733. What are the angular velocity and acceleration at 5 seconds of a wheel rotating under constant moment according to the observed law $\theta = 2t^2$?

Given $\theta = 2t^2$, where θ = radians and t = time in seconds. Then, at 5 sec., the angular velocity

$$\omega = \frac{d\theta}{dt} = 4t = 4 \cdot 5 = 20 \quad \text{radians/sec.}$$

And, at 5 sec., the angular acceleration

$$\alpha = \frac{d\omega}{dt} = \frac{d\left(\frac{d\theta}{dt}\right)}{dt} = \frac{d^2\theta}{dt^2} = 4 \text{ radians/sec.}^2$$

734. What is the velocity at the end of **5** seconds of a bullet fired so that its components are $x = 800t$, $y = 800t - 16.1t^2$?

Given $\qquad\qquad x = 800t,$

Then $\dfrac{dx}{dt} = 800$ ft./sec. = the x component of the velocity

Given $\qquad\qquad y = 800t - 16.1t^2$

Then $\dfrac{dy}{dt} = 800 - 32.2t = 800 - 32.2 \cdot 5$

$\qquad\qquad = 639$ ft./sec. = the y component of the velocity

Then $\dfrac{ds}{dt} = \sqrt{800^2 + 639^2} = 1{,}024$ ft./sec. = the actual velocity

735. What is the relation of angular to linear velocities?

The linear velocity of a point on a body rotating about

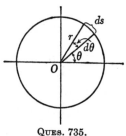

QUES. 735.

a fixed axis is the product of the distance of the point from the axis and the angular velocity of the body.

$$ds = r \, d\theta$$

Now divide both sides by dt. Then

Linear velocity $v = \dfrac{ds}{dt} = r \cdot \dfrac{d\theta}{dt} = r$

$\times$ angular velocity $= r\omega$

736. What is the relation of tangential to angular accelerations?

The tangential acceleration of a point on a body rotating about a fixed axis is the distance of the point from the axis times the angular acceleration about the axis.

From the previous question,

$$v = r\omega$$

Then $\quad \dfrac{dv}{dt} = r\dfrac{d\omega}{dt} = r \times$ angular acceleration

$$= \text{tangential acceleration}$$

737. What is the distance of a body from the surface of the earth 4 seconds after it starts to fall from a height of 300 feet?

Now 32.2t ft./sec. = the velocity of a falling body for any time t sec. after it starts to fall. Let H = the height in feet of the body above earth at any time. H decreases as t increases.

Then $\dfrac{dH}{dt} = -32.2t =$ the velocity of the body

or $\quad dH = -32.2t \cdot dt =$ the differential-equation form

Now sum up all the differentials of height for every differential of time elapsed to get the total height or distance above the surface of the earth.

Then $\quad\quad \int dH = -32.2\int t \cdot dt$

or $\quad\quad\quad H = -32.2 \cdot \dfrac{t^2}{2} = -16.1t^2 + C$

To determine C, we know $H = 300$ when $t = 0$. Substitute these values.

$$300 = -16.1 \cdot 0 + C \quad\text{or}\quad C = 300$$

Then at any time t sec. after the body starts to fall

$$H = -16.1t^2 + 300$$

Now, when $t = 4$,

$$H = -16.1 \cdot 16 + 300 = 42.4 \text{ ft.}$$

At the end of 5 sec. the body is 42.4 ft. above the earth's surface.

738. What distance s will a body fall in 6 seconds if it started downward with an initial velocity $v_o = 30$ feet

per second when the velocity v feet per second at any time t seconds after it starts to fall is $v = gt + v_o$?

Since
$$v = \frac{ds}{dt}$$

then
$$\frac{ds}{dt} = gt + v_o$$

or $ds = (gt + v_o)\, dt =$ the differential-equation form

Now sum up all the ds distances of the fall. Then

$$\int ds = \int (gt + v_o)\, dt = \int gt \cdot dt + v_o \int dt$$

or
$$s = g\frac{t^2}{2} + v_o t + C$$

Now $s = 0$ when $t = 0$.

$$\therefore C = 0$$

and
$$s = g\frac{t^2}{2} + v_o t$$

For $t = 6$ sec.,

$$s = 32.2 \cdot \tfrac{36}{2} + 30 \cdot 6 = 760^- \text{ ft.}$$

739. What is the total distance s a body will have moved at the end of t seconds if the acceleration is proportional to the time and the initial velocity $v_o = v$ and the initial distance $s_o = s$ when $t = 0$?

$$\frac{dv}{dt} = \frac{d^2s}{dt^2} = \text{acceleration}$$

and $\dfrac{dv}{dt} = kt =$ acceleration proportional to time (given)

or $dv = kt \cdot dt =$ the differential-equation form

Now sum up all dv's.

$$\int dv = \int kt \cdot dt = k \int t \cdot dt$$

or
$$v = \frac{kt^2}{2} + C_1$$

Now $v = v_o$ when $t = 0$.

$$\therefore v_o = C_1$$

and
$$v = \frac{kt^2}{2} + v_o$$

But
$$\frac{ds}{dt} = v$$

$$\therefore \frac{ds}{dt} = \frac{kt^2}{2} + v_o$$

and
$$ds = \frac{kt^2}{2} \cdot dt + v_o \cdot dt$$

Now sum up all ds's.

$$\int ds = \frac{k}{2} \int t^2 \cdot dt + v_o \int dt$$

or
$$s = \frac{kt^3}{6} + v_o t + C_2$$

Now $s = s_o$ when $t = 0$.

$$\therefore C_2 = s_o$$

Finally,
$$s = \frac{kt^3}{6} + v_o t + s_o$$

which is the distance the body will have moved in t sec.

740. How long will it take a flywheel to make one revolution if its angular velocity in radians per second is $\omega = 0.03t^2$?

Let θ = the number of radians the wheel turns through.

Then $\dfrac{d\theta}{dt} = \omega = 0.03t^2$ (given)

or $d\theta = 0.03t^2 \cdot dt$ = the differential-equation form

And $\int d\theta = 0.03 \int t^2 \cdot dt$

or $\theta = \dfrac{0.03t^3}{3} + C$

Now $\theta = 0$ when $t = 0$.

$$\therefore C = 0$$

Then $\theta = 0.01t^3$ or $t^3 = \dfrac{\theta}{0.01}$

For one revolution, $\theta = 2\pi$. Then

$$t^3 = \frac{2 \cdot 3.1416}{0.01} = \frac{6.2832}{0.01} = 628.32$$

$$\therefore t = 8.57 \text{ sec. for one revolution}$$

741. At what rate is a man approaching the top of a pole 75 feet high if he is walking at the rate of 3 miles per hour toward the foot when he is 100 feet from the foot?

$$z^2 = x^2 + y^2$$

Differentiate with respect to time.

$$2z \frac{dz}{dt} = 2x \frac{dx}{dt} + 2y \frac{dy}{dt}$$

or.
$$z \frac{dz}{dt} = x \frac{dx}{dt} + y \frac{dy}{dt} \qquad (1)$$

Given
$$x = 100 \text{ ft.}$$

$$\frac{dx}{dt} = -3 \text{ m.p.h.} = -3 \cdot 5{,}280 \text{ ft./hr.}$$

$$y = 75 = \text{constant}$$

Then
$$\frac{dy}{dt} = 0$$

Now $z = \sqrt{x^2 + y^2} = \sqrt{100^2 + 75^2} = \sqrt{15{,}625} = 125$ ft.

Substitute in Eq. (1).

$$125 \frac{dz}{dt} = 100(-3 \cdot 5{,}280)$$

or
$$\frac{dz}{dt} = -\frac{100}{125} \cdot 3 \cdot 5{,}280 = -2.4 \text{ m.p.h.}$$

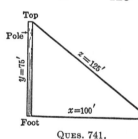

Ques. 741.

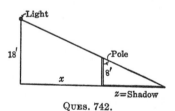

Ques. 742.

742. How fast is the shadow of an 8-foot pole lengthening if it is moved away at the rate of 100 feet per minute from a source of light that is suspended 18 feet from the ground?

Let x = horizontal distance of the pole from the light

z = length of shadow of pole

Then
$$\frac{8}{18} = \frac{z}{x + z} \text{ from similar triangles}$$

$$8x + 8z = 18z$$
$$8x = 10z$$
or
$$z = 0.8x$$

Now differentiate with respect to time.

$$\frac{dz}{dt} = 0.8 \frac{dx}{dt}$$

The rate of change of the shadow is 0.8 the rate of change of the pole position.   But

$$\frac{dx}{dt} = 100 \text{ ft./min.}$$

$$\therefore \frac{dz}{dt} = 0.8 \cdot 100 = 80 \text{ ft./min.}$$

The shadow is lengthening at the rate of 80 ft./min.

743. At what rate are trains A and B separating from each other when each is 12 miles from the station at the same instant and A is approaching the station at 40 miles per hour and B is going away from the station at 50 miles per hour on tracks that make an angle of 45 degrees?

Let x = distance of A from station at any instant

y = distance of B from station at the same instant

s = distance apart of trains

Given $\quad \dfrac{dx}{dt} = -40 \quad$ and $\quad \dfrac{dy}{dt} = 50$

By the cosine law of trigonometry,

$$s = \sqrt{x^2 + y^2 - 2xy \cos 45°} = \sqrt{x^2 + y^2 - \sqrt{2}\, xy}$$

Then $\dfrac{ds}{dt} = \dfrac{1}{2} \cdot \dfrac{1}{\sqrt{x^2 + y^2 - \sqrt{2}\,xy}} \cdot \dfrac{d}{dt}(x^2 + y^2 - \sqrt{2}\,xy)$

or $\quad \dfrac{ds}{dt} = \dfrac{2x\dfrac{dx}{dt} + 2y\dfrac{dy}{dt} - \sqrt{2}\,x\dfrac{dy}{dt} - \sqrt{2}\,y\dfrac{dx}{dt}}{2\sqrt{x^2 + y^2 - \sqrt{2}\,xy}}$

Now, when $x = y = 12$,

$\dfrac{ds}{dt}$

$= \dfrac{2 \cdot 12 \cdot (-40) + 2 \cdot 12 \cdot 50 - \sqrt{2} \cdot 12 \cdot 50 - \sqrt{2} \cdot 12 \cdot (-40)}{2\sqrt{144 + 144 - \sqrt{2} \cdot 144}}$

$= \dfrac{70.3}{2 \cdot 9.19} = 3.8 \text{ m.p.h.} = $ rate at which the trains are sepa-

rating when each is 12 miles from the station

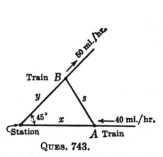

Ques. 743.

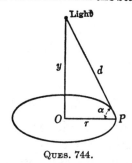

Ques. 744.

744. At what height should a light be placed over the center of a circle to provide a maximum illumination to the circumference if the intensity varies as the sine of the angle at which the rays strike the illuminated surface, divided by the square of the distance from the light?

Given $I = k\dfrac{\sin \alpha}{d^2} = $ illumination at P, any point on the

circumference

Now $\qquad\qquad d = \sqrt{y^2 + r^2}$

and $\qquad\qquad \sin \alpha = \dfrac{y}{\sqrt{y^2 + r^2}}$

Then $I = k \cdot \dfrac{y}{\sqrt{y^2 + r^2}} \cdot \dfrac{1}{(y^2 + r^2)} = k \cdot \dfrac{y}{(y^2 + r^2)^{\frac{3}{2}}}$

and $\dfrac{dI}{dy} = k\left[\dfrac{(y^2 + r^2)^{\frac{3}{2}} \cdot 1 - y \cdot \frac{3}{2}(y^2 + r^2)^{\frac{1}{2}} \cdot 2y}{(y^2 + r^2)^3}\right]$

$\qquad = k\left[\dfrac{(y^2 + r^2)^{\frac{3}{2}} - 3y^2(y^2 + r^2)^{\frac{1}{2}}}{(y^2 + r^2)^3}\right]$

$\qquad\qquad = \dfrac{k(y^2 + r^2)^{\frac{1}{2}}(y^2 + r^2 - 3y^2)}{(y^2 + r^2)^3}$

$\qquad = \dfrac{k(r^2 - 2y^2)}{(y^2 + r^2)^{\frac{5}{2}}}$

Now $\dfrac{dI}{dy} = 0$ when the numerator $= 0$. Then

$$r^2 - 2y^2 = 0 \qquad \text{or} \qquad y = \frac{r}{2}\sqrt{2} \quad \text{for a maximum}$$

and this is indicated by $\dfrac{d^2I}{dy^2}$, where both terms of the resulting numerator would be minus.

Now, as $y = \dfrac{r}{2}\sqrt{2}$ is positive, this would give a minus result for $\dfrac{d^2I}{dy^2}$.

745. How do you arrive at the relation between the pressure developed by a centrifugal fan and the fan speed?

Let Δx = thickness of layer of air between two vanes
$\qquad A$ = area of the layer
$\qquad x$ = its distance from the axis
$\qquad r$ = radius of the tip of the blade
$\qquad D$ = density of air

Now $\quad A \cdot \Delta x$ = volume of an elementary layer of air
and $\quad A \cdot \Delta x \cdot D$ = mass of an elementary layer of air
Assume that the fan outlet is closed and the wheel revolves at ω radians/sec.

Now $\quad C$ = centrifugal force = $\dfrac{m\omega^2 r}{g}$, for a mass m at

$\qquad\qquad\qquad\qquad\qquad\qquad\qquad\qquad$ radius r

Then ΔC = an element of pressure = $\dfrac{A \cdot \Delta x \cdot D}{g} \cdot \omega^2 \cdot x$

and $\Delta p = \dfrac{\Delta C}{A}$ = unit pressure   (dividing by the area)

But $\Delta h = \dfrac{\Delta p}{D} = \dfrac{\Delta C}{A \cdot D}$ = equivalent head for pressure Δp

$\therefore \Delta h = \dfrac{A \cdot \Delta x \cdot D}{g} \cdot \omega^2 x \cdot \dfrac{1}{A \cdot D} = \dfrac{\omega^2 x \cdot \Delta x}{g}$

Now passing to the limit and integrating between 0 and r,

$$h = \int_0^r \frac{\omega^2 x \cdot dx}{g} = \frac{\omega^2}{g} \left[\frac{x^2}{2} \right]_0^r = \frac{\omega^2 r^2}{2g}$$

Now, if v = linear tip velocity, then

$$v = \omega r \qquad \text{or} \qquad \omega = \frac{v}{r}$$

$\therefore h = \dfrac{v^2}{r^2} \cdot \dfrac{r^2}{2g} = \dfrac{v^2}{2g}$ from the first source of pressure due to centrifugal force of air between vanes

Also, $h' = \dfrac{v^2}{2g}$ from the second source of pressure due to linear velocity of air as it leaves the tip of the blades

$\therefore h + h' = 2 \cdot \dfrac{v^2}{2g} = \dfrac{v^2}{g}$ = total pressure head

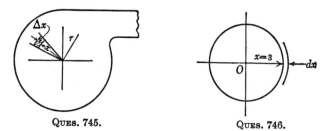

QUES. 745.   QUES. 746.

746. At what rate is the surface of a 6-inch-diameter metal plate increasing if it is heated so that the radius increases at the rate of 0.02 inch per second?

Let x = radius of the plate

Then y = area of plate = πx^2

and $\dfrac{dy}{dt} = 2\pi x \dfrac{dx}{dt}$

Now at any instant the area is increasing in square inches $2\pi x$ times as fast as the radius is increasing in linear inches.

But when $x = 3$ in., then $\dfrac{dx}{dt} = 0.02$ in./sec.

$\therefore \dfrac{dy}{dt} = 2\pi \cdot 3 \cdot 0.02 = 0.12\pi = 0.377$ sq. in./sec. = rate of increase of surface of the plate

747. How fast is the volume of a metal cube increasing per hour if the edge of the cube is expanding at the rate of 0.06 inch per hour when the edge is 20 inches?

Let x = edge of cube

Then $v = x^3$ = volume   and $\dfrac{dv}{dx} = 3x^2$

But x and v are also functions of time t.   Now

$\dfrac{dv}{dt} = \dfrac{dv}{dx} \cdot \dfrac{dx}{dt}$ and $\dfrac{dx}{dt} = 0.06$ in./hr. (given)

$\therefore \dfrac{dv}{dt} = 3x^2 \cdot 0.06 = 0.18x^2 = 0.18 \cdot \overline{20}^2 = \mathbf{72}$ cu. in./hr.

which is the rate of increase of the volume when the edge is 20 in.

748. What dimensions will make the surface of a steel cylindrical tank open at the top a minimum if it is to have a capacity of 10,000 cubic feet and the walls and bottom are to be of uniform thickness?

Let D = diameter

                 h = height

                 V = volume

                 A = area of surface

Then $A = \dfrac{\pi D^2}{4} + \pi Dh$ = area of bottom and walls   (1)

Given $V = \dfrac{\pi D^2}{4} \cdot h = 10{,}000$ cu. ft. = constant.

$$\therefore h = \frac{4 \cdot 10{,}000}{\pi D^2} \tag{2}$$

Substitute in Eq. (1).

$$A = \frac{\pi D^2}{4} + \frac{\pi D \cdot 4 \cdot 10{,}000}{\pi D^2} = \frac{\pi D^2}{4} + \frac{40{,}000}{D}$$

Differentiating,

$$\frac{dA}{dD} = \frac{\pi D}{2} - \frac{40{,}000}{D^2} = \text{rate of change of surface with respect}$$

to diameter

For a minimum,

$$\frac{dA}{dD} = \frac{\pi D}{2} - \frac{40{,}000}{D^2} = 0$$

or

$$\frac{\pi D}{2} = \frac{40{,}000}{D^2}$$

and

$$\pi D^3 = 80{,}000$$

$$\therefore D^3 = 25{,}464.7 \quad \text{or} \quad D = 29.42 \text{ ft.} = \text{diameter}$$

Substitute this in Eq. (2). Then

$$h = \frac{4 \cdot 10{,}000}{\pi D^2} = \frac{40{,}000}{3.1416 \cdot \overline{29.42}^2}$$

$$\therefore h = \frac{40{,}000}{2{,}719.2} = 14.71 \text{ ft.}$$

For a minimum, this shows that the diameter should be twice the height.

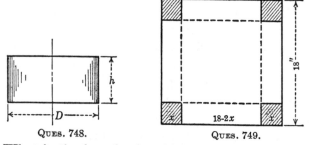

QUES. 748.   QUES. 749.

749. What is the length of a side of one of the four small corner squares that must be cut out to form an open-

top box of maximum volume from a square piece of
metal whose side is 18 inches?

Let x = side of small square

           = depth of box

Then $18 - 2x$ = side of the square to form the bottom of
the box

and the volume

$$V = (18 - 2x)^2 \cdot x$$

or $V = (324 - 72x + 4x^2)x = 324x - 72x^2 + 4x^3$

Then $\dfrac{dV}{dx} = 324 - 144x + 12x^2$

For maximum or minimum,

$$324 - 144x + 12x^2 = 0$$

or $12(x^2 - 12x + 27) = 0$

$$(x - 3)(x - 9) = 0$$

$$\therefore x = 3 \quad \text{and} \quad x = 9$$

Now $\dfrac{d^2V}{dx^2} = -144 + 24x = 0$

For $x = 3$, $\dfrac{d^2V}{dx^2} = -144 + 24 \cdot 3 = (-)$ minus, which indicates clearly a *maximum*.

For $x = 9$, $\dfrac{d^2V}{dx^2} = -144 + 24 \cdot 9 = (+)$ plus, which indicates clearly a *minimum*.

For $x = 3$, $V = (18 - 2x)^2 \cdot x = (18 - 2 \cdot 3)^2 \cdot 3 = 432$
cu. in., which is the maximum value.

750. Show that, when a square sheet of metal is heated,
     the relative rate of increase of its area equals twice
     the coefficient of expansion of the material.

Relative rate of increase = rate of increase of a function
per unit value of the function.

Coefficient of expansion = increase in length per unit
length for 1° rise in temperature =, therefore, the relative
rate of increase = $\dfrac{1}{x} \cdot \dfrac{dx}{dT}$, when T = temperature.

Now let x = one side of the square

Then $\qquad\qquad y = x^2$ = area

and $\qquad\qquad \log_e y = 2 \log_e x$

and, differentiating,

$$\frac{1}{y} \cdot \frac{dy}{dT} = 2\frac{1}{x} \cdot \frac{dx}{dT}$$

Therefore, relative rate of increase of area = two times the coefficient of expansion of the material.

751. What is the approximate volume of a spherical shell having an outside diameter of 6 inches and $\frac{1}{64}$ inch thick, using differentials?

Now $V = \frac{1}{6}\pi x^3$ = volume of a sphere of diameter x

Then $dV = \dfrac{\pi}{6} \cdot 3x^2 \cdot dx = \dfrac{\pi x^2}{2} \cdot dx$ = a differential of volume

Now $x = 6$ in. and $dx = \frac{1}{32}$ in.

As $\frac{1}{32}$ in. can be considered small as compared with 6 in., the differential will approximate the increment.

$$\therefore\ dV = \frac{3.1416}{2} \cdot 36 \cdot \frac{1}{32} = \frac{28.2744}{16} = 1.7671 \text{ cu. in.}$$

$\qquad\qquad$ = approximate volume of the shell

The exact volume $= \dfrac{\pi}{6}\left(\overline{6}^3 - \overline{5\tfrac{31}{32}}^{\,3}\right)$

$$= 113.10 - 111.34 = 1.76 \text{ cu. in.}$$

752. What is the approximate value of the tan of 46°, using differentials when given tan 45° = 1, sec 45° = $\sqrt{2}$, and 1° = 0.01745 radian?

Let $y = \tan x = \tan 45° = 1$ $\qquad$ and $\qquad dx = 0.0175$

Then $\quad dy = \sec^2 x \cdot dx = (\sqrt{2})^2 \cdot 0.0175 = 0.0350$

Now, when x changes to $x + dx$ or when 45° becomes 46°, y becomes

$y + dy = 1 + 0.0350 = \mathbf{1.0350} = \tan 46°$ **approximately**

The tables give tan 46° as 1.0355.

753. What is the approximate maximum error in the area of a circle when the diameter is measured to be 6.3 inches with a maximum error of 0.04 inch?

Let $\qquad A = \text{area} = \dfrac{\pi}{4}\, x^2$ (where x = diameter)

and finding the differential of this,

then the error $= dA = \dfrac{\pi}{4} \cdot 2x \cdot dx = \dfrac{\pi}{2} \cdot x \cdot dx$

Or the error in the area $= \dfrac{3.1416}{2} \cdot 6.3 \cdot 0.04 = 0.3958$ sq. in.

754. What are the relative error and the percentage of error of the above?

If dA is the error in A, then

$$\text{Ratio } \frac{dA}{A} = \text{the relative error}$$

and $\qquad 100 \cdot \dfrac{dA}{A} = \text{percentage of error}$

Now $\qquad A = \dfrac{\pi x^2}{4}$ (from above) $= \text{area of circle}$

and $\qquad \log_e A = \log_e \dfrac{\pi}{4} + 2 \log_e x$

and, differentiating, we get

$$\frac{1}{A} \cdot \frac{dA}{dx} = \frac{2}{x}$$

$\therefore \dfrac{dA}{A} = \dfrac{2 \cdot dx}{x} = \dfrac{2 \cdot 0.04}{6.3} = 0.0127 = \text{the relative error}$

and $100 \cdot 0.0127 = 1.27$ per cent $= \text{percentage of error}$

755. What is the approximate volume of a cylindrical thin copper cup of inside diameter 4 inches, height 6 inches, and thickness $\frac{1}{16}$ inch?

Let x = diameter, y = height, and V = volume. Then

$$V = \frac{\pi x^2}{4} \cdot y = \text{volume of cylinder}$$

Take the sum of the partial derivatives, or

$$dV = \frac{\pi x}{2} \cdot y \cdot dx + \frac{\pi x^2}{4} \cdot dy$$

Substitute $x = 4$, $y = 6$, $dx = \frac{1}{8}$, $dy = \frac{1}{16}$. Then

$$dV = \frac{\pi \cdot 4}{2} \cdot 6 \cdot \frac{1}{8} + \frac{\pi \cdot 4^2}{4} \cdot \frac{1}{16} = \frac{3\pi}{2} + \frac{\pi}{4} = \frac{7\pi}{4} = 5.4978$$

cu. in.

∴ Approximate volume = 5.50 cu. in.

The exact volume $= \dfrac{\pi}{4} \cdot \overline{4.125}^2 \cdot 6.0625 - \dfrac{\pi}{4} \cdot \overline{4}^2 \cdot 6$

$$= \frac{\pi}{4} \,(103.16 - 96.00) = 0.7854 \cdot 7.16$$

$$= 5.62 \text{ cu. in.}$$

756. **What are the approximate error and percentage error in calculating the third side of an oblique plane triangle two sides of which were measured to be 50 feet and 80 feet with the included angle 60 degrees, the measurements being subject to a maximum error of 0.05 foot in each length and 30 seconds in the angle?**

Let $z =$ the calculated side and x and y the other sides
   $\alpha =$ the included angle
Then, by the cosine law

$$z^2 = x^2 + y^2 - 2xy \cos \alpha \qquad \text{or}$$

$$z = \sqrt{x^2 + y^2 - 2xy \cos \alpha}$$

By partial differentiation,

$$\frac{\partial z}{\partial x} = \frac{2x - 2y \cos \alpha}{2(x^2 + y^2 - 2xy \cos \alpha)^{\frac{1}{2}}} = \frac{x - y \cos \alpha}{z} \;\; (y \text{ and } \alpha \text{ are}$$

considered constant.)

$$\frac{\partial z}{\partial y} = \frac{y - x \cos \alpha}{z} \;(\text{similarly } x \text{ and } \alpha \text{ are considered constant.})$$

$$\frac{\partial z}{\partial \alpha} = \frac{xy \sin \alpha}{z} \;\; (x \text{ and } y \text{ are constant.})$$

Then $dz = \dfrac{\partial z}{\partial x} \cdot dx + \dfrac{\partial z}{\partial y} \cdot dy + \dfrac{\partial z}{\partial \alpha} \cdot d\alpha$

$\quad = \dfrac{(x - y \cos \alpha)\ dx + (y - x \cos \alpha)\ dy + (xy \sin \alpha)\ d\alpha}{z}$

Substituting values from above

$$\cos 60° = \tfrac{1}{2}$$
$$\sin 60° = 0.866$$

$$dz = \dfrac{\begin{array}{c}(50 - 80 \cos 60°) \cdot 0.05 + (80 - 50 \cos 60°) \cdot 0.05 \\ + (50 \cdot 80 \sin 60°)0.000145\end{array}}{(\overline{50}^2 + \overline{80}^2 - 2 \cdot 50 \cdot 80 \cos 60°)^{\frac{1}{2}}}$$

$$= \dfrac{\begin{array}{c}(50 - 80 \cdot \tfrac{1}{2})0.05 + (80 - 50 \cdot \tfrac{1}{2})0.05 \\ + (50 \cdot 80 \cdot 0.866)0.000145\end{array}}{(2{,}500 + 6{,}400 - 2 \cdot 50 \cdot 80 \cdot \tfrac{1}{2})^{\frac{1}{2}}}$$

$$= \dfrac{0.5 + 2.75 + 0.502}{(8{,}900 - 4{,}000)^{\frac{1}{2}}} = \dfrac{3.752}{70.0} = 0.0536 \text{ ft. error}$$

$$100 \cdot \dfrac{0.0536}{70.0} = 0.077 \text{ per cent} = \text{percentage error}$$

APPENDIX A

ANSWERS TO PROBLEMS OF SECTIONS I AND II

Page 6

1. $C = 2\sqrt{\pi A}$ **2.** $A = \dfrac{d^2}{2}$ **3.** $S = \sqrt[3]{36\pi V^2}$

4. $f = \dfrac{M}{s}$ **5.** $h = \dfrac{p}{w}$ **6.** $\text{Hp.} = \dfrac{EI}{746}$

7. (a) 0; (b) $\frac{1}{2}$; (c) 1; (d) $\dfrac{\sqrt{3}}{2}$; (e) $\dfrac{\sqrt{2}}{2}$

8. (a) $-5x^3y - 3x^2y - 6y^2$; (b) $-5x^3y + 3x^2y - 6y^2$;
  (c) $5x^3y - 3x^2y - 6y^2$

9. (a) 2^y; (b) 2^{x+y} **10.** $x = \cos^{-1} y$

11. $y = \log_4 x$ **12.** $x = \sqrt{8 - y^2}$; $y = \sqrt{8 - x^2}$

Page 26

1. (a) $8 \cdot \Delta x$; (b) 16; (c) 16; (d) No
2. (a) 3.72; (b) 0.3612; (c) 0.036012
3. (a) 112; (b) 97.6; (c) 96.16; (d) 96.016; (e) 96.0016; (f) 96
5. (a) $\Delta A = \frac{3}{4}$; (b) No   **7.** $\Delta A = 4\frac{1}{4}$ **9.** $\Delta y = 0.00004$
11. (a) Average rate of speed in miles per hour for s miles;
  (b) Average rate of speed in miles per hour for Δs miles;
  (c) Instantaneous rate of speed at any instant
13. (a) 30; (b) $6\sqrt{5}$; (c) $-6\sqrt{5}$
14. (a) $\dfrac{dy}{dx} = 12x^2$; (b) $\dfrac{dy}{dx} = 2x - \dfrac{1}{x^2}$; (c) $\dfrac{ds}{dt} = 2 - 2t$

Pages 39, 40, 41

1. $\dfrac{dy}{dx} = 17x^{16}$ **2.** $\dfrac{dy}{dx} = -\dfrac{7}{5\sqrt[5]{x^{12}}}$

3. $\dfrac{dv}{du} = \dfrac{1}{4\sqrt[4]{u^3}}$ **4.** $\dfrac{dy}{dx} = 4cx^{c-1}$

5. $\dfrac{dy}{dx} = -\dfrac{7}{2\sqrt{x^9}}$ **6.** $\dfrac{dy}{dx} = \dfrac{2}{a}x^{\frac{2}{a}-1}$

7. $\dfrac{dy}{dx} = 3b \cdot x^{3b-1}$ **8.** $\dfrac{dv}{du} = 3.8u^{2.8}$

9. $\dfrac{dy}{dx} = \dfrac{7}{9} \cdot x^6$ **10.** $\dfrac{dy}{dy} = \dfrac{b}{2\sqrt{y}}$

11. $\dfrac{dy}{dx} = -\dfrac{4b + 2a - \sqrt{a-b}}{b - 3a - {}^4\!2\sqrt{a-b}}$ **13.** $\dfrac{dx}{dy} = \dfrac{9b}{2}\sqrt{ay} - \dfrac{6a\sqrt[4]{b}}{y^2}$

15. $\dfrac{dv}{dt} = 64t^3 + 55.2t^2 + 90.58t + 23$

17. $\dfrac{dy}{d\alpha} = \dfrac{4}{3}\sqrt[3]{\alpha} + \dfrac{2}{3\sqrt[3]{\alpha}} - \dfrac{2}{\sqrt[3]{\alpha^5}} - \dfrac{4}{\sqrt[3]{\alpha^7}}$ **19.** $\dfrac{dx}{dy} = \dfrac{y^2(9 - y^4)}{(y^4 + 3)^2}$

21. $\dfrac{d\theta}{dt} = \dfrac{\dfrac{2}{3} \cdot \dfrac{b}{\sqrt[3]{t}} - \dfrac{1}{3} \cdot b^2 \sqrt[3]{t^2} + b}{(1 - b\sqrt[3]{t^2})^2}$ **23.** $h = 27.64$ in., $r = 13.82$ in.

25. $\left(0 = 50 \cdot 1.41 \cdot 6^{.41} \cdot 0.2 + 6^{1.41} \cdot \dfrac{dp}{dt}\right); \dfrac{dp}{dt} = -2.35$ lb./sec.; the
pressure is decreasing at the rate of 2.35 lb./sec.

27. At 5 sec., $v = \dfrac{dx}{dt} = 4$ ft./sec., $\alpha = 0.8$ ft./sec.^{2}

At 80 ft., $v = \dfrac{dx}{dt} = 10.62$ ft./sec., $\alpha = 0.8$ ft./sec.^{2}

29. Rising at the rate of 7.11 m.p.h.
31. Increasing at 9,218.7 ft./min.

33. $\dfrac{dy}{dx} = -\dfrac{b^2x}{a^2y}$ **35.** $\dfrac{dy}{dx} = -\dfrac{(x+y)^{\frac{1}{3}} + (x-y)^{\frac{1}{3}}}{(x+y)^{\frac{1}{3}} - (x-y)^{\frac{1}{3}}}$ obtained

from $\dfrac{4}{3}(x+y)^{\frac{1}{3}} \cdot \dfrac{d}{dx}(x+y) + \dfrac{4}{3}(x-y)^{\frac{1}{3}} \cdot \dfrac{d}{dx}(x-y)$

37. $\dfrac{dx}{dy} = \dfrac{6x^{\frac{1}{3}}}{3x^{\frac{2}{3}} + 2}$

39. $\dfrac{dy}{dx} = -\dfrac{2x^3 + 6yx^2}{2x^3 + 3y}$ **41.** $\dfrac{d^2y}{dx^2} = 6$

43. $f''(x) = \dfrac{2a^2}{(a+x)^3}$ **45.** $\phi''(x) = -\dfrac{4x}{y^5}$

Page 46

1. $\dfrac{dx}{dy} = \dfrac{y}{\sqrt{y^2 + a^2}}$ **3.** $\dfrac{dy}{dx} = \dfrac{4a^2 - x^2}{3x^3\sqrt{x^2 - 2a^2}}$ **5.** $\dfrac{dy}{dx} = \dfrac{2x}{\sqrt{2x^2 + 4a^2}}$

7. $\dfrac{dy}{dx} = \dfrac{3x^2(x^2 + a) - 4x(x^3 - a)}{4(x^2 + a)^{\frac{3}{2}}(x^3 - a)^{\frac{1}{2}}}$ **9.** $\dfrac{dy}{dx} = 10x(x^2 + 4)^4$

11. $\dfrac{dy}{dx} = \dfrac{x(3a^2 - x^2)}{(a^2 - x^2)^{\frac{1}{2}}}$ **13.** $\dfrac{dw}{du} = -\dfrac{8\sqrt{3x}(x+1)}{25(x^2 + 2x)^3}$

Page 50

1. (a) $y = \sqrt{x - a^2}$; **(b)** $y = \log_a x$; **(c)** $y = \arcsin x$

Page 55

1. $\dfrac{dy}{d\theta} = 2 \sin \theta \cdot \cos \theta$ **3.** $\dfrac{dy}{d\theta} = 2 \cos 2\theta$ **5.** $\dfrac{dy}{d\theta} = \dfrac{1}{1 + \theta^2}$

7. $\dfrac{dy}{d\theta} = \dfrac{1}{\theta\sqrt{\theta^2 - 1}}$ **9.** $\dfrac{dy}{d\theta} = 2(\theta + 1) \cos (\theta^2 + 2\theta - 3)$

11. $\dfrac{dy}{d\theta} = \dfrac{2 \sin \theta}{(1 + \cos \theta)^2}$ **13.** $\dfrac{dy}{d\theta} = a \cos \left(\theta - \dfrac{\pi}{2}\right)$

15. $\dfrac{dx}{dt} = -n \sin 2\pi nt$

Pages 66, 67

1. $\dfrac{dy}{dx} = 27$ **3.** $x = \pm 2.83$

5. $y = 24x - 48$ **7.** $x = 8.12, y = 264.74$

9. $\dfrac{dy}{dx}$ is never negative and the function is never decreasing. It is always increasing, except when $x = 0$.

11. (a) Positive in the second and fourth quadrants; (b) negative in the first and third quadrants; (c) zero when $x = 0$.

13. A maximum at $(0, 6)$; a minimum at $(\tfrac{4}{3}, 4.81)$.

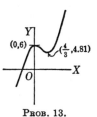

PROB. 13.

Pages 76, 77

1. $24x - y - 48 = 0$, tangent; $x + 24y - 1156 = 0$, normal

3. Tangents, $0.85x + y - 4.53 = 0$; $0.85x - y - 4.53 = 0$; normals, $1.18x - y - 1.55 = 0$, $1.18x + y - 1.55 = 0$

5. $1.73x + y + 6.19 = 0$

7. Tangent $= 7.74$, normal $= 6.32$, subtangent $= 6.0$, subnormal $= 4.0$

9. Tangent $= -2.82$, normal $= 2.82$, subtangent $= -2.0$, subnormal $= -2.0$

11. Horizontal tangents, $(3, 7), (3,1)$; vertical tangents $(-1, 4), (7, 4)$

Pages 87, 88

1. Maximum at $(0, 0)$; minimum at $(\tfrac{2}{3}, -1.24)$

3. Maximum at $-0.25, -0.195$

5. $\tfrac{2}{3}R$.   **7.** $r = h = 12.41$ ft.   **9.** $396

11. $I = 6.67$ amp.   **13.** Minimum at $(1.82, 8.46)$

15. 17.32 by 24.49 in.   **17.** 18.39 ft.

19. 6.25 tons of structural steel

Pages 97, 98

1. $dy = (5x^4 - 10x + 3)\,dx$ **3.** (a) $gt \cdot dt$; (b) 96

5. $dy = -0.075$, $ds = 0.125$ **7.** $dy = \dfrac{(4 - 2x^2)\,dx}{\sqrt{4 - x^2}}$

9. (a) $dA = 2\pi r \cdot dr$; (b) $\Delta A = \pi\,\Delta r(2r + \Delta r)$ **11.** 8.34

13. $\Delta y = 2.002$, $dy = 1.9$

Page 108

1. $\dfrac{7}{5(x - 2)} + \dfrac{3}{5(x + 3)}$ **3.** $\dfrac{3}{2(x - 1)} + \dfrac{1}{2(x - 3)}$

5. $\dfrac{9}{13(x + 3)} - \dfrac{23}{13(4x - 1)}$

7. $\dfrac{5}{16(4x - 3)} + \dfrac{3}{16(4x + 5)}$

9. $x + \dfrac{2}{3(x + 1)} + \dfrac{1 - 2x}{3(x^2 - x + 1)}$ **11.** $\dfrac{1}{x + 2} - \dfrac{1}{x + 1} + \dfrac{2}{(x + 2)^2}$

13. $\dfrac{1}{6(x - 2)} + \dfrac{1}{3(x - 2)^2} - \dfrac{x}{6(x^2 + 2x + 4)}$

Pages 113, 114

1. (a) \$1,050; (b) \$1,100; (c) \$1,150   **2.** 20 years

3. (a) \$1,050; (b) \$1,102.50; (c) 1,157.63

4. 14.21 years; 20.5 years   **5.** \$7,204.50

6. 5.745 as fast as at 0°C., or 0.843 total transformation

Page 120

1. $\dfrac{dy}{dx} = 2a(\epsilon^{2x} + \epsilon^{-2x})$ **2.** $\dfrac{dy}{dx} = \dfrac{8\epsilon^{-\frac{2x}{x-1}}}{(x - 1)^2}$

3. $\dfrac{dy}{dx} = -4a\epsilon^{-4ax}$ **4.** $\dfrac{dy}{dx} = \dfrac{15x^2}{9}\epsilon^{\frac{5x^3}{9}}$

5. $\dfrac{dy}{dx} = \epsilon^{\frac{6x}{x+a}} \cdot \dfrac{6a}{(x + a)^2}$ **6.** $\dfrac{dx}{dy} = \dfrac{\epsilon^{\sqrt{y^2+4}}}{\sqrt{1 + \dfrac{4}{y^2}}}$

7. $\dfrac{dx}{dy} = -\dfrac{12\epsilon^{-\frac{3y}{x-5}}(x - 5)}{(x - 5)^2 - 12y\epsilon^{-\frac{3y}{x-5}}}$

Pages 130, 131

1. $\dfrac{dy}{dx} = \dfrac{4x}{a + 2x^2}$ **3.** $\dfrac{dy}{dx} = \dfrac{(x + 1)(5x - 11)}{2(x - 3)^{\frac{1}{2}}}$

5. $\dfrac{dy}{dx} = -\dfrac{x(x + 1)}{\sqrt{x^2 + 1} \cdot \sqrt[3]{(x^3 - 1)^4}}$ **7.** $\dfrac{dx}{dy} = \dfrac{1}{4y\sqrt[4]{(\log_\epsilon y)^3}}$

9. $\dfrac{dy}{dx} = b^{ax}$

11. $\dfrac{dx}{dy} = \dfrac{10y^2 + 24y^{\frac{1}{2}} + 1}{y^{\frac{1}{2}}}$

13. $\dfrac{dy}{dx} = \dfrac{x+1}{x}$

15. $\dfrac{dy}{dx} = \dfrac{1 - \log_\epsilon (x+2)}{(x+2)^2}$

17. $\dfrac{1}{y} \cdot \dfrac{dy}{dx} = 3$

19. $\dfrac{dy}{dx} = -\dfrac{y(1-x)}{x(y+1)}$

21. Ratio $= -\cot^2 \cdot x$

23. Relative rate of increase $= 0.35$, percentage rate $= 35.0$

Page 139

1. $\dfrac{dx}{dy} = 2y^{2v}(1 + \log_\epsilon y)$

3. $\dfrac{dy}{dx} = 2x^{2x} \cdot \epsilon^{x^{2x}}(1 + \log_\epsilon x)$

5. Minimum for $y = \dfrac{1}{\epsilon}$

7. $t = 0.1,\ I = 4.46;\ t = 0.2,\ I = 1.00;\ t = 0.3,\ I = 0.22;\ t = 0.4,\ I = 0.05$

9. $E = 16.06$ volts

Page 150

1. $dz = \left(\dfrac{x}{3} \cdot y^{\frac{x}{3}-1}\right) dy + \left(\dfrac{y^{\frac{x}{3}}}{3} \log_\epsilon y\right) dx$

3. $dy = x(\sin \theta)^{x-1} \cdot \cos \theta \cdot d\theta + (\sin \theta)^x \cdot \log_\epsilon \sin \theta \cdot dx$

5. $\dfrac{6x^2}{5} + 9x^2y - 5y^2(x \text{ alone}); 3x^3 - 10xy + \dfrac{1}{6} (y \text{ alone})$

7. Maximum $z = -4.23$ for $x = \frac{5}{3}, y = \frac{3}{2}$

9. $\dfrac{\partial y}{\partial \theta} = 3 \cos (3\theta + 4\alpha); \dfrac{\partial y}{\partial \alpha} = 4 \cos (3\theta + 4\alpha);$

$d_\theta y = 3 \cos (3\theta + 4\alpha)\, d\theta; d_\alpha y = 4 \cos (3\theta + 4\alpha)\, d\alpha;$
$dy = 3 \cos (3\theta + 4\alpha)\, d\theta + 4 \cos (3\theta + 4\alpha)\, d\alpha$

11. $\dfrac{\partial y}{\partial u} = \log_\epsilon v \cdot u^{\log_\epsilon v - 1}; \dfrac{\partial y}{\partial v} = \dfrac{u^{\log_\epsilon v} \cdot \log_\epsilon u}{v}$

13. $\dfrac{\partial z}{\partial x} = 9x^2y^3 + 4y^2 - 5y; \dfrac{\partial z}{\partial y} = 9y^2x^3 + 8yx - 5x$

15. $\dfrac{\partial^2 z}{\partial y\, \partial x} = \dfrac{\partial^2 z}{\partial x \cdot \partial y} = -\dfrac{xy}{(x^2 + y^2)^{\frac{3}{2}}}$

17. $\dfrac{\partial z}{\partial x} = \sec \theta; \dfrac{\partial z}{\partial \theta} = x \sec \theta \cdot \tan \theta$

Page 162

1. $R = 6.13, y_1 = -2.07, x_1 = 6.71$

3. $R = 2.64; x_1 = -1\frac{1}{2}; y_1 = \frac{5}{8}$

5. $R = 580.8; x_1 = -578.12; y_1 = 46.62$; point of inflection at $(-\frac{5}{8}, \frac{27}{16})$

7. $R = 1$; $x_1 = \frac{\pi}{2}$; $y_1 = 0$; point of inflection at $\theta = \pi$

9. $K = \dfrac{1}{a}$ **11.** $R = 18.52$

Page 167

1. $x + C$ **2.** $y + C$ **3.** $\dfrac{y^2}{2} + C$ **4.** $\dfrac{5x^2}{2} + C$ **5.** $4x^2 + C$ **6.** $\dfrac{5x^3}{3} + C$

7. $2x^3 + C$ **8.** $\frac{1}{4}x^4 + C$ **9.** $x^6 + C$ **10.** $(x+1)^2 + C$ **11.** $\dfrac{3t^2}{2} + C$

12. $(x+1)^3 + C$ **13.** $(x^2+1)^2 + C$ **14.** $(x^2-1)^3 + C$

15. $y = 3x^2 + C$ **16.** $y = -x^4 + C$ **17.** $-\dfrac{5x^2}{2} + C$

18. $y = (x^2-3)^3 + C$ **19.** $y = -\dfrac{x}{2} + \dfrac{7}{2}$ **20.** $y = \frac{3}{2}x^2 - 3$

21. 400 ft. above earth

Pages 175, 176

1. $-\dfrac{2}{\sqrt{x}}$ **3.** $-\dfrac{5}{3x^3}$ **5.** $2x^4 - 7x$

7. $-\dfrac{4.3}{x}$ **9.** $5 \log_\epsilon x - \dfrac{3x^2}{2}$ **11.** $2 \log_\epsilon (x+3) + 4cx$

13. $-2\epsilon^{-x}$ **15.** $-8\epsilon^{-\frac{x}{2}}$ **17.** $-\dfrac{1}{2} \cos x - \dfrac{3x^2}{2}$

19. $-\cos 2\pi nt$ **21.** $\sqrt{2} \tan x$ **23.** $3 \arcsin x$

25. $2 \tan 4x$ **27.** $\tan (4x)$ **29.** $3 \sin (x+4)$

31. $3 \sin x \ \cos x$ **33.** $\dfrac{3a^x}{\log_\epsilon a}$ **35.** $\sin 2x + 2x$

37. 500 ft. **39.** 31.55 sec.

Pages 184, 185

1. $6\sqrt[3]{x^2} + \dfrac{3}{5x} - \dfrac{16a\sqrt{x^3}}{3}$ **3.** $\frac{1}{9}\sqrt[4]{(3x^2-7)^3}$

5. $\dfrac{3a}{2\sqrt{5} \cdot b} \cdot \tan^{-1} \dfrac{\sqrt{5}}{2} \ bx$ **7.** $3x - \frac{3}{2}\log_\epsilon (4x - 3)$

9. $\dfrac{1}{\sqrt{19}} \cdot \tan^{-1} \dfrac{x+1}{\sqrt{19}}$ **11.** $\dfrac{1}{10}\log_\epsilon \dfrac{3x-4}{3x+6}$

13. $\dfrac{2}{\sqrt{75}} \log_\epsilon (\sqrt{75}\ x^2 + \sqrt{75x^4 + 12})$

15. $\dfrac{x}{2}\sqrt{36x^2 - 100} - \frac{25}{3}\log_\epsilon (6x + \sqrt{36x^2 - 100})$

17. $\dfrac{1}{6}\sin^{-1} \dfrac{9x^2 - 4}{4}$

Page 186

1. $\dfrac{6}{5} \cdot \dfrac{c^{5x}}{\log_\epsilon c}$

2. $\dfrac{(cd)^{3x}}{3(\log_\epsilon c + \log_\epsilon d)}$

3. $\frac{1}{3}\epsilon^{3x} - \frac{1}{2}\epsilon^{2x} + \epsilon^x - \log_\epsilon (\epsilon^x + 1)$

4. $\epsilon^{\sin x}$

5. $\dfrac{1}{4} \dfrac{a^{4x}}{\log_\epsilon a}$

6. $\epsilon^{\tan x}$

7. $\dfrac{1}{b} \cdot \dfrac{a^{bx+c}}{\log_\epsilon a}$

8. $\frac{1}{4}\epsilon^{x^4}$

9. $-\frac{1}{2}\epsilon^{-x^2}$

10. $\frac{1}{3}\epsilon^{\sin 3x}$

Page 188

1. $-\dfrac{1}{5a} \cdot \cos 5ax$

3. $\frac{1}{3} \cdot \sin 3x$

5. $-\frac{1}{5}\cos (5x + 3)$

7. $-\frac{1}{4}\log_\epsilon \cos 4x$

9. $\frac{1}{3}\log_\epsilon (\sec 3x + \tan 3x)$

11. $-\cot x - x$

13. $\frac{1}{2}\log_\epsilon (3 + \sin^2 x)$

15. $\frac{1}{3}\sec 3x$

17. $-\frac{1}{3}\csc 3x$

19. $\frac{1}{5}\cos^5 x - \frac{1}{3}\cos^3 x$

Page 194

1. $\dfrac{(x^5 + 9x)^2}{2}$

3. $\frac{2}{15}(a^2 + x^3)^{\frac{5}{2}}$

5. $-\dfrac{x^{-2.62}}{2.62}$

7. $\dfrac{(3x^2 + 4)^4}{24}$

9. $-\dfrac{1}{2 \sin^2 x}$

11. $\dfrac{1}{2 \cos^2 x}$

13. $\log_\epsilon (7 + x^4)$

15. $\frac{1}{25}\log_\epsilon (5x^5 + 2)$

17. $-\frac{1}{4}\log_\epsilon \cos 4\theta$

19. $\dfrac{1}{4}\log_\epsilon \tan \dfrac{4\theta}{2}$

Page 202

1. $\frac{1}{24}\sqrt{(3 - 4x)^3} - \frac{3}{8}\sqrt{3 - 4x}$

3. $\frac{4}{7}x^{\frac{7}{2}} - \frac{2}{3}x^{\frac{9}{2}} + \frac{4}{5}x^{\frac{5}{2}} - x + \frac{4}{3}x^{\frac{3}{2}} - 2x^{\frac{1}{2}} + 4x^{\frac{1}{2}} - 4\log_\epsilon (x^{\frac{1}{2}} + 1)$

5. $\dfrac{1}{\sqrt{5}}\log_\epsilon \dfrac{\sqrt{3x + 5} - \sqrt{5}}{\sqrt{3x + 5} + \sqrt{5}}$

7. $2 \sin^{-1} \dfrac{x}{2} + \dfrac{x\sqrt{4 - x^2}}{2}$

9. $\dfrac{1}{2}\log_\epsilon \left(\dfrac{x + \sqrt{x^2 - 4}}{x - \sqrt{x^2 - 4}}\right) - \dfrac{\sqrt{x^2 - 4}}{x}$

11. $-\frac{3}{2}\sqrt[3]{x}(\sqrt[3]{x} + 2) - 3\log_\epsilon (\sqrt[3]{x} - 1)$

13. $\dfrac{2}{\sqrt{3}}\tan^{-1} \dfrac{\sqrt{\epsilon^x}}{\sqrt{3}}$

15. $\dfrac{3a^2x - 2x^3}{3a^4(a^2 - x^2)^{\frac{3}{2}}}$

17. $\dfrac{1}{18}\left(\dfrac{1}{3}\cos^{-1}\dfrac{3}{x} + \dfrac{\sqrt{x^2 - 9}}{x^2}\right)$

Page 210

1. $-\dfrac{\cos^3 \theta}{3} + \dfrac{2 \cos^5 \theta}{5} - \dfrac{\cos^7 \theta}{7}$

3. $\frac{1}{2}\tan^2 \theta + \log_\epsilon \cos \theta$

5. $\dfrac{\tan^5 2\theta}{10} + \dfrac{\tan^3 2\theta}{3} + \dfrac{\tan 2\theta}{2}$

7. $\dfrac{\tan^7 \theta}{7} + \dfrac{\tan^5 \theta}{5}$

9. $\dfrac{\sec^7 \theta}{7} - \dfrac{2 \sec^5 \theta}{5} + \dfrac{\sec^3 \theta}{3}$ **11.** $\dfrac{\cos 2\theta}{4} - \dfrac{\cos 8\theta}{16}$

13. $\dfrac{\sin 5\theta}{10} + \dfrac{\sin \theta}{2}$ **15.** $\dfrac{9\theta}{2} + 4 \cos \theta - \dfrac{\sin 2\theta}{4}$

Page 216

1. $\dfrac{\epsilon^{ax}}{a} \left(x^3 - \dfrac{3x^2}{a} + \dfrac{6x}{a^2} - \dfrac{6}{a^3} \right)$ **3.** $(x^2 - 2) \cdot \sin x + 2x \cdot \cos x$

5. $\frac{1}{2} \cdot \epsilon^x (\sin x + \cos x)$ **7.** $\dfrac{\epsilon^x}{17} (\sin 4x - 4 \cos 4x)$

9. $x \cos^{-1} x - \sqrt{1 - x^2}$ **11.** $\frac{1}{4}(\cos 2x + 2x \cdot \sin 2x - 2x^2 \cdot \cos 2x)$

13. $\dfrac{\epsilon^{ax}(3 \sin 3x + a \cos 3x)}{a^2 + 9}$ **15.** $\dfrac{\cos 3x}{9} + \dfrac{x \sin 3x}{3}$

17. $\dfrac{x}{1 + x} \log_\epsilon x - \log_\epsilon (1 + x)$ **19.** $\dfrac{\epsilon^x}{1 + x}$

Page 225

1. $4 \log_\epsilon (x + 1) + \dfrac{2}{x + 1}$ **3.** $\frac{3}{2} \log_\epsilon (x - 2) + \frac{5}{2} \log_\epsilon (x + 2)$

5. $-\dfrac{3}{4} \log_\epsilon x + \dfrac{3}{8} \log_\epsilon (x^2 + 2) + \dfrac{1}{\sqrt{2}} \tan^{-1} \dfrac{x}{\sqrt{2}} - \dfrac{5}{4(x^2 + 2)}$

$+ 2 \displaystyle\int \dfrac{dx}{(x^2 + 2)^2}$ (Look up a table of integrals for this.)

7. $\log_\epsilon (x - 3) + \dfrac{2}{(x - 3)}$

9. $\frac{1}{2} \log_\epsilon x + \log_\epsilon (x - 2) + \frac{1}{2} \log_\epsilon (x + 2)$

11. $-\dfrac{\dfrac{\sqrt{3}}{x} + x \tan^{-1} \dfrac{x}{\sqrt{3}}}{\sqrt{27}}$

Pages 241, 242

1. 86.0   **3.** 0.51   **5.** 0.866   **7.** 2.45
9. $\frac{2}{3}a^2\pi$ **11.** 0.1168   **13.** 1.7183   **15.** $18\frac{3}{4}$ square units
17. $\frac{16}{3}\pi^3$ **19.** 31.6   **21.** $0.866a^2$ **23.** a^2

Page 249

1. Mean $y = \frac{12}{5}$ **3.** $y = 3l^{4a} \sqrt{\dfrac{1}{8a + 1}}$, $y = 16$

5. $A = 93\frac{1}{3}$, mean $y = 18\frac{2}{3}$ **7.** Mean $y = 1.386$
9. Arithmetical mean = 4.33, quadratic mean = 4.92

Pages 257, 258

1. $8a$ **3.** $6a$ **5.** $21.26a$ **7.** πa **9.** 2.086
11. 27.17 units

APPENDIX A

Pages 269, 270, 271

1. $\dfrac{\pi}{4}$ **3.** 3π **5.** 20π **7.** $1\frac{1}{3}\pi$ **9.** $5\frac{1}{3}\pi$

11. $2\frac{1}{3}\pi$ cubic units **13.** 15π cubic units **15.** 120π cubic units
17. 3.6π cubic units **19.** $2,250$ cubic in. **21.** 9π cubic units
23. $5\pi^2 a^3$

Pages 279, 280

1. $y = \frac{1}{6}x^3 + C_1 x + C_2$ **3.** $y = \dfrac{x^6}{15} + \dfrac{x^4}{12} + C_1 x + C_2$

5. $y = \dfrac{1}{8}\epsilon^{2x} + \dfrac{C_1 x^2}{2} + C_2 x + C_3$ **7.** $u = 3x^2 - 5yx - 3x + \phi(y)$

9. $10\frac{2}{3}$ **11.** $\frac{1}{8}\epsilon^4 - \frac{3}{4}\epsilon^2 + \epsilon - \frac{3}{8} = 3.63$ **13.** $\frac{2}{3}$

Page 287

1. $A = 4 \displaystyle\int_0^3 \int_0^{\sqrt{9-x^2}} dy \cdot dx = 9\pi$ **3.** 14.07 square units

5. $\dfrac{3\pi a^2}{4}$ **7.** 6π **9.** 7.135 square units

Pages 291, 292

1. $1706\frac{2}{3}$ cubic units **3.** 4 cubic units **5.** $13\frac{1}{2}$ cubic units

Page 303

1. $\left(\dfrac{4r}{3\pi}, 0\right)$ **3.** $(1.5, 0)$ **5.** $\left(\dfrac{4a}{3\pi}, \dfrac{4b}{3\pi}\right)$

7. $(\bar{x} = \frac{3}{10}a^{\frac{1}{3}} \cdot b^{\frac{2}{3}}, \bar{y} = \frac{3}{10}a^{\frac{2}{3}} \cdot b^{\frac{1}{3}})$ **9.** $\dfrac{81a\sqrt{3}}{80\pi}$

11. $(1, 3\frac{2}{3})$ **13.** $\bar{x} = \dfrac{\pi^2 - 4\pi + 8}{8(\pi - 2)}$ **15.** $\bar{x} = \frac{2}{5}a$

Pages 311, 312

1. $\frac{1}{12}bh^3$ **3.** $\left(I_x = \dfrac{71A}{8}, I_y = 10A\right)$ **5.** $I_x = I_y = \dfrac{7A}{64}$

7. $\left(\dfrac{16A}{5} = I_x, \dfrac{48A}{7} = I_y\right)$ **9.** $\dfrac{A}{48}(3\pi + 8)a^2$

Pages 324, 325

1. $2x^2 + C$ **2.** $xy = C$ **3.** $\epsilon^{-2x} + C_1 x + C_2$ **4.** $y^2(x^2+2)^{\frac{3}{2}} = \epsilon^C$
5. $2x^5 y = 3xy + C$ **6.** $\log_\epsilon (3+x^2)(3+4y^2)^3 = \log_\epsilon Cx^2$

7. $\dfrac{3y}{4x} = 2x + C$ **8.** $y' = 4 \sin 2\theta \cdot \cos 2\theta$ **9.** $dy = y \cdot dx$

10. $x = t \cdot \dfrac{dx}{dt} + \left(\dfrac{dx}{dt}\right)^2$ **11.** $x(y')^2 = 2y(y') + x$

12. $x^2 \, dy + y^2 \, dx = 0$ **13.** $y = C\epsilon^{x^2}$

14. $x = \theta + \frac{1}{2} \cos 2\theta + C$ **15.** $t^2x + Cx + 2 = 0$

16. $x^2y = C(x^2 - xy - ay)$ **17.** $\cos \beta = C \cos \theta$

18. $x + y = C(1 - xy)$

APPENDIX B

USEFUL FORMULAS, NOTATIONS, AND TABLES

Greek-letter Symbols Most Commonly Used

α = alpha	θ = theta
β = beta	π = pi
γ = gamma	ρ = rho]
δ = delta = Δ	τ = tau
ϵ = epsilon	ϕ = phi
λ = lambda	ψ = psi
μ = mu	ω = omega

Solution of the Quadratic Equation $Ax^2 + Bx + C = 0$

$$x = \frac{-B \pm \sqrt{B^2 - 4AC}}{2A}$$

If $B^2 - 4AC$ = $\begin{cases} > \\ 0 \\ < \end{cases}$ $\begin{cases} \text{The roots are real and unequal.} \\ \text{The roots are real and equal.} \\ \text{The roots are imaginary.} \end{cases}$

Binomial Expansions

$$(a \pm b)^2 = a^2 \pm 2ab + b^2$$
$$(a \pm b)^3 = a^3 \pm 3a^2b + 3ab^2 \pm b^3$$
$$(a \pm b)^4 = a^4 \pm 4a^3b + 6a^2b^2 \pm 4ab^3 + b^4$$

$$(a \pm b)^n = a^n \pm \frac{n}{1} a^{n-1}b + \frac{n(n-1)}{1 \cdot 2} a^{n-2}b^2$$
$$\pm \frac{n(n-1)(n-2)}{1 \cdot 2 \cdot 3} a^{n-3}b^3$$
$$+ \frac{n(n-1)(n-2)(n-3)}{1 \cdot 2 \cdot 3 \cdot 4} a^{n-4}b^4 \pm \cdots *$$

* n may be positive or negative, integral or fractional. When n is a positive integer, the series has $(n + 1)$ terms; otherwise, the number of terms is infinite.

Circle. Circumference $= 2\pi r$. Area $= \pi r^2$.

Circular Sector.
Area $= \frac{1}{2}r^2\alpha$, where $\alpha = $ central angle of sector measured in radians.

Prism. Volume $= $ area of base $\times$ altitude.

Pyramid. Volume $= \frac{1}{3}$ area of base $\times$ altitude.

Right Circular Cylinder. Volume $= \pi r^2 \times$ altitude. Lateral surface $= 2\pi r \times$ altitude. Total surface $= 2\pi r$ $(r + $ altitude$)$.

Right Circular Cone. Volume $= \frac{1}{3}\pi r^2 \times$ altitude. Lateral surface $= \pi r \times$ slant height. Total surface $= \pi r$ $(r + $ slant height$)$.

Sphere. Volume $= \frac{4}{3}\pi r^3$. Surface $= 4\pi r^2$.

Frustum of a Right Circular Cone
Volume $= \frac{1}{3}\pi \times$ height $(R^2 + r^2 + Rr)$ $(r = $ small radius; $R = $ large radius$)$

Lateral surface $= \pi \times$ slant height $(R + r)$

Relation between Unit Angles

180 deg. $= \pi$ radians $(\pi = 3.14159 \cdots)$

$$1 \text{ deg.} = \frac{\pi}{180} = 0.0174 \cdots \text{ radian}$$

$$1 \text{ radian} = \frac{180}{\pi} = 57.29 \cdots \text{ deg.}$$

$\therefore$ Number of radians in an angle $= \dfrac{\text{subtending arc}}{\text{radius}}$

Operations with Zero and Infinity

$a \cdot 0 = 0$	$a \cdot \infty = \infty$
$\dfrac{0}{a} = 0$	$\dfrac{a}{0} = \infty$
$\dfrac{a}{\infty} = 0$	$\dfrac{\infty}{a} = \infty$
$a^{-\infty} = 0$ if $a^2 > 1$	$\infty^a = \infty$
$0^a = 0$	$a^{\infty} = \infty$ if $a^2 > 1$
$a^{\infty} = 0$ if $a^2 < 1$	$a^{-\infty} = \infty$ if $a^2 < 1$
	$\infty - a = \infty$

$$0 \cdot \infty = \text{indeterminate}$$

$$\frac{0}{0} = \text{indeterminate}$$

$$\frac{\infty}{\infty} = \text{indeterminate}$$

$$0^0 = \text{indeterminate}$$

$$\infty^0 = \text{indeterminate}$$

$$\infty - \infty = \text{indeterminate}$$

Factors

$$(a^2 - b^2) = (a + b)(a - b)$$

$$(a^3 + b^3) = (a + b)(a^2 - ab + b^2)$$

$$(a^3 - b^3) = (a - b)(a^2 + ab + b^2)$$

$$(a^4 + b^4) = (a^2 + ab\sqrt{2} + b^2)(a^2 - ab\sqrt{2} + b^2)$$

$$(a^{2n} - b^{2n}) = (a^n + b^n)(a^n - b^n)$$

$$(a^n - b^n) = (a - b)(a^{n-1} + a^{n-2}b + a^{n-3}b^2 + \cdots + b^{n-1})$$

$$(a^n - b^n) = (a + b)(a^{n-1} - a^{n-2}b + a^{n-3}b^2 - \cdots - b^{n-1}) \text{ if } n \text{ is even}$$

$$(a^n + b^n) = (a + b)(a^{n-1} - a^{n-2}b + a^{n-3}b^2 - \cdots + b^{n-1}) \text{ if } n \text{ is odd}$$

Powers and Roots

$$a^n = a \cdot a \cdot a \cdot a \rightarrow \text{to } n \text{ factors} \qquad \frac{a^m}{a^n} = a^{m-n}$$

$$a^{-n} = \frac{1}{a^n} \qquad\qquad (a^m)^n = a^{mn}$$

$$a^m \cdot a^n = a^{m+n} \qquad\qquad (ab)^n = a^n \cdot b^n$$

$$\left(\frac{a}{b}\right)^n = \frac{a^n}{b^n}$$

$$(\sqrt[n]{a})^n = \sqrt[n]{a^n} = a \qquad\qquad \frac{a^{-m}}{b^{-n}} = \frac{\dfrac{1}{a^m}}{\dfrac{1}{b^n}} = \frac{1}{a^m} \cdot \frac{b^n}{1} = \frac{b^n}{a^m}$$

$$a^{\frac{1}{n}} = \sqrt[n]{a} \qquad\qquad \sqrt[n]{\frac{1}{a^m}} = \sqrt[n]{a^{-m}} = a^{-\frac{m}{n}}$$

$$\frac{m}{a^n} = \sqrt[n]{a^m} = (\sqrt[n]{a})^m$$

$$\sqrt[n]{ab} = \sqrt[n]{a} \cdot \sqrt[n]{b}$$

$$\sqrt[n]{\frac{a}{b}} = \frac{\sqrt[n]{a}}{\sqrt[n]{b}}$$

$$\sqrt[n]{\sqrt[m]{a}} = \sqrt[mn]{a} = a^{\frac{1}{mn}}$$

$$a^0 = a^{n-n} = \frac{a^n}{a^n} = 1$$

$$\sqrt[v]{\sqrt[n]{\frac{1}{a^{ms}}}} = \sqrt[v]{\sqrt[n]{a^{-ms}}}$$

$$= \sqrt[v]{a^{-\frac{ms}{n}}} = a^{-\frac{ms}{nv}}$$

$$\sqrt[n]{a^m} = \sqrt[ns]{a^{ms}}$$

$$\sqrt[n]{a} \cdot \sqrt[n]{a} = \sqrt[n]{a \cdot a}$$

$$= \sqrt[n]{a^2}$$

$$(a + b)^n = a^n\left(1 + \frac{b}{a}\right)^n$$

$$a^{-n} \cdot a^n = a^{n-n} = a^0 = 1$$

Logarithms

$$\log (a \cdot b) = \log a + \log b \qquad \log \sqrt[x]{a} = \frac{1}{x} \log a$$

$$\log \frac{a}{b} = \log a - \log b \qquad \log_a a = 1$$

$$\log \frac{1}{a} = - \log a \qquad \log 1 = 0$$

$$\log a^x = x \log a$$

Relations among Functions of an Angle

$$\sin \theta = \frac{1}{\operatorname{cosec} \theta}, \qquad \cos \theta = \frac{1}{\sec \theta}, \qquad \tan \theta = \frac{1}{\cot \theta}$$

$$= \frac{\sin \theta}{\cos \theta}$$

$$\sin^2 \theta + \cos^2 \theta = 1, \qquad \sec^2 \theta - \tan^2 \theta = 1, \qquad \operatorname{cosec}^2 \theta$$

$$- \cot^2 \theta = 1$$

Functions of Multiple Angles

$$\sin 2\theta = 2 \sin \theta \cdot \cos \theta$$

$$\cos 2\theta = \cos^2 \theta - \sin^2 \theta = 2 \cos^2 \theta - 1 = 1 - 2 \sin^2 \theta$$

$$\tan 2\theta = \frac{2 \tan \theta}{1 - \tan^2 \theta}$$

Functions of Half Angles

$$\sin \frac{\theta}{2} = \pm \sqrt{\frac{1 - \cos \theta}{2}}$$

$$\cos \frac{\theta}{2} = \pm \sqrt{\frac{1 + \cos \theta}{2}}$$

$$\tan \frac{\theta}{2} = \pm \sqrt{\frac{1 - \cos \theta}{1 + \cos \theta}} = \frac{1 - \cos \theta}{\sin \theta} = \frac{\sin \theta}{1 + \cos \theta}$$

Powers of Functions

$$\sin^2 \theta = \tfrac{1}{2} - \tfrac{1}{2} \cos 2\theta$$
$$\cos^2 \theta = \tfrac{1}{2} + \tfrac{1}{2} \cos 2\theta$$

Functions of Sum or Difference of Two Angles

$$\sin (A \pm B) = \sin A \cos B \pm \cos A \sin B$$
$$\cos (A \pm B) = \cos A \cos B \mp \sin A \sin B$$

$$\tan (A \pm B) = \frac{\tan A \pm \tan B}{1 \mp \tan A \tan B}$$

Sums, Differences, and Products of Two Functions

$$\sin A + \sin B = 2 \sin \tfrac{1}{2}(A + B) \cos \tfrac{1}{2}(A - B)$$
$$\sin A - \sin B = 2 \cos \tfrac{1}{2}(A + B) \sin \tfrac{1}{2}(A - B)$$
$$\cos A + \cos B = 2 \cos \tfrac{1}{2}(A + B) \cos \tfrac{1}{2}(A - B)$$
$$\cos A - \cos B = -2 \sin \tfrac{1}{2}(A + B) \sin \tfrac{1}{2}(A - B)$$

$$\tan A \pm \tan B = \frac{\sin (A \pm B)}{\cos A \cos B}$$

$$\sin^2 A - \sin^2 B = \sin (A + B) \sin (A - B)$$
$$\cos^2 A - \cos^2 B = -\sin (A + B) \sin (A - B)$$
$$\cos^2 A - \sin^2 B = \cos (A + B) \cos (A - B)$$
$$\sin A \sin B = \tfrac{1}{2} \cos (A - B) - \tfrac{1}{2} \cos (A + B)$$
$$\cos A \cos B = \tfrac{1}{2} \cos (A - B) + \tfrac{1}{2} \cos (A + B)$$
$$\sin A \cos B = \tfrac{1}{2} \sin (A + B) + \tfrac{1}{2} \sin (A - B)$$

Law of Sines

$$\frac{a}{\sin A} = \frac{b}{\sin B} = \frac{c}{\sin C}$$

Law of Cosines

$$a^2 = b^2 + c^2 - 2bc \cos A$$

Area of Any Triangle

$K = \frac{1}{2}bc \sin A$ Given two sides and included angle.

$K = \dfrac{\frac{1}{2}a^2 \sin B \sin C}{\sin (B + C)}$ Given one side and two angles.

$K = \sqrt{s(s - a)(s - b)(s - c)}$ where $s = \frac{1}{2}(a + b + c)$ Given three sides.

VALUES OF ϵ^x FROM $x = 0$ TO $x = 4.9$

x	0.0	0.1	0.2	0.3	0.4	0.5	0.6	0.7	0.8	0.9
0	1.00	1.11	1.22	1.35	1.49	1.65	1.82	2.01	2.23	2.46
1	2.72	3.00	3.32	3.67	4.06	4.48	4.95	5.47	6.05	6.69
2	7.39	8.17	9.03	9.97	11.0	12.2	13.5	14.9	16.4	18.2
3	20.1	22.2	24.5	27.1	30.0	33.1	36.6	40.4	44.7	49.4
4	54.6	60.3	66.7	73.7	81.5	90.0	99.5	109.9	121.5	134.3

VALUES OF ϵ^{-x} FROM $x = 0$ TO $x = 4.9$

x	0.0	0.1	0.2	0.3	0.4	0.5	0.6	0.7	0.8	0.9
0	1.00	0.90	0.82	0.74	0.67	0.61	0.55	0.50	0.45	0.41
1	0.37	0.33	0.30	0.27	0.25	0.22	0.20	0.18	0.17	0.15
2	0.14	0.12	0.11	0.10	0.09	0.08	0.07	0.07	0.06	0.06
3	0.05	0.05	0.04	0.04	0.03	0.03	0.03	0.02	0.02	0.02
4	0.02	0.02	0.01	0.01	0.01	0.01	0.01	0.01	0.01	0.01

FREQUENTLY USED TRIGONOMETRIC FUNCTIONS

$\theta°$	θ radians	$\sin\theta$	$\cos\theta$	$\tan\theta$	$\cot\theta$	$\sec\theta$	cosec θ
0°	0	0	1	0	∞	1	∞
30°	$\frac{\pi}{6}$	$\frac{1}{2}$	$\frac{\sqrt{3}}{2}$	$\frac{\sqrt{3}}{3}$	$\sqrt{3}$	$\frac{2\sqrt{3}}{3}$	2
45°	$\frac{\pi}{4}$	$\frac{\sqrt{2}}{2}$	$\frac{\sqrt{2}}{2}$	1	1	$\sqrt{2}$	$\sqrt{2}$
60°	$\frac{\pi}{3}$	$\frac{\sqrt{3}}{2}$	$\frac{1}{2}$	$\sqrt{3}$	$\frac{\sqrt{3}}{3}$	2	$\frac{2\sqrt{3}}{3}$
90°	$\frac{\pi}{2}$	1	0	∞	0	∞	1
120°	$\frac{2\pi}{3}$	$\frac{\sqrt{3}}{2}$	$-\frac{1}{2}$	$-\sqrt{3}$	$-\frac{\sqrt{3}}{3}$	-2	$\frac{2\sqrt{3}}{3}$
135°	$\frac{3\pi}{4}$	$\frac{\sqrt{2}}{2}$	$-\frac{\sqrt{2}}{2}$	-1	-1	$-\sqrt{2}$	$\sqrt{2}$
150°	$\frac{5\pi}{6}$	$\frac{1}{2}$	$-\frac{\sqrt{3}}{2}$	$-\frac{\sqrt{3}}{3}$	$-\sqrt{3}$	$-\frac{2\sqrt{3}}{3}$	2
180°	π	0	-1	0	∞	-1	∞
210°	$\frac{7\pi}{6}$	$-\frac{1}{2}$	$-\frac{\sqrt{3}}{2}$	$\frac{\sqrt{3}}{3}$	$\sqrt{3}$	$-\frac{2\sqrt{3}}{3}$	-2
225°	$\frac{5\pi}{4}$	$-\frac{\sqrt{2}}{2}$	$-\frac{\sqrt{2}}{2}$	1	1	$-\sqrt{2}$	$-\sqrt{2}$
240°	$\frac{4\pi}{3}$	$-\frac{\sqrt{3}}{2}$	$-\frac{1}{2}$	$\sqrt{3}$	$\frac{\sqrt{3}}{3}$	-2	$-\frac{2\sqrt{3}}{3}$
270°	$\frac{3\pi}{2}$	-1	0	∞	0	∞	-1
300°	$\frac{5\pi}{3}$	$-\frac{\sqrt{3}}{2}$	$\frac{1}{2}$	$-\sqrt{3}$	$-\frac{\sqrt{3}}{3}$	2	$-\frac{2\sqrt{3}}{3}$
315°	$\frac{7\pi}{4}$	$-\frac{\sqrt{2}}{2}$	$\frac{\sqrt{2}}{2}$	-1	-1	$\sqrt{2}$	$-\sqrt{2}$
330°	$\frac{11\pi}{6}$	$-\frac{1}{2}$	$\frac{\sqrt{3}}{2}$	$-\frac{\sqrt{3}}{3}$	$-\sqrt{3}$	$\frac{2\sqrt{3}}{3}$	-2
360°	2π	0	1	0	∞	1	∞

ANY TRIGONOMETRIC FUNCTION IN TERMS OF EACH OF THE OTHERS

	$\sin\theta$	$\cos\theta$	$\tan\theta$	$\cot\theta$	$\sec\theta$	$\operatorname{cosec}\theta$
$\sin\theta$	$\sin\theta$	$\sqrt{1-\cos^2\theta}$	$\dfrac{\tan\theta}{\sqrt{1+\tan^2\theta}}$	$\dfrac{1}{\sqrt{1+\cot^2\theta}}$	$\dfrac{\sqrt{\sec^2\theta-1}}{\sec\theta}$	$\dfrac{1}{\operatorname{cosec}\theta}$
$\cos\theta$	$\sqrt{1-\sin^2\theta}$	$\cos\theta$	$\dfrac{1}{\sqrt{1+\tan^2\theta}}$	$\dfrac{\cot\theta}{\sqrt{1+\cot^2\theta}}$	$\dfrac{1}{\sec\theta}$	$\dfrac{\sqrt{\operatorname{cosec}^2\theta-1}}{\operatorname{cosec}\theta}$
$\tan\theta$	$\dfrac{\sin\theta}{\sqrt{1-\sin^2\theta}}$	$\dfrac{\sqrt{1-\cos^2\theta}}{\cos\theta}$	$\tan\theta$	$\dfrac{1}{\cot\theta}$	$\sqrt{\sec^2\theta-1}$	$\dfrac{1}{\sqrt{\operatorname{cosec}^2\theta-1}}$
$\cot\theta$	$\dfrac{\sqrt{1-\sin^2\theta}}{\sin\theta}$	$\dfrac{\cos\theta}{\sqrt{1-\cos^2\theta}}$	$\dfrac{1}{\tan\theta}$	$\cot\theta$	$\dfrac{1}{\sqrt{\sec^2\theta-1}}$	$\sqrt{\operatorname{cosec}^2\theta-1}$
$\sec\theta$	$\dfrac{1}{\sqrt{1-\sin^2\theta}}$	$\dfrac{1}{\cos\theta}$	$\sqrt{1+\tan^2\theta}$	$\dfrac{\sqrt{1+\cot^2\theta}}{\cot\theta}$	$\sec\theta$	$\dfrac{\sec\theta}{\sqrt{\sec^2\theta-1}}$
$\operatorname{cosec}\theta$	$\dfrac{1}{\sin\theta}$	$\dfrac{1}{\sqrt{1-\cos^2\theta}}$	$\dfrac{\sqrt{1+\tan^2\theta}}{\tan\theta}$	$\sqrt{1+\cot^2\theta}$	$\dfrac{\sec\theta}{\sqrt{\sec^2\theta-1}}$	$\operatorname{cosec}\theta$

Formulas Used in the Text

$$\frac{d(x^n)}{dx} = n \cdot x^{n-1}$$

$$\frac{d(c)}{dx} = 0$$

$$\frac{d(x)}{dx} = 1$$

$$\frac{d(cy)}{dx} = c\,\frac{dy}{dx}$$

$$\frac{d\left(\dfrac{y}{c}\right)}{dx} = \frac{1}{c} \cdot \frac{dy}{dx}$$

$$\frac{d(u + v + w + \cdots)}{dx} = \frac{du}{dx} + \frac{dv}{dx} + \frac{dw}{dx} + \cdots$$

$$\frac{d(uv)}{dx} = u\,\frac{dv}{dx} + v\,\frac{du}{dx}$$

$$\frac{d\left(\dfrac{u}{v}\right)}{dx} = \frac{v \cdot \dfrac{du}{dx} - u\,\dfrac{dv}{dx}}{v^2}$$

$$\frac{dy}{dx} = \frac{1}{\dfrac{dx}{dy}} \qquad \text{or} \qquad \frac{dy}{dx} \cdot \frac{dx}{dy} = 1$$

$$\frac{d(\sin \theta)}{d\theta} = \cos \theta$$

$$\frac{d(\cos \theta)}{d\theta} = -\sin \theta$$

$$\frac{d(\tan \theta)}{d\theta} = \sec^2 \theta$$

$$\frac{d(\cot \theta)}{d\theta} = -\operatorname{cosec}^2 \theta$$

$$\frac{d^2(\sin \theta)}{d\theta^2} = -\sin \theta$$

$$\frac{d^2(\cos \theta)}{d\theta^2} = -\cos \theta$$

$$\frac{d(\arcsin x)}{dx} = \frac{1}{\sqrt{1 - x^2}}$$

$$\frac{d(\sin\theta \cdot \cos\theta)}{d\theta} = \cos^2\theta - \sin^2\theta$$

$$\frac{d(\sec\theta)}{d\theta} = \tan\theta \cdot \sec\theta$$

$$m = \tan\theta = \frac{y_2 - y_1}{x_2 - x_1} = \text{slope}$$

$$y - y_1 = \frac{dy}{dx}(x - x_1) = \text{point slope form of}$$

a line

$$y - y_1 = -\frac{1}{\dfrac{dy}{dx}}(x - x_1) = \text{equation of the}$$

normal at a point

$$y_1 \frac{\sqrt{1 + \left(\dfrac{dy}{dx}\right)^2}}{\dfrac{dy}{dx}} = \text{length of tangent}$$

$$y_1 \sqrt{1 + \left(\frac{dy}{dx}\right)^2} = \text{length of normal}$$

$$\frac{y_1}{\tan\theta} = \frac{y_1}{\dfrac{dy}{dx}} = \text{length of subtangent}$$

$$y_1 \frac{dy}{dx} = \text{length of subnormal}$$

Maximum or a minimum at $\dfrac{dy}{dx} = 0$,   or $\dfrac{dy}{dx} = \infty$

$$\frac{d^2y}{dx^2} = + \text{ positive for a minimum}$$

$$\frac{d^2y}{dx^2} = - \text{ negative for a maximum}$$

$$P_t = C + \frac{C}{n} \cdot t = \text{simple-interest growth}$$

$$C_x = C\left(1 + \frac{1}{n}\right)^x = \text{compound-interest growth}$$

$$\epsilon = \left(1 + \frac{1}{n}\right)^n = 2.71828$$

$$\frac{d(\epsilon^x)}{dx} = \epsilon^x$$

$$\frac{d(\epsilon^{ax})}{dx} = a\epsilon^{ax}$$

$$\log_\epsilon \epsilon = 1$$

$$\log x = 0.4343 \log_\epsilon x$$

$$\log_\epsilon x = 2.3026 \log x$$

$$\frac{d(\log_\epsilon x)}{dx} = \frac{1}{x}$$

$$d_v y = \frac{\partial y}{\partial v} \cdot dv \quad \text{and} \quad d_u y = \frac{\partial y}{\partial u} \cdot du$$

$$dy = \frac{\partial y}{\partial v} \cdot dv + \frac{\partial y}{\partial u} \cdot du = \text{total differential}$$

$$R = \frac{\left[1 + \left(\frac{dy}{dx}\right)^2\right]^{\frac{3}{2}}}{\frac{d^2y}{dx^2}} = \text{radius of curvature,} \quad K = \frac{1}{R}$$

$$y_1 = y + \frac{1 + \left(\frac{dy}{dx}\right)^2}{\frac{d^2y}{dx^2}} = y \text{ coordinate of center}$$

of curvature

$$x_1 = x - \frac{\left[1 + \left(\frac{dy}{dx}\right)^2\right]\frac{dy}{dx}}{\frac{d^2y}{dx^2}} = x \text{ coordinate of}$$

center of curvature

$$\int x^n \cdot dx = \frac{x^{n+1}}{n+1} + C$$

$$\int bx^n = \frac{bx^{n+1}}{n+1} + C$$

$$\int (u \pm v)\, dx = \int u \cdot dx \pm \int v \cdot dx$$

$$\int (x^n \cdot dx \pm a \cdot dx) = \int x^n \cdot dx \pm a\int dx$$

$$\int x^{-1} \cdot dx = \log_\epsilon x + C$$

$$\int \epsilon^x \cdot dx = \epsilon^x + C$$

$\int \sin x \cdot dx = - \cos x + C$

$\int \cos x \cdot dx = \sin x + C$

$\int \sec^2 x \cdot dx = \tan x + C$

$\int \text{cosec}^2 x \cdot dx = - \cot x + C$

$\int \dfrac{dx}{\sqrt{1 - x^2}} = \text{arc } \sin x + C$

$\int \log_\epsilon x = x(\log_\epsilon x - 1) + C$

(I) $\displaystyle\int u^n \cdot du = \dfrac{u^{n+1}}{n+1} + C$

(II) $\displaystyle\int u^{-1} \cdot du = \int \dfrac{du}{u} = \log_\epsilon u + C$

(III) $\displaystyle\int \dfrac{du}{u^2 + a^2} = \dfrac{1}{a} \tan^{-1} \dfrac{u}{a}$ or $- \dfrac{1}{a} \cot^{-1} \dfrac{u}{a} + C$

(IV) $\displaystyle\int \dfrac{du}{u^2 - a^2} = \dfrac{1}{2a} \log_\epsilon \dfrac{u - a}{u + a} + C$ when

$$u^2 > a^2$$

(V) $\displaystyle\int \dfrac{du}{\sqrt{a^2 - u^2}} = \sin^{-1} \dfrac{u}{a}$ or $- \cos^{-1} \dfrac{u}{a} + C$

(VI) $\displaystyle\int \dfrac{du}{\sqrt{u^2 \pm a^2}} = \log_\epsilon (u + \sqrt{u^2 \pm a^2}) + C$

(VII) $\displaystyle\int \sqrt{a^2 - u^2} \cdot du = \dfrac{u}{2} \sqrt{a^2 - u^2} + \dfrac{a^2}{2} \sin^{-1} \dfrac{u}{a} + C$

(VIII) $\displaystyle\int \sqrt{u^2 \pm a^2} \cdot du = \dfrac{u}{2} \sqrt{u^2 \pm a^2}$

$$\pm \dfrac{a^2}{2} \log_\epsilon (u + \sqrt{u^2 \pm a^2}) + C$$

(IX) $\displaystyle\int \dfrac{du}{u \sqrt{u^2 - a^2}} = \dfrac{1}{a} \sec^{-1} \dfrac{u}{a}$ or

$$- \dfrac{1}{a} \text{cosec}^{-1} \dfrac{u}{a} + C$$

(X) $\displaystyle\int \dfrac{du}{\sqrt{2au - u^2}} = \sin^{-1} \dfrac{u - a}{a}$ or $\text{vers}^{-1} \dfrac{u}{a}$

(XI) $\displaystyle\int a^u \cdot du = \dfrac{a^u}{\log_\epsilon a} + C$

(XII) $\int \epsilon^u \cdot du = \epsilon^u + C$

(XIII) $\int \sin u \cdot du = - \cos u + C$

(XIV) $\int \cos u \cdot du = \sin u + C$

(XV) $\int \tan u \cdot du = - \log_e \cos u = \log_e \sec u + C$

(XVI) $\int \cot u \cdot du = \log_e \sin u = - \log_e \operatorname{cosec} u + C$

(XVII) $\int \sec u \cdot du = \log_e (\sec u + \tan u)$ or

$$\log_e \tan \left(\frac{u}{2} + \frac{\pi}{4} \right) + C$$

(XVIII) $\int \operatorname{cosec} u \cdot du = \log_e (\operatorname{cosec} u - \cot u)$ or

$$\log_e \tan \frac{u}{2} + C$$

(XIX) $\int \sec^2 u \cdot du = \tan u + C$

(XX) $\int \operatorname{cosec}^2 u \cdot du = - \cot u + C$

(XXI) $\int \sec u \cdot \tan u \cdot du = \sec u + C$

(XXII) $\int \operatorname{cosec} u \cdot \cot u \cdot du = - \operatorname{cosec} u + C$

$\int u \cdot dv = u \cdot v - \int v \cdot du$

$\int_{x_1}^{x_2} y \cdot dx = \text{area}$

$\frac{1}{2} \int_{\theta_1}^{\theta_2} \rho^2 \cdot d\theta = \text{area polar coordinates}$

$y = \dfrac{1}{x_1} \displaystyle\int_0^{x_1} y \cdot dx = \text{mean value, limits 0 to } x_1$

$y = \dfrac{1}{x_2 - x_1} \displaystyle\int_{x_1}^{x_2} y \cdot dx = \text{mean value, limits } x_1 \text{ to } x_2$

$\sqrt{\dfrac{1}{x_1} \displaystyle\int_0^{x_1} y^2 \cdot dx} = \text{quadratic mean}$

$\dfrac{\sqrt{\dfrac{1}{x_1} \displaystyle\int_0^{x_1} y^2 \cdot dx}}{\dfrac{1}{x_1} \displaystyle\int_0^{x_1} y \cdot dx} = \text{form factor}$

$\int \pi y^2 \cdot dx = \text{volume—single integration}$

$\int\int dy \cdot dx = \text{area—double integration}$

$\int\int \rho \cdot d\rho \cdot d\theta = \text{area—polar coordinates}$

$\int\int\int dz \cdot dy \cdot dx = \text{volume—triple integration}$

$s = \displaystyle\int \sqrt{1 + \left(\frac{dy}{dx} \right)^2} \cdot dx = \text{length of an arc}$

$$s = \int \sqrt{1 + \left(\frac{dx}{dy}\right)^2} \cdot dy = \text{length of an arc}$$

$$s = \int \sqrt{\rho^2 + \left(\frac{d\rho}{d\theta}\right)^2} \cdot d\theta = \text{length of an arc polar coordinates}$$

$$\bar{x} = \frac{\iint x \cdot dy \cdot dx}{\iint dy \cdot dx} \qquad \bar{y} = \frac{\iint y \cdot dy \cdot dx}{\iint dy \cdot dx} \text{ c.g. of an area}$$

$$\bar{x} = \frac{\iint \rho^2 \cos \theta \cdot d\rho \cdot d\theta}{\iint \rho \cdot d\rho \cdot d\theta}, \qquad \bar{y} = \frac{\iint \rho^2 \cdot \sin \theta \cdot d\rho \cdot d\theta}{\iint \rho \cdot d\rho \cdot d\theta}$$

c.g. polar coordinates

Moments of Inertia of a Body

$I_x = \iint \mu y^2 \, dy \, dx$, with respect to the x axis

$I_y = \iint \mu x^2 \, dy \, dx$, with respect to the y axis

$I_0 = \iint \mu(x^2 + y^2) \, dy \, dx$, with respect to the origin

Polar Moment of Inertia

$$I_0 = I_x + I_y \qquad \text{or} \qquad I_0 = \iint \mu \rho^2 \, dy \, dx.$$

Moments of Inertia in Polar Coordinates

$$M_x = \iint \rho^2 \sin \theta \, d\rho \, d\theta$$
$$M_y = \iint \rho^2 \cos \theta \, d\rho \, d\theta$$
$$I_x = \iint \rho^3 \sin^2 \theta \, d\rho \, d\theta$$
$$I_y = \iint \rho^3 \cos^2 \theta \, d\rho \, d\theta$$
$$I_0 = \iint \rho^3 \, d\rho \, d\theta$$

Radius of Gyration

$$\sqrt{\frac{I}{A}}$$

Integration Formulas and Tables of Integrals

In the following formulas the base ϵ of all log terms and the constant of integration of all integrals have been omitted for convenience.

I. General Formulas

(u, v, and w are any functions of x.)

1. $\int c \, du = c \int du$

2. $\int (du \pm dv \pm dw \pm \cdots) = \int du \pm \int dv \pm \int dw$
$$+ \cdots$$

3. $\int u^n \, du = \dfrac{u^{n+1}}{n+1} \cdot \quad n \neq -1$

4. $\int u^{-1} \, du = \int \dfrac{du}{u} = \log u$

5. $\int u \, dv = uv - \int v \, du$ (Integration by parts)

II. Algebraic Forms

A. *Integrand Containing* $(a + bx)$

6. $\int (a + bx)^n \, dx = \dfrac{(a + bx)^{n+1}}{b(n+1)} \cdot \quad n \neq -1$

7. $\int \dfrac{dx}{a + bx} = \dfrac{1}{b} \log (a + bx)$

8. $\int \dfrac{x \, dx}{a + bx} = \dfrac{1}{b^2} [a + bx - a \log (a + bx)]$

9. $\int \dfrac{x^2 \, dx}{a + bx} = \dfrac{1}{b^3} [\tfrac{1}{2}(a + bx)^2 - 2a(a + bx)$
$$+ \, a^2 \log (a + bx)]$$

10. $\int \dfrac{dx}{x(a + bx)} = -\dfrac{1}{a} \log \dfrac{a + bx}{x}$

11. $\int \dfrac{dx}{x^2(a + bx)} = -\dfrac{1}{ax} + \dfrac{b}{a^2} \log \dfrac{a + bx}{x}$

12. $\int \dfrac{dx}{(a + bx)^2} = -\dfrac{1}{b(a + bx)}$

13. $\int \dfrac{dx}{(a + bx)^3} = -\dfrac{1}{2b(a + bx)^2}$

14. $\int \dfrac{x \, dx}{(a + bx)^2} = \dfrac{1}{b^2} \left[\log (a + bx) + \dfrac{a}{a + bx} \right]$

15. $\int \dfrac{x \, dx}{(a + bx)^3} = \dfrac{1}{b^2} \left[-\dfrac{1}{a + bx} + \dfrac{a}{2(a + bx)^2} \right]$

16. $\int \dfrac{x^2 \, dx}{(a + bx)^2} = \dfrac{1}{b^3} \Bigg[a + bx - 2a \log (a + bx)$
$$- \, \dfrac{a^2}{a + bx} \Bigg]$$

17. $\int \dfrac{dx}{x(a + bx)^2} = \dfrac{1}{a(a + bx)} - \dfrac{1}{a^2} \log \dfrac{a + bx}{x}$

B. Integrand Containing $(a^2 \pm x^2)$ or $(a + bx^2)$

18. $\displaystyle \int \frac{dx}{a^2 + x^2} = \frac{1}{a} \tan^{-1} \frac{x}{a}$

19. $\displaystyle \int \frac{dx}{a^2 - x^2} = \frac{1}{2a} \log \frac{a + x}{a - x}$, or $\frac{1}{2a} \log \frac{x + a}{x - a}$

20. $\displaystyle \int \frac{dx}{x^2 - a^2} = \frac{1}{2a} \log \frac{x - a}{x + a}$, or $\frac{1}{2a} \log \frac{a - x}{a + x}$

21. $\displaystyle \int \frac{dx}{a + bx^2} = \frac{1}{\sqrt{ab}} \tan^{-1} x \sqrt{\frac{b}{a}}$, when $a > 0$ and
$$b > 0$$

22. $\displaystyle \int \frac{dx}{a + bx^2} = \frac{1}{2\sqrt{-ab}} \log \frac{\sqrt{a} + x\sqrt{-b}}{\sqrt{a} - x\sqrt{-b}}$, when
$$a > 0 \text{ and } b < 0$$

23. $\displaystyle \int \frac{x\,dx}{a + bx^2} = \frac{1}{2b} \log \left(x^2 + \frac{a}{b} \right)$

24. $\displaystyle \int \frac{x^2\,dx}{a + bx^2} = \frac{x}{b} - \frac{a}{b} \int \frac{dx}{a + bx^2}$.   Use 21 or 22.

25. $\displaystyle \int \frac{dx}{x(a + bx^2)} = \frac{1}{2a} \log \frac{x^2}{a + bx^2}$

26. $\displaystyle \int \frac{dx}{x^2(a + bx^2)} = -\frac{1}{ax} - \frac{b}{a} \int \frac{dx}{a + bx^2}$.   Use 21
$$\text{or } 22.$$

27. $\displaystyle \int \frac{dx}{(a + bx^2)^2} = \frac{x}{2a(a + bx^2)} + \frac{1}{2a} \int \frac{dx}{a + bx^2}$.
$$\text{Use 21 or 22.}$$

C. Integrand Containing $x^m(a + bx^n)^p$

(Reduction formulas)

28. $\displaystyle \int x^m(a + bx^n)^p \, dx = \frac{x^{m-n+1}(a + bx^n)^{p+1}}{(np + m + 1)b}$
$$- \frac{(m - n + 1)a}{(np + m + 1)b} \int x^{m-n}(a + bx^n)^p \, dx$$

29. $\displaystyle \int x^m(a + bx^n)^p \, dx = \frac{x^{m+1}(a + bx^n)^p}{np + m + 1}$
$$+ \frac{anp}{np + m + 1} \int x^m(a + bx^n)^{p-1} \, dx$$

30. $\displaystyle\int x^m(a + bx^n)^p\, dx = \frac{x^{m+1}(a + bx^n)^{p+1}}{(m + 1)a}$

$$- \frac{(np + m + n + 1)b}{(m + 1)a} \int x^{m+n}(a + bx^n)^p\, dx$$

31. $\displaystyle\int x^m(a + bx^n)^p\, dx = - \frac{x^{m+1}(a + bx^n)^{p+1}}{n(p + 1)a}$

$$+ \frac{np + m + n + 1}{n(p + 1)a} \int x^m(a + bx^n)^{p+1}\, dx$$

D. Integrand Containing $\sqrt{a + bx}$

32. $\displaystyle\int \sqrt{a + bx}\, dx = \frac{2\sqrt{(a + bx)^3}}{3b}$

33. $\displaystyle\int x\sqrt{a + bx}\, dx = -\frac{2(2a - 3bx)\sqrt{(a + bx)^3}}{15b^2}$

34. $\displaystyle\int \frac{dx}{\sqrt{a + bx}} = \frac{2}{b}\sqrt{a + bx}$

35. $\displaystyle\int \frac{x\, dx}{\sqrt{a + bx}} = -\frac{2(2a - bx)\sqrt{a + bx}}{3b^2}$

36. $\displaystyle\int \frac{x^2\, dx}{\sqrt{a + bx}} = \frac{2(8a^2 - 4abx + 3b^2x^2)\sqrt{a + bx}}{15b^3}$

37. $\displaystyle\int \frac{dx}{x\sqrt{a + bx}} = \frac{1}{\sqrt{a}}\log\frac{\sqrt{a + bx} - \sqrt{a}}{\sqrt{a + bx} + \sqrt{a}},$

when $a > 0$

38. $\displaystyle\int \frac{dx}{x\sqrt{a + bx}} = \frac{2}{\sqrt{-a}}\tan^{-1}\frac{\sqrt{a + bx}}{\sqrt{-a}},$

when $a < 0$

39. $\displaystyle\int \frac{dx}{x^2\sqrt{a + bx}} = -\frac{\sqrt{a + bx}}{ax} - \frac{b}{2a}\int \frac{dx}{x\sqrt{a + bx}}.$

Use 37 or 38.

40. $\displaystyle\int \frac{\sqrt{a + bx}\, dx}{x} = 2\sqrt{a + bx} + a\int \frac{dx}{x\sqrt{a + bx}}.$

Use 37 or 38.

E. Integrand Containing $\sqrt{a^2 + x^2}$

41. $\int \sqrt{a^2 + x^2}\, dx = \frac{1}{2}x\,\sqrt{a^2 + x^2}$
$$+ \tfrac{1}{2}a^2 \log\,(x + \sqrt{a^2 + x^2})$$

42. $\int \sqrt{(a^2 + x^2)^3}\, dx = \frac{1}{8}x(2x^2 + 5a^2)\,\sqrt{a^2 + x^2}$
$$+ \tfrac{3}{8}a^4 \log\,(x + \sqrt{a^2 + x^2})$$

43. $\displaystyle \int \sqrt{(a^2 + x^2)^n}\, dx = \frac{x\,\sqrt{(a^2 + x^2)^n}}{n + 1}$
$$+ \frac{a^2 n}{n + 1} \int \sqrt{(a^2 + x^2)^{n-2}}\, dx$$

44. $\displaystyle \int x\,\sqrt{(a^2 + x^2)^n}\, dx = \frac{\sqrt{(a^2 + x^2)^{n+2}}}{n + 2}$

45. $\int x^2 \sqrt{a^2 + x^2}\, dx = \frac{1}{8}x(2x^2 + a^2)\,\sqrt{a^2 + x^2}$
$$- \tfrac{1}{8}a^4 \log\,(x + \sqrt{a^2 + x^2})$$

46. $\displaystyle \int \frac{dx}{\sqrt{a^2 + x^2}} = \log\,(x + \sqrt{a^2 + x^2})$

47. $\displaystyle \int \frac{x\, dx}{\sqrt{a^2 + x^2}} = \sqrt{a^2 + x^2}$

48. $\displaystyle \int \frac{x^2\, dx}{\sqrt{a^2 + x^2}} = \frac{1}{2}x\,\sqrt{a^2 + x^2}$
$$- \tfrac{1}{2}a^2 \log\,(x + \sqrt{a^2 + x^2})$$

49. $\displaystyle \int \frac{x^2\, dx}{\sqrt{(a^2 + x^2)^3}} = -\frac{x}{\sqrt{a^2 + x^2}}$
$$+ \log\,(x + \sqrt{a^2 + x^2})$$

50. $\displaystyle \int \frac{dx}{x\,\sqrt{a^2 + x^2}} = \frac{1}{a}\log \frac{x}{a + \sqrt{a^2 + x^2}}$

51. $\displaystyle \int \frac{dx}{x^2\,\sqrt{a^2 + x^2}} = -\frac{\sqrt{a^2 + x^2}}{a^2 x}$

52. $\displaystyle \int \frac{dx}{x^3\,\sqrt{a^2 + x^2}} = -\frac{\sqrt{a^2 + x^2}}{2a^2 x^2}$
$$+ \frac{1}{2a^3} \log \frac{a + \sqrt{a^2 + x^2}}{x}$$

53. $\displaystyle\int \frac{\sqrt{a^2 + x^2}\,dx}{x} = \sqrt{a^2 + x^2}$

$$- a \log \frac{a + \sqrt{a^2 + x^2}}{x}$$

54. $\displaystyle\int \frac{\sqrt{a^2 + x^2}\,dx}{x^2} = -\frac{\sqrt{a^2 + x^2}}{x}$

$$+ \log (x + \sqrt{a^2 + x^2})$$

F. Integrand Containing $\sqrt{a^2 - x^2}$

55. $\displaystyle\int \sqrt{a^2 - x^2}\,dx = \tfrac{1}{2}\left(x\sqrt{a^2 - x^2} + a^2 \sin^{-1} \frac{x}{a} \right)$

56. $\displaystyle\int \sqrt{(a^2 - x^2)^3}\,dx = \tfrac{1}{8}x(5a^2 - 2x^2)\sqrt{a^2 - x^2}$

$$+ \tfrac{3}{8}a^4 \sin^{-1} \frac{x}{a}$$

57. $\displaystyle\int \sqrt{(a^2 - x^2)^n}\,dx = \frac{x\sqrt{(a^2 - x^2)^n}}{n + 1}$

$$+ \frac{a^2 n}{n + 1}\int \sqrt{(a^2 - x^2)^{n-2}}\,dx$$

58. $\displaystyle\int x\sqrt{(a^2 - x^2)^n}\,dx = -\frac{\sqrt{(a^2 - x^2)^{n+2}}}{n + 2}$

59. $\displaystyle\int x^2 \sqrt{a^2 - x^2}\,dx = \tfrac{1}{8}x(2x^2 - a^2)\sqrt{a^2 - x^2}$

$$+ \tfrac{1}{8}a^4 \sin^{-1} \frac{x}{a}$$

60. $\displaystyle\int \frac{dx}{\sqrt{a^2 - x^2}} = \sin^{-1} \frac{x}{a}$

61. $\displaystyle\int \frac{dx}{\sqrt{(a^2 - x^2)^3}} = \frac{x}{a^2 \sqrt{a^2 - x^2}}$

62. $\displaystyle\int \frac{x\,dx}{\sqrt{a^2 - x^2}} = -\sqrt{a^2 - x^2}$

63. $\displaystyle\int \frac{x^2\,dx}{\sqrt{a^2 - x^2}} = -\tfrac{1}{2}x\sqrt{a^2 - x^2} + \tfrac{1}{2}a^2 \sin^{-1} \frac{x}{a}$

64. $\displaystyle\int \frac{x^2\,dx}{\sqrt{(a^2 - x^2)^3}} = \frac{x}{\sqrt{a^2 - x^2}} - \sin^{-1} \frac{x}{a}$

65. $\displaystyle\int \frac{dx}{x\sqrt{a^2-x^2}} = \frac{1}{a}\log\frac{x}{a+\sqrt{a^2-x^2}}$

66. $\displaystyle\int \frac{dx}{x^2\sqrt{a^2-x^2}} = -\frac{\sqrt{a^2-x^2}}{a^2 x}$

67. $\displaystyle\int \frac{dx}{x^3\sqrt{a^2-x^2}} = -\frac{\sqrt{a^2-x^2}}{2a^2x^2}$
$$+ \frac{1}{2a^3}\log\frac{x}{a+\sqrt{a^2-x^2}}$$

68. $\displaystyle\int \frac{\sqrt{a^2-x^2}\,dx}{x} = \sqrt{a^2-x^2}$
$$- a\log\frac{a+\sqrt{a^2-x^2}}{x}$$

69. $\displaystyle\int \frac{\sqrt{a^2-x^2}\,dx}{x^2} = -\frac{\sqrt{a^2-x^2}}{x} - \sin^{-1}\frac{x}{a}$

G. Integrand Containing $\sqrt{x^2-a^2}$

70. $\displaystyle\int \sqrt{x^2-a^2}\,dx = \frac{1}{2}[x\sqrt{x^2-a^2}$
$$- a^2\log(x+\sqrt{x^2-a^2})]$$

71. $\displaystyle\int \sqrt{(x^2-a^2)^3}\,dx = \frac{1}{8}x(2x^2-5a^2)\sqrt{x^2-a^2}$
$$+ \frac{3}{8}a^4\log(x+\sqrt{x^2-a^2})$$

72. $\displaystyle\int \sqrt{(x^2-a^2)^n}\,dx = \frac{x\sqrt{(x^2-a^2)^n}}{n+1}$
$$- \frac{a^2 n}{n+1}\int \sqrt{(x^2-a^2)^{n-2}}\,dx$$

73. $\displaystyle\int x\sqrt{(x^2-a^2)^n}\,dx = \frac{\sqrt{(x^2-a^2)^{n+2}}}{n+2}$

74. $\displaystyle\int x^2\sqrt{x^2-a^2}\,dx = \frac{1}{8}x(2x^2-a^2)\sqrt{x^2-a^2}$
$$- \frac{1}{8}a^4\log(x+\sqrt{x^2-a^2})$$

75. $\displaystyle\int \frac{dx}{\sqrt{x^2-a^2}} = \log(x+\sqrt{x^2-a^2})$

76. $\displaystyle\int \frac{dx}{\sqrt{(x^2-a^2)^3}} = -\frac{x}{a^2\sqrt{x^2-a^2}}$

77. $\int \dfrac{x\,dx}{\sqrt{x^2 - a^2}} = \sqrt{x^2 - a^2}$

78. $\int \dfrac{x^2\,dx}{\sqrt{x^2 - a^2}} = x\sqrt{x^2 - a^2}$
$$+ \tfrac{1}{2}a^2 \log\,(x + \sqrt{x^2 - a^2})$$

79. $\int \dfrac{x^2\,dx}{\sqrt{(x^2 - a^2)^3}} = -\dfrac{x}{\sqrt{x^2 - a^2}}$
$$+ \log\,(x + \sqrt{x^2 - a^2})$$

80. $\int \dfrac{dx}{x\sqrt{x^2 - a^2}} = \dfrac{1}{a}\cos^{-1}\dfrac{a}{x}$

81. $\int \dfrac{dx}{x^2\sqrt{x^2 - a^2}} = \dfrac{\sqrt{x^2 - a^2}}{a^2 x}$

82. $\int \dfrac{dx}{x^3\sqrt{x^2 - a^2}} = \dfrac{\sqrt{x^2 - a^2}}{2a^2 x^2} + \dfrac{1}{2a^3}\cos^{-1}\dfrac{a}{x}$

83. $\int \dfrac{\sqrt{x^2 - a^2}\,dx}{x} = \sqrt{x^2 - a^2} - a\cos^{-1}\dfrac{a}{x}$

84. $\int \dfrac{\sqrt{x^2 - a^2}\,dx}{x^2} = -\dfrac{\sqrt{x^2 - a^2}}{x}$
$$+ \log\,(x + \sqrt{x^2 - a^2})$$

H. Integrand Containing $\sqrt{2ax - x^2}$ *or* $\sqrt{2ax + x^2}$

85. $\int \sqrt{2ax - x^2}\,dx = \tfrac{1}{2}(x - a)\sqrt{2ax - x^2}$
$$+ \tfrac{1}{2}a^2\sin^{-1}\dfrac{x - a}{a}$$

86. $\int \sqrt{2ax + x^2}\,dx = \tfrac{1}{2}(x + a)\sqrt{2ax + x^2}$
$$- \tfrac{1}{2}a^2\log\,(x + a + \sqrt{2ax + x^2})$$

87. $\int x\sqrt{2ax - x^2}\,dx = -\tfrac{1}{6}(3a^2 + ax$
$$- 2x^2)\sqrt{2ax - x^2} + \tfrac{1}{2}a^3\sin^{-1}\dfrac{x - a}{a}$$

88. $\int x\sqrt{2ax + x^2}\,dx = \tfrac{1}{3}\sqrt{(2ax + x^2)^3}$
$$- a\int \sqrt{2ax + x^2}\,dx$$

89. $\int x^n\sqrt{2ax - x^2}\,dx = -\dfrac{x^{n-1}\sqrt{(2ax - x^2)^3}}{n + 2}$
$$+ \dfrac{(2n + 1)a}{n + 2}\int x^{n-1}\sqrt{2ax - x^2}\,dx$$

90. $\displaystyle\int \frac{dx}{\sqrt{2ax - x^2}} = \text{vers}^{-1}\frac{x}{a}, \text{ or } \sin^{-1}\frac{x - a}{a}$

91. $\displaystyle\int \frac{dx}{\sqrt{2ax + x^2}} = \log\left(x + a + \sqrt{2ax + x^2}\right)$

92. $\displaystyle\int \frac{dx}{\sqrt{(2ax - x^2)^3}} = \frac{x - a}{a^2\sqrt{2ax - x^2}}$

93. $\displaystyle\int \frac{dx}{\sqrt{(2ax + x^2)^3}} = -\frac{x + a}{a^2\sqrt{2ax + x^2}}$

94. $\displaystyle\int \frac{x\,dx}{\sqrt{2ax - x^2}} = -\sqrt{2ax - x^2} + a\sin^{-1}\frac{x - a}{a}$

95. $\displaystyle\int \frac{x\,dx}{\sqrt{2ax + x^2}} = \sqrt{2ax + x^2} - a\log\left(x + a\right.$
$$\left. + \sqrt{2ax + x^2}\right)$$

96. $\displaystyle\int \frac{x\,dx}{\sqrt{(2ax \pm x^2)^3}} = \frac{x}{a\sqrt{2ax \pm x^2}}$

97. $\displaystyle\int \frac{x^2\,dx}{\sqrt{2ax - x^2}} = -\tfrac{1}{2}(x + 3a)\sqrt{2ax - x^2}$
$$+ \tfrac{3}{2}a^2\sin^{-1}\frac{x - a}{a}$$

98. $\displaystyle\int \frac{x^n\,dx}{\sqrt{2ax - x^2}} = -\frac{x^{n-1}\sqrt{2ax - x^2}}{n}$
$$-\frac{a(1 - 2n)}{n}\int \frac{x^{n-1}\,dx}{\sqrt{2ax - x^2}}$$

99. $\displaystyle\int \frac{dx}{x\sqrt{2ax \pm x^2}} = -\frac{\sqrt{2ax \pm x^2}}{ax}$

100. $\displaystyle\int \frac{dx}{x^n\sqrt{2ax - x^2}} = \frac{\sqrt{2ax - x^2}}{a(1 - 2n)x^n}$
$$+ \frac{n - 1}{(2n - 1)a}\int \frac{dx}{x^{n-1}\sqrt{2ax - x^2}}$$

101. $\displaystyle\int \frac{\sqrt{2ax - x^2}\,dx}{x} = \sqrt{2ax - x^2} + a\sin^{-1}\frac{x - a}{a}$

102. $\displaystyle\int \frac{\sqrt{2ax + x^2}\,dx}{x} = \sqrt{2ax + x^2}$
$$+ a\log\left(x + a + \sqrt{2ax + x^2}\right)$$

103. $\displaystyle\int \frac{\sqrt{2ax - x^2}\,dx}{x^2} = -\frac{2\sqrt{2ax - x^2}}{x} - \sin^{-1}\frac{x - a}{a}$

104. $\displaystyle\int \frac{\sqrt{2ax + x^2}\,dx}{x^2} = -\frac{2\sqrt{2ax + x^2}}{x}$
$$+ \log\left(x + a + \sqrt{2ax + x^2}\right)$$

105. $\displaystyle\int \frac{\sqrt{2ax - x^2}\,dx}{x^n} = \frac{\sqrt{(2ax - x^2)^3}}{(3 - 2n)ax^n}$
$$+ \frac{n - 3}{(2n - 3)a}\int \frac{\sqrt{2ax - x^2}\,dx}{x^{n-1}}$$

I. Integrand Containing $ax^2 + bx + c$

106. $\displaystyle\int \frac{dx}{ax^2 + bx + c} = \frac{2}{\sqrt{4ac - b^2}}\tan^{-1}\frac{2ax + b}{\sqrt{4ac - b^2}},$
$$\text{when } b^2 - 4ac < 0$$

107. $\displaystyle\int \frac{dx}{ax^2 + bx + c} = \frac{1}{\sqrt{b^2 - 4ac}}$
$$\log \frac{2ax + b - \sqrt{b^2 - 4ac}}{2ax + b + \sqrt{b^2 - 4ac}}, \text{ when } b^2 - 4ac > 0$$

108. $\displaystyle\int \frac{x\,dx}{ax^2 + bx + c} = \frac{1}{2a}\log(ax^2 + bx + c)$
$$- \frac{b}{2a}\int \frac{dx}{ax^2 + bx + c}$$

109. $\displaystyle\int \frac{x^2\,dx}{ax^2 + bx + c} = \frac{x}{a} - \frac{b}{2a^2}\log(ax^2 + bx + c)$
$$+ \frac{b^2 - 4ac}{2a^2}\int \frac{dx}{ax^2 + bx + c}$$

110. $\displaystyle\int \frac{dx}{x(ax^2 + bx + c)} = \frac{1}{2c}\log \frac{x^2}{ax^2 + bx + c}$
$$- \frac{b}{2c}\int \frac{dx}{ax^2 + bx + c}$$

111. $\displaystyle\int \frac{dx}{x^2(ax^2 + bx + c)} = \frac{b}{2c^2}\log \frac{ax^2 + bx + c}{x^2} - \frac{1}{cx}$
$$+ \left(\frac{b^2}{2c^2} - \frac{a}{c}\right)\int \frac{dx}{ax^2 + bx + c}$$

J. Integrand Containing $\sqrt{ax^2 + bx + c}$

112. $\displaystyle \int \sqrt{ax^2 + bx + c}\, dx = \frac{2ax + b}{4a} \sqrt{ax^2 + bx + c}$
$$- \frac{b^2 - 4ac}{8a} \int \frac{dx}{\sqrt{ax^2 + bx + c}}$$

113. $\displaystyle \int \frac{dx}{\sqrt{ax^2 + bx + c}} = \frac{1}{\sqrt{a}} \log\,(2ax + b$
$$+ 2\sqrt{a}\,\sqrt{ax^2 + bx + c}),\ \text{when } a > 0$$

114. $\displaystyle \int \frac{dx}{\sqrt{ax^2 + bx + c}} = \frac{1}{\sqrt{-a}} \sin^{-1} \frac{-2ax - b}{\sqrt{b^2 - 4ac}},$
$$\text{when } a < 0$$

115. $\displaystyle \int \frac{x\, dx}{\sqrt{ax^2 + bx + c}} = \frac{1}{a} \sqrt{ax^2 + bx + c}$
$$- \frac{b}{2a} \int \frac{dx}{\sqrt{ax^2 + bx + c}}$$

116. $\displaystyle \int \frac{x^2\, dx}{\sqrt{ax^2 + bx + c}} = \left(\frac{x}{2a} - \frac{3b}{4a^2} \right) \sqrt{ax^2 + bx + c}$
$$+ \frac{3b^2 - 4ac}{8a^2} \int \frac{dx}{\sqrt{ax^2 + bx + c}}$$

117. $\displaystyle \int \frac{dx}{x\,\sqrt{ax^2 + bx + c}} = -\frac{1}{\sqrt{c}}$
$$\log \left(\frac{\sqrt{ax^2 + bx + c} + \sqrt{c}}{x} + \frac{b}{2\sqrt{c}} \right),\ \text{when } c > 0$$

118. $\displaystyle \int \frac{dx}{x\,\sqrt{ax^2 + bx + c}} = \frac{1}{\sqrt{-c}} \sin^{-1} \left(\frac{bx + 2c}{x\,\sqrt{b^2 - 4ac}} \right),$
$$\text{when } c < 0$$

119. $\displaystyle \int \frac{dx}{x^2\,\sqrt{ax^2 + bx + c}} = -\frac{\sqrt{ax^2 + bx + c}}{cx}$
$$- \frac{b}{2c} \int \frac{dx}{x\,\sqrt{ax^2 + bx + c}}$$

K. Miscellaneous Algebraic Forms

120. $\displaystyle \int \sqrt{\frac{a + x}{b + x}}\, dx = \sqrt{(a + x)(b + x)}$
$$+ (a - b) \log\,(\sqrt{a + x} + \sqrt{b + x})$$

121. $\int \sqrt{\dfrac{a+x}{b-x}}\, dx = -\sqrt{(a+x)(b-x)}$

$$-(a+b)\sin^{-1}\sqrt{\dfrac{b-x}{a+b}}$$

122. $\int \sqrt{\dfrac{a-x}{b+x}}\, dx = \sqrt{(a-x)(b+x)}$

$$+(a+b)\sin^{-1}\sqrt{\dfrac{b+x}{a+b}}$$

123. $\int \dfrac{dx}{x\sqrt{x^n-a^2}} = \dfrac{2}{an}\sec^{-1}\dfrac{\sqrt{x^n}}{a}$

III. Exponential and Logarithmic Forms

124. $\int \epsilon^x\, dx = \epsilon^x$ **125.** $\int a^x\, dx = \dfrac{a^x}{\log a}$

126. $\int \epsilon^{ax}\, dx = \dfrac{\epsilon^{ax}}{a}$ **127.** $\int a^{bx}\, dx = \dfrac{a^{bx}}{b\log a}$

128. $\int \log x\, dx = x\log x - x$

129. $\int \dfrac{dx}{\log x} = \log(\log x) + \log x + \dfrac{(\log x)^2}{2\cdot 2!}$

$$+\dfrac{(\log x)^3}{3\cdot 3!} + \cdots$$

130. $\int \dfrac{dx}{x\log x} = \log(\log x)$

131. $\int x\epsilon^{ax}\, dx = \dfrac{\epsilon^{ax}}{a^2}(ax-1)$

132. $\int x^n\epsilon^{ax}\, dx = \dfrac{x^n\epsilon^{ax}}{a} - \dfrac{n}{a}\int x^{n-1}\epsilon^{ax}\, dx$

133. $\int \dfrac{\epsilon^{ax}\, dx}{x} = \log x + ax + \dfrac{a^2x^2}{2\cdot 2!} + \dfrac{a^3x^3}{3\cdot 3!} + \cdots$

134. $\int \dfrac{\epsilon^{ax}\, dx}{x^n} = \dfrac{1}{n-1}\left(-\dfrac{\epsilon^{ax}}{x^{n-1}} + a\int \dfrac{\epsilon^{ax}\, dx}{x^{n-1}}\right)$

135. $\int xa^x\, dx = \dfrac{a^x x}{\log a} - \dfrac{a^x}{(\log a)^2}$

136. $\int \dfrac{a^x\, dx}{x} = \log x + x\log a + \dfrac{(x\log a)^2}{2\cdot 2!} + \dfrac{(x\log a)^3}{3\cdot 3!}$

$$+\cdots$$

137. $\int x \log x \, dx = \frac{1}{2}x^2(\log x - \frac{1}{2})$

138. $\int x^2 \log x \, dx = \frac{1}{3}x^3(\log x - \frac{1}{3})$

139. $\displaystyle\int x^n \log x \, dx = \frac{x^{n+1}}{n+1}\left(\log x - \frac{1}{n+1}\right)$

140. $\displaystyle\int \epsilon^{ax} \log x \, dx = \frac{\epsilon^{ax}\log x}{a} - \frac{1}{a}\int \frac{\epsilon^{ax}}{x}\, dx$

141. $\displaystyle\int \frac{dx}{1+\epsilon^x} = \log \frac{\epsilon^x}{1+\epsilon^x}$

142. $\displaystyle\int \frac{dx}{a+b\epsilon^{nx}} = \frac{1}{an}\left[nx - \log(a+b\epsilon^{nx})\right]$

143. $\displaystyle\int \frac{dx}{a\epsilon^{nx}+b\epsilon^{-nx}} = \frac{1}{n\sqrt{ab}}\tan^{-1}\left(\epsilon^{nx}\sqrt{\frac{a}{b}}\right)$

144. $\int \log(a^2+x^2)\, dx = x \log(a^2+x^2) - 2x$
$$+ 2a \tan^{-1}\frac{x}{a}$$

IV. Trigonometric Forms

145. $\int \sin x \, dx = -\cos x$

146. $\int \sin^2 x \, dx = \frac{1}{2}x - \frac{1}{2}\cos x \sin x = \frac{1}{2}x - \frac{1}{4}\sin 2x$

147. $\displaystyle\int \sin^n x \, dx = -\frac{\sin^{n-1} x \cos x}{n}$
$$+ \frac{n-1}{n}\int \sin^{n-2} x \, dx$$

148. $\int \cos x \, dx = \sin x$

149. $\int \cos^2 x \, dx = \frac{1}{2}x + \frac{1}{2}\sin x \cos x = \frac{1}{2}x + \frac{1}{4}\sin 2x$

150. $\displaystyle\int \cos^n x \, dx = \frac{\cos^{n-1} x \sin x}{n} + \frac{n-1}{n}\int \cos^{n-2} x \, dx$

151. $\int \tan x \, dx = -\log \cos x$

152. $\int \tan^2 x \, dx = \tan x - x$

153. $\displaystyle\int \tan^n x \, dx = \frac{\tan^{n-1} x}{n-1} - \int \tan^{n-2} x \, dx$

154. $\int \cot x \, dx = \log \sin x$

155. $\int \cot^2 x \, dx = -\cot x - x$

156. $\displaystyle\int \cot^n x \, dx = -\frac{\cot^{n-1} x}{n-1} - \int \cot^{n-2} x \, dx$

157. $\displaystyle\int \sec x \, dx = \log \tan(\frac{1}{4}\pi + \frac{1}{2}x) = \frac{1}{2}\log \frac{1+\sin x}{1-\sin x}$

158. $\int \sec^2 x \, dx = \tan x$

159. $\int \sec^n x \, dx = \dfrac{\sin x \sec^{n-1} x}{n-1} + \dfrac{n-2}{n-1} \int \sec^{n-2} x \, dx$

160. $\int \operatorname{cosec} x \, dx = \log \tan \tfrac{1}{2}x$

161. $\int \operatorname{cosec}^2 x \, dx = -\cot x$

162. $\int \operatorname{cosec}^n x \, dx = -\dfrac{\cos x \operatorname{cosec}^{n-1} x}{n-1}$

$$+ \dfrac{n-2}{n-1} \int \operatorname{cosec}^{n-2} x \, dx$$

163. $\int \sin x \cos x \, dx = \tfrac{1}{2}\sin^2 x \text{ or } -\tfrac{1}{2}\cos^2 x$

164. $\int \sin^n x \cos x \, dx = \dfrac{\sin^{n+1} x}{n+1}$

165. $\int \cos^n x \sin x \, dx = -\dfrac{\cos^{n+1} x}{n+1}$

166. $\int \sin^2 x \cos^2 x \, dx = \tfrac{1}{8}(x - \tfrac{1}{4}\sin 4x)$

167. $\int \sin^n x \cos^m x \, dx = -\dfrac{\sin^{n-1} x \cos^{m+1} x}{m+n}$

$$+ \dfrac{n-1}{m+n} \int \sin^{n-2} x \cos^m x \, dx$$

or $\quad = \dfrac{\sin^{n+1} x \cos^{m-1} x}{m+n} + \dfrac{m-1}{m+n} \int \sin^n x \cos^{m-2} x \, dx$

168. $\displaystyle\int \dfrac{\sin^n x \, dx}{\cos^m x} = \dfrac{1}{n-m}\left[-\dfrac{\sin^{n-1} x}{\cos^{m-1} x} \right.$

$$\left. + (n-1) \int \dfrac{\sin^{n-2} x \, dx}{\cos^m x} \right]$$

or $\quad = \dfrac{1}{m-1}\left[\dfrac{\sin^{n+1} x}{\cos^{m-1} x} - (n-m+2) \int \dfrac{\sin^n x \, dx}{\cos^{m-2} x} \right]$

169. $\displaystyle\int \dfrac{\cos^n x \, dx}{\sin^m x} = -\dfrac{1}{m-1}\left[\dfrac{\cos^{n+1} x}{\sin^{m-1} x} \right.$

$$\left. + (n-m+2) \int \dfrac{\cos^n x \, dx}{\sin^{m-2} x} \right]$$

or $\quad = \dfrac{1}{n-m}\left[\dfrac{\cos^{n-1} x}{\sin^{m-1} x} + (n-1) \int \dfrac{\cos^{n-2} x \, dx}{\sin^m x} \right]$

170. $\displaystyle\int \dfrac{dx}{1+\sin x} = -\tan(\tfrac{1}{4}\pi - \tfrac{1}{2}x)$

171. $\int \dfrac{dx}{1 - \sin x} = \cot \left(\tfrac{1}{4}\pi - \tfrac{1}{2}x\right) = \tan \left(\tfrac{1}{4}\pi + \tfrac{1}{2}x\right)$

172. $\int \dfrac{dx}{1 + \cos x} = \tan \tfrac{1}{2}x$

173. $\int \dfrac{dx}{1 - \cos x} = -\cot \tfrac{1}{2}x$

174. $\int \dfrac{dx}{a + b \sin x} = \dfrac{2}{\sqrt{a^2 - b^2}}$
$$\tan^{-1} \dfrac{a \tan \tfrac{1}{2}x + b}{\sqrt{a^2 - b^2}}, \text{ when } a > b$$

175. $\int \dfrac{dx}{a + b \sin x} = \dfrac{1}{\sqrt{b^2 - a^2}}$
$$\log \dfrac{a \tan \tfrac{1}{2}x + b - \sqrt{b^2 - a^2}}{a \tan \tfrac{1}{2}x + b + \sqrt{b^2 - a^2}}, \text{ when } a < b$$

176. $\int \dfrac{dx}{a + b \cos x} = \dfrac{2}{\sqrt{a^2 - b^2}} \tan^{-1} \left(\sqrt{\dfrac{a - b}{a + b}} \tan \tfrac{1}{2}x\right),$
$$\text{when } a > b$$

177. $\int \dfrac{dx}{a + b \cos x} = \dfrac{1}{\sqrt{b^2 - a^2}}$
$$\log \dfrac{\sqrt{b - a} \tan \tfrac{1}{2}x + \sqrt{b + a}}{\sqrt{b - a} \tan \tfrac{1}{2}x - \sqrt{b + a}}, \text{ when } a < b$$

178. $\int \sec x \tan x \, dx = \sec x$

179. $\int \operatorname{cosec} x \cot x \, dx = -\operatorname{cosec} x$

180. $\int \dfrac{dx}{a^2 \cos^2 x + b^2 \sin^2 x} = \dfrac{1}{ab} \tan^{-1} \left(\dfrac{b \tan x}{a}\right)$

181. $\int \sin mx \sin nx \, dx = \dfrac{\sin (m - n)x}{2(m - n)}$
$$- \dfrac{\sin (m + n)x}{2(m + n)}$$

182. $\int \cos mx \cos nx \, dx = \dfrac{\sin (m - n)x}{2(m - n)}$
$$+ \dfrac{\sin (m + n)x}{2(m + n)}$$

183. $\int \sin mx \cos nx \, dx = -\dfrac{\cos (m - n)x}{2(m - n)}$
$$- \dfrac{\cos (m + n)x}{2(m + n)}$$

184. $\int \sin^{-1} x \, dx = x \sin^{-1} x + \sqrt{1 - x^2}$

185. $\int \cos^{-1} x \, dx = x \cos^{-1} x - \sqrt{1 - x^2}$

V. Algebraic and Trigonometric Forms

186. $\int x \sin x \, dx = \sin x - x \cos x$

187. $\int x^2 \sin x \, dx = 2x \sin x - (x^2 - 2) \cos x$

188. $\int x^n \sin x \, dx = -x^n \cos x + n \int x^{n-1} \cos x \, dx$

189. $\int x \cos x \, dx = \cos x + x \sin x$

190. $\int x^2 \cos x \, dx = 2x \cos x + (x^2 - 2) \sin x$

191. $\int x^n \cos x \, dx = x^n \sin x - n \int x^{n-1} \sin x \, dx$

192. $\displaystyle\int \frac{\sin x}{x} \, dx = x - \frac{x^3}{3 \cdot 3!} + \frac{x^5}{5 \cdot 5!} - \frac{x^7}{7 \cdot 7!} + \cdots$

193. $\displaystyle\int \frac{\cos x}{x} \, dx = \log x - \frac{x^2}{2 \cdot 2!} + \frac{x^4}{4 \cdot 4!} - \frac{x^6}{6 \cdot 6!}$
$$+ \cdots$$

194. $\displaystyle\int \frac{x \, dx}{1 + \sin x} = 2 \log \cos \tfrac{1}{2}(\tfrac{1}{2}\pi - x)$
$$- x \tan \tfrac{1}{2}(\tfrac{1}{2}\pi - x)$$

195. $\displaystyle\int \frac{x \, dx}{1 - \sin x} = 2 \log \sin \tfrac{1}{2}(\tfrac{1}{2}\pi - x)$
$$+ x \cot \tfrac{1}{2}(\tfrac{1}{2}\pi - x)$$

196. $\displaystyle\int \frac{x \, dx}{1 + \cos x} = 2 \log \cos \tfrac{1}{2}x + x \tan \tfrac{1}{2}x$

197. $\displaystyle\int \frac{x \, dx}{1 - \cos x} = 2 \log \sin \tfrac{1}{2}x - x \cot \tfrac{1}{2}x$

VI. Exponential and Trigonometric Forms

198. $\displaystyle\int \epsilon^{ax} \sin x \, dx = \frac{\epsilon^{ax}(a \sin x - \cos x)}{1 + a^2}$

199. $\displaystyle\int \epsilon^{ax} \cos x \, dx = \frac{\epsilon^{ax}(\sin x + a \cos x)}{1 + a^2}$

200. $\displaystyle\int \epsilon^{ax} \sin nx \, dx = \frac{\epsilon^{ax}(a \sin nx - n \cos nx)}{a^2 + n^2}$

201. $\displaystyle\int \epsilon^{ax} \cos nx \, dx = \frac{\epsilon^{ax}(n \sin nx + a \cos nx)}{a^2 + n^2}$

COMMON LOGARITHMS

N.	0	1	2	3	4	5	6	7	8	9
10	0000	0043	0086	0128	0170	0212	0253	0294	0334	0374
11	0414	0453	0492	0531	0569	0607	0645	0682	0719	0755
12	0792	0828	0864	0899	0934	0969	1004	1038	1072	1106
13	1139	1173	1206	1239	1271	1303	1335	1367	1399	1430
14	1461	1492	1523	1553	1584	1614	1644	1673	1703	1732
15	1761	1790	1818	1847	1875	1903	1931	1959	1987	2014
16	2041	2068	2095	2122	2148	2175	2201	2227	2253	2279
17	2304	2330	2355	2380	2405	2430	2455	2480	2504	2529
18	2553	2577	2601	2625	2648	2672	2695	2718	2742	2765
19	2788	2810	2833	2856	2878	2900	2923	2945	2967	2989
20	3010	3032	3054	3075	3096	3118	3139	3160	3181	3201
21	3222	3243	3263	3284	3304	3324	3345	3365	3385	3404
22	3424	3444	3464	3483	3502	3522	3541	3560	3579	3598
23	3617	3636	3655	3674	3692	3711	3729	3747	3766	3784
24	3802	3820	3838	3856	3874	3892	3909	3927	3945	3962
25	3979	3997	4014	4031	4048	4065	4082	4099	4116	4133
26	4150	4166	4183	4200	4216	4232	4249	4265	4281	4298
27	4314	4330	4346	4362	4378	4393	4409	4425	4440	4456
28	4472	4487	4502	4518	4533	4548	4564	4579	4594	4609
29	4624	4639	4654	4669	4683	4698	4713	4728	4742	4757
30	4771	4786	4800	4814	4829	4843	4857	4871	4886	4900
31	4914	4928	4942	4955	4969	4983	4997	5011	5024	5038
32	5051	5065	5079	5092	5105	5119	5132	5145	5159	5172
33	5185	5198	5211	5224	5237	5250	5263	5276	5289	5302
34	5315	5328	5340	5353	5366	5378	5391	5403	5416	5428
35	5441	5453	5465	5478	5490	5502	5514	5527	5539	5551
36	5563	5575	5587	5599	5611	5623	5635	5647	5658	5670
37	5682	5694	5705	5717	5729	5740	5752	5763	5775	5786
38	5798	5809	5821	5832	5843	5855	5866	5877	5888	5899
39	5911	5922	5933	5944	5955	5966	5977	5988	5999	6010
40	6021	6031	6042	6053	6064	6075	6085	6096	6107	6117
41	6128	6138	6149	6160	6170	6180	6191	6201	6212	6222
42	6232	6243	6253	6263	6274	6284	6294	6304	6314	6325
43	6335	6345	6355	6365	6375	6385	6395	6405	6415	6425
44	6435	6444	6454	6464	6474	6484	6493	6503	6513	6522
45	6532	6542	6551	6561	6571	6580	6590	6599	6609	6618
46	6628	6637	6646	6656	6665	6675	6684	6693	6702	6712
47	6721	6730	6739	6749	6758	6767	6776	6785	6794	6803
48	6812	6821	6830	6839	6848	6857	6866	6875	6884	6893
49	6902	6911	6920	6928	6937	6946	6955	6964	6972	6981
50	6990	6998	7007	7016	7024	7033	7042	7050	7059	7067
51	7076	7084	7093	7101	7110	7118	7126	7135	7143	7152
52	7160	7168	7177	7185	7193	7202	7210	7218	7226	7235
53	7243	7251	7259	7267	7275	7284	7292	7300	7308	7316
54	7324	7332	7340	7348	7356	7364	7372	7380	7388	7396
N.	0	1	2	3	4	5	6	7	8	9

COMMON LOGARITHMS.—(*Continued*)

N.	0	1	2	3	4	5	6	7	8	9
55	7404	7412	7419	7427	7435	7443	7451	7459	7466	7474
56	7482	7490	7497	7505	7513	7520	7528	7536	7543	7551
57	7559	7566	7574	7582	7589	7597	7604	7612	7619	7627
58	7634	7642	7649	7657	7664	7672	7679	7686	7694	7701
59	7709	7716	7723	7731	7738	7745	7752	7760	7767	7774
60	7782	7789	7796	7803	7810	7818	7825	7832	7839	7846
61	7853	7860	7868	7875	7882	7889	7896	7903	7910	7917
62	7924	7931	7938	7945	7952	7959	7966	7973	7980	7987
63	7993	8000	8007	8014	8021	8028	8035	8041	8048	8055
64	8062	8069	8075	8082	8089	8096	8102	8109	8116	8122
65	8129	8136	8142	8149	8156	8162	8169	8176	8182	8189
66	8195	8202	8209	8215	8222	8228	8235	8241	8248	8254
67	8261	8267	8274	8280	8287	8293	8299	8306	8312	8319
68	8325	8331	8338	8344	8351	8357	8363	8370	8376	8382
69	8388	8395	8401	8407	8414	8420	8426	8432	8439	8445
70	8451	8457	8463	8470	8476	8482	8488	8494	8500	8506
71	8513	8519	8525	8531	8537	8543	8549	8555	8561	8567
72	8573	8579	8585	8591	8597	8603	8609	8615	8621	8627
73	8633	8639	8645	8651	8657	8663	8669	8675	8681	8686
74	8692	8698	8704	8710	8716	8722	8727	8733	8739	8745
75	8751	8756	8762	8768	8774	8779	8785	8791	8797	8802
76	8808	8814	8820	8825	8831	8837	8842	8848	8854	8859
77	8865	8871	8876	8882	8887	8893	8899	8904	8910	8915
78	8921	8927	8932	8938	8943	8949	8954	8960	8965	8971
79	8976	8982	8987	8993	8998	9004	9009	9015	9020	9025
80	9031	9036	9042	9047	9053	9058	9063	9069	9074	9079
81	9085	9090	9096	9101	9106	9112	9117	9122	9128	9133
82	9138	9143	9149	9154	9159	9165	9170	9175	9180	9186
83	9191	9196	9201	9206	9212	9217	9222	9227	9232	9238
84	9243	9248	9253	9258	9263	9269	9274	9279	9284	9289
85	9294	9299	9304	9309	9315	9320	9325	9330	9335	9340
86	9345	9350	9355	9360	9365	9370	9375	9380	9385	9390
87	9395	9400	9405	9410	9415	9420	9425	9430	9435	9440
88	9445	9450	9455	9460	9465	9469	9474	9479	9484	9489
89	9494	9499	9504	9509	9513	9518	9523	9528	9533	9538
90	9542	9547	9552	9557	9562	9566	9571	9576	9581	9586
91	9590	9595	9600	9605	9609	9614	9619	9624	9628	9633
92	9638	9643	9647	9652	9657	9661	9666	9671	9675	9680
93	9685	9689	9694	9699	9703	9708	9713	9717	9722	9727
94	9731	9736	9741	9745	9750	9754	9759	9763	9768	9773
95	9777	9782	9786	9791	9795	9800	9805	9809	9814	9818
96	9823	9827	9832	9836	9841	9845	9850	9854	9859	9863
97	9868	9872	9877	9881	9886	9890	9894	9899	9903	9908
98	9912	9917	9921	9926	9930	9934	9939	9943	9948	9952
99	9956	9961	9965	9969	9974	9978	9983	9987	9991	9996
N.	0	1	2	3	4	5	6	7	8	9

Trigonometric Functions

Angles	Sines		Cosines		Tangents		Cotangents		Angles
	Nat.	Log.	Nat.	Log.	Nat.	Log.	Nat.	Log.	
0°00′	.0000	∞	1.0000	0.0000	.0000	∞	∞	∞	90°00′
10	.0029	7.4637	1.0000	0000	.0029	7.4637	343.77	2.5363	50
20	.0058	7648	1.0000	0000	.0058	7648	171.89	2352	40
30	.0087	9408	1.0000	0000	.0087	9409	114.59	0591	30
40	.0116	8.0658	.9999	0000	.0116	8.0658	85.940	1.9342	20
50	.0145	1627	.9999	0000	.0145	1627	68.750	8373	10
1°00′	.0175	8.2419	.9998	9.9999	.0175	8.2419	57.290	1.7581	89°00′
10	.0204	3088	.9998	9999	.0204	3089	49.104	6911	50
20	.0233	3668	.9997	9999	.0233	3669	42.964	6331	40
30	.0262	4179	.9997	9999	.0262	4181	38.188	5819	30
40	.0291	4637	.9996	9998	.0291	4638	34.368	5362	20
50	.0320	5050	.9995	9998	.0320	5053	31.242	4947	10
2°00′	.0349	8.5428	.9994	9.9997	.0349	8.5431	28.636	1.4569	88°00′
10	.0378	5776	.9993	9997	.0378	5779	26.432	4221	50
20	.0407	6097	.9992	9996	.0407	6101	24.542	3899	40
30	.0436	6397	.9990	9996	.0437	6401	22.904	3599	30
40	.0465	6677	.9989	9995	.0466	6682	21.470	3318	20
50	.0494	6940	.9988	9995	.0495	6945	20.206	3055	10
3°00′	.0523	8.7188	.9986	9.9994	.0524	8.7194	19.081	1.2806	87°00′
10	.0552	7423	.9985	9993	.0553	7429	18.075	2571	50
20	.0581	7645	.9983	9993	.0582	7652	17.169	2348	40
30	.0610	7857	.9981	9992	.0612	7865	16.350	2135	30
40	.0640	8059	.9980	9991	.0641	8067	15.605	1933	20
50	.0669	8251	.9978	9990	.0670	8261	14.924	1739	10
4°00′	.0698	8.8436	.9976	9.9989	.0699	8.8446	14.301	1.1554	86°00′
10	.0727	8613	.9974	9989	.0729	8624	13.727	1376	50
20	.0756	8783	.9971	9988	.0758	8795	13.197	1205	40
30	.0785	8946	.9969	9987	.0787	8960	12.706	1040	30
40	.0814	9104	.9967	9986	.0816	9118	12.251	0882	20
50	.0843	9256	.9964	9985	.0846	9272	11.826	0728	10
5°00′	.0872	8.9403	.9962	9.9983	.0875	8.9420	11.430	1.0580	85°00′
10	.0901	9545	.9959	9982	.0904	9563	11.059	0437	50
20	.0929	9682	.9957	9981	.0934	9701	10.712	0299	40
30	.0958	9816	.9954	9980	.0963	9836	10.385	0164	30
40	.0987	9945	.9951	9979	.0992	9966	10.078	0034	20
50	.1016	9.0070	.9948	9977	.1022	9.0093	9.7882	0.9907	10
6°00′	.1045	9.0192	.9945	9.9976	.1051	9.0216	9.5144	0.9784	84°00′
10	.1074	0311	.9942	9975	.1080	0336	9.2553	9664	50
20	.1103	0426	.9939	9973	.1110	0453	9.0098	9547	40
30	.1132	0539	.9936	9972	.1139	0567	8.7769	9433	30
40	.1161	0648	.9932	9971	.1169	0678	8.5555	9322	20
50	.1190	0755	.9929	9969	.1198	0786	8.3450	9214	10
7°00′	.1219	9.0859	.9925	9.9968	.1228	9.0891	8.1443	0.9109	83°00′
10	.1248	0961	.9922	9966	.1257	0995	7.9530	9005	50
20	.1276	1060	.9918	9964	.1287	1096	7.7704	8904	40
30	.1305	1157	.9914	9963	.1317	1194	7.5958	8806	30
40	.1334	1252	.9911	9961	.1346	1291	7.4287	8709	20
50	.1363	1345	.9907	9959	.1376	1385	7.2687	8615	10
8°00′	.1392	9.1436	.9903	9.9958	.1405	9.1478	7.1154	0.8522	82°00′
10	.1421	1525	.9899	9956	.1435	1569	6.9682	8431	50
20	.1449	1612	.9894	9954	.1465	1658	6.8269	8342	40
30	.1478	1697	.9890	9952	.1495	1745	6.6912	8255	30
40	.1507	1781	.9886	9950	.1524	1831	6.5606	8169	20
50	.1536	1863	.9881	9948	.1554	1915	6.4348	8085	10
9°00′	.1564	9.1943	.9877	9.9946	1584	9.1997	6.3138	0.8003	81°00′
	Nat.	Log.	Nat.	Log.	Nat.	Log.	Nat.	Log.	
Angles	Cosines		Sines		Cotangents		Tangents		Angles

TRIGONOMETRIC FUNCTIONS.—(Continued)

Angles	Sines Nat.	Sines Log.	Cosines Nat.	Cosines Log.	Tangents Nat.	Tangents Log.	Cotangents Nat.	Cotangents Log.	Angles
9°00′	.1564	9.1943	.9877	9.9946	.1584	9.1997	6.3138	0.8003	81°00′
10	.1593	2022	.9872	9944	.1614	2078	6.1970	7922	50
20	.1622	2100	.9868	9942	.1644	2158	6.0844	7842	40
30	.1650	2176	.9863	9940	.1673	2236	5.9758	7764	30
40	.1679	2251	.9858	9938	.1703	2313	5.8708	7687	20
50	.1708	2324	.9853	9936	.1733	2389	5.7694	7611	10
10°00′	.1736	9.2397	.9848	9.9934	.1763	9.2463	5.6713	0.7537	80°00′
10	.1765	2468	.9843	9931	.1793	2536	5.5764	7464	50
20	.1794	2538	.9838	9929	.1823	2609	5.4845	7391	40
30	.1822	2606	.9833	9927	.1853	2680	5.3955	7320	30
40	.1851	2674	.9827	9924	.1883	2750	5.3093	7250	20
50	.1880	2740	.9822	9922	.1914	2819	5.2257	7181	10
11°00′	.1908	9.2806	.9816	9.9919	.1944	9.2887	5.1446	0.7113	79°00′
10	.1937	2870	.9811	9917	.1974	2953	5.0658	7047	50
20	.1965	2934	.9805	9914	.2004	3020	4.9894	6980	40
30	.1994	2997	.9799	9912	.2035	3085	4.9152	6915	30
40	.2022	3058	.9793	9909	.2065	3149	4.8430	6851	20
50	.2051	3119	.9787	9907	.2095	3212	4.7729	6788	10
12°00′	.2079	9.3179	.9781	9.9904	.2126	9.3275	4.7046	0.6725	78°00′
10	.2108	3238	.9775	9901	.2156	3336	4.6382	6664	50
20	.2136	3296	.9769	9899	.2186	3397	4.5736	6603	40
30	.2164	3353	.9763	9896	.2217	3458	4.5107	6542	30
40	.2193	3410	.9757	9893	.2247	3517	4.4494	6483	20
50	.2221	3466	.9750	9890	.2278	3576	4.3897	6424	10
13°00′	.2250	9.3521	.9744	9.9887	.2309	9.3634	4.3315	0.6366	77°00′
10	.2278	3575	.9737	9884	.2339	3691	4.2747	6309	50
20	.2306	3629	.9730	9881	.2370	3748	4.2193	6252	40
30	.2334	3682	.9724	9878	.2401	3804	4.1653	6196	30
40	.2363	3734	.9717	9875	.2432	3859	4.1126	6141	20
50	.2391	3786	.9710	9872	.2462	3914	4.0611	6086	10
14°00′	.2419	9.3837	.9703	9.9869	.2493	9.3968	4.0108	0.6032	76°00′
10	.2447	3887	.9696	9866	.2524	4021	3.9617	5979	50
20	.2476	3937	.9689	9863	.2555	4074	3.9136	5926	40
30	.2504	3986	.9681	9859	.2586	4127	3.8667	5873	30
40	.2532	4035	.9674	9856	.2617	4178	3.8208	5822	20
50	.2560	4083	.9667	9853	.2648	4230	3.7760	5770	10
15°00′	.2588	9.4130	.9659	9.9849	.2679	9.4281	3.7321	0.5719	75°00′
10	.2616	4177	.9652	9846	.2711	4331	3.6891	5669	50
20	.2644	4223	.9644	9843	.2742	4381	3.6470	5619	40
30	.2672	4269	.9636	9839	.2773	4430	3.6059	5570	30
40	.2700	4314	.9628	9836	.2805	4479	3.5656	5521	20
50	.2728	4359	.9621	9832	.2836	4527	3.5261	5473	10
16°00′	.2756	9.4403	.9613	9.9828	.2867	9.4575	3.4874	0.5425	74°00
10	.2784	4447	.9605	9825	.2899	4622	3.4495	5378	50
20	.2812	4491	.9596	9821	.2931	4669	3.4124	5331	40
30	.2840	4533	.9588	9817	.2962	4716	3.3759	5284	30
40	.2868	4576	.9580	9814	.2994	4762	3.3402	5238	20
50	.2896	4618	.9572	9810	.3026	4808	3.3052	5192	10
17°00′	.2924	9.4659	.9563	9.9806	.3057	9.4853	3.2709	0.5147	73°00′
10	.2952	4700	.9555	9802	.3089	4898	3.2371	5102	50
20	.2979	4741	.9546	9798	.3121	4943	3.2041	5057	40
30	.3007	4781	.9537	9794	.3153	4987	3.1716	5013	30
40	.3035	4821	.9528	9790	.3185	5031	3.1397	4969	20
50	.3062	4861	.9520	9786	.3217	5073	3.1084	4925	10
18°00′	.3090	9.4900	.9511	9.9782	.3249	9.5118	3.0777	0.4882	72°00′
	Nat.	Log.	Nat.	Log.	Nat.	Log.	Nat.	Log.	
Angles	Cosines		Sines		Cotangents		Tangents		Angles

TRIGONOMETRIC FUNCTIONS.—(*Continued*)

Angles	Sines		Cosines		Tangents		Cotangents		Angles
	Nat.	Log.	Nat.	Log.	Nat.	Log.	Nat.	Log.	
18°00′	.3090	9.4900	.9511	9.9782	.3249	9.5118	3.0777	0.4882	72°00′
10	.3118	4939	.9502	9778	.3281	5161	3.0475	4839	50
20	.3145	4977	.9492	9774	.3314	5203	3.0178	4797	40
30	.3173	5015	.9483	9770	.3346	5245	2.9887	4755	30
40	.3201	5052	.9474	9765	.3378	5287	2.9600	4713	20
50	.3228	5090	.9465	9761	.3411	5329	2.9319	4671	10
19°00′	.3256	9.5126	.9455	9.9757	.3443	9.5370	2.9042	0.4630	71°00′
10	.3283	5163	.9446	9752	.3476	5411	2.8770	4589	50
20	.3311	5199	.9436	9748	.3508	5451	2.8502	4549	40
30	.3338	5235	.9426	9743	.3541	5491	2.8239	4509	30
40	.3365	5270	.9417	9739	.3574	5531	2.7980	4469	20
50	.3393	5306	.9407	9734	.3607	5571	2.7725	4429	10
20°00′	.3420	9.5341	.9397	9.9730	.3640	9.5611	2.7475	0.4389	70°00′
10	.3448	5375	.9387	9725	.3673	5650	2.7228	4350	50
20	.3475	5409	.9377	9721	.3706	5689	2.6985	4311	40
30	.3502	5443	.9367	9716	.3739	5727	2.6746	4273	30
40	.3529	5477	.9356	9711	.3772	5766	2.6511	4234	20
50	.3557	5510	.9346	9706	.3805	5804	2.6279	4196	10
21°00′	.3584	9.5543	.9336	9.9702	.3839	9.5842	2.6051	0.4158	69°00′
10	.3611	5576	.9325	9697	.3872	5879	2.5826	4121	50
20	.3638	5609	.9315	9692	.3906	5917	2.5605	4083	40
30	.3665	5641	.9304	9687	.3939	5954	2.5386	4046	30
40	.3692	5673	.9293	9682	.3973	5991	2.5172	4009	20
50	.3719	5704	.9283	9677	.4006	6028	2.4960	3972	10
22°00′	.3746	9.5736	.9272	9.9672	.4040	9.6064	2.4751	0.3936	68°00′
10	.3773	5767	.9261	9667	.4074	6100	2.4545	3900	50
20	.3800	5798	.9250	9661	.4108	6136	2.4342	3864	40
30	.3827	5828	.9239	9656	.4142	6172	2.4142	3828	30
40	.3854	5859	.9228	9651	.4176	6208	2.3945	3792	20
50	.3881	5889	.9216	9646	.4210	6243	2.3750	3757	10
23°00′	.3907	9.5919	.9205	9.9640	.4245	9.6279	2.3559	0.3721	67°00′
10	.3934	5948	.9194	9635	.4279	6314	2.3369	3686	50
20	.3961	5978	.9182	9629	.4314	6348	2.3183	3652	40
30	.3987	6007	.9171	9624	.4348	6383	2.2998	3617	30
40	.4014	6036	.9159	9618	.4383	6417	2.2817	3583	20
50	.4041	6065	.9147	9613	.4417	6452	2.2637	3548	10
24°00′	.4067	9.6093	.9135	9.9607	.4452	9.6486	2.2460	0.3514	66°00′
10	.4094	6121	.9124	9602	.4487	6520	2.2286	3480	50
20	.4120	6149	.9112	9596	.4522	6553	2.2113	3447	40
30	.4147	6177	.9100	9590	.4557	6587	2.1943	3413	30
40	.4173	6205	.9088	9584	.4592	6620	2.1775	3380	20
50	.4200	6232	.9075	9579	.4628	6654	2.1609	3346	10
25°00′	.4226	9.6259	.9063	9.9573	.4663	9.6687	2.1445	0.3313	65°00′
10	.4253	6286	.9051	9567	.4699	6720	2.1283	3280	50
20	.4279	6313	.9038	9561	.4734	6752	2.1123	3248	40
30	.4305	6340	.9026	9555	.4770	6785	2.0965	3215	30
40	.4331	6366	.9013	9549	.4806	6817	2.0809	3183	20
50	.4358	6392	.9001	9543	.4841	6850	2.0655	3150	10
26°00′	.4384	9.6418	.8988	9.9537	.4877	9.6882	2.0503	0.3118	64°00′
10	.4410	6444	.8975	9530	.4913	6914	2.0353	3086	50
20	.4436	6470	.8962	9524	.4950	6946	2.0204	3054	40
30	.4462	6495	.8949	9518	.4986	6977	2.0057	3023	30
40	.4488	6521	.8936	9512	.5022	7009	1.9912	2991	20
50	.4514	6546	.8923	9505	.5059	7040	1.9768	2960	10
27°00′	.4540	9.6570	.8910	9.9499	.5095	9.7072	1.9626	0.2928	63°00′
	Nat.	Log.	Nat.	Log.	Nat.	Log.	Nat.	Log.	
Angles	Cosines		Sines		Cotangents		Tangents		Angles

TRIGONOMETRIC FUNCTIONS.—(Continued)

Angles	Sines		Cosines		Tangents		Cotangents		Angles
	Nat.	Log.	Nat.	Log.	Nat.	Log.	Nat.	Log.	
27°00'	.4540	9.6570	.8910	9.9499	.5095	9.7072	1.9626	0.2928	63°00'
10	.4566	6595	.8897	9492	.5132	7103	1.9486	2897	50
20	.4592	6620	.8884	9486	.5169	7134	1.9347	2866	40
30	.4617	6644	.8870	9479	.5206	7165	1.9210	2835	30
40	.4643	6668	.8857	9473	.5243	7196	1.9074	2804	20
50	.4669	6692	.8843	9466	.5280	7226	1.8940	2774	10
28°00'	.4695	9.6716	.8829	9.9459	.5317	9.7257	1.8807	0.2743	62°00'
10	.4720	6740	.8816	9453	.5354	7287	1.8676	2713	50
20	.4746	6763	.8802	9446	.5392	7317	1.8546	2683	40
30	.4772	6787	.8788	9439	.5430	7348	1.8418	2652	30
40	.4797	6810	.8774	9432	.5467	7378	1.8291	2622	20
50	.4823	6833	.8760	9425	.5505	7408	1.8165	2592	10
29°00'	.4848	9.6856	.8746	9.9418	.5543	9.7438	1.8040	0.2562	61°00'
10	.4874	6878	.8732	9411	.5581	7467	1.7917	2533	50
20	.4899	6901	.8718	9404	.5619	7497	1.7796	2503	40
30	.4924	6923	.8704	9397	.5658	7526	1.7675	2474	30
40	.4950	6946	.8689	9390	.5696	7556	1.7556	2444	20
50	.4975	6968	.8675	9383	.5735	7585	1.7437	2415	10
30°00'	.5000	9.6990	.8660	9.9375	.5774	9.7614	1.7321	0.2386	60°00'
10	.5025	7012	.8646	9368	.5812	7644	1.7205	2356	50
20	.5050	7033	.8631	9361	.5851	7673	1.7090	2327	40
30	.5075	7055	.8616	9353	.5890	7701	1.6977	2299	30
40	.5100	7076	.8601	9346	.5930	7730	1.6864	2270	20
50	.5125	7097	.8587	9338	.5969	7759	1.6753	2241	10
31°00'	.5150	9.7118	.8572	9.9331	.6009	9.7788	1.6643	0.2212	59°00'
10	.5175	7139	.8557	9323	.6048	7816	1.6534	2184	50
20	.5200	7160	.8542	9315	.6088	7845	1.6426	2155	40
30	.5225	7181	.8526	9308	.6128	7873	1.6319	2127	30
40	.5250	7201	.8511	9300	.6168	7902	1.6212	2098	20
50	.5275	7222	.8496	9292	.6208	7930	1.6107	2070	10
32°00'	.5299	9.7242	.8480	9.9284	.6249	9.7958	1.6003	0.2042	58°00'
10	.5324	7262	.8465	9276	.6289	7986	1.5900	2014	50
20	.5348	7282	.8450	9268	.6330	8014	1.5798	1986	40
30	.5373	7302	.8434	9260	.6371	8042	1.5697	1958	30
40	.5398	7322	.8418	9252	.6412	8070	1.5597	1930	20
50	.5422	7342	.8403	9244	.6453	8097	1.5497	1903	10
33°00'	.5446	9.7361	.8387	9.9236	.6494	9.8125	1.5399	0.1875	57°00'
10	.5471	7380	.8371	9228	.6536	8153	1.5301	1847	50
20	.5495	7400	.8355	9219	.6577	8180	1.5204	1820	40
30	.5519	7419	.8339	9211	.6619	8208	1.5108	1792	30
40	.5544	7438	.8323	9203	.6661	8235	1.5013	1765	20
50	.5568	7457	.8307	9194	.6703	8263	1.4919	1737	10
34°00'	.5592	9.7476	.8290	9.9186	.6745	9.8290	1.4826	0.1710	56°00'
10	.5616	7494	.8274	9177	.6787	8317	1.4733	1683	50
20	.5640	7513	.8258	9169	.6830	8344	1.4641	1656	40
30	.5664	7531	.8241	9160	.6873	8371	1.4550	1629	30
40	.5688	7550	.8225	9151	.6916	8398	1.4460	1602	20
50	.5712	7568	.8208	9142	.6959	8425	1.4370	1575	10
35°00'	.5736	9.7586	.8192	9.9134	.7002	9.8452	1.4281	0.1548	55°00'
10	.5760	7604	.8175	9125	.7046	8479	1.4193	1521	50
20	.5783	7622	.8158	9116	.7089	8506	1.4106	1494	40
30	.5807	7640	.8141	9107	.7133	8533	1.4019	1467	30
40	.5831	7657	.8124	9098	.7177	8559	1.3934	1441	20
50	.5854	7675	.8107	9089	.7221	8586	1.3848	1414	10
36°00'	.5878	9.7692	.8090	9.9080	.7265	9.8613	1.3764	0.1387	54°00'
	Nat.	Log.	Nat.	Log.	Nat.	Log.	Nat.	Log.	
Angles	Cosines		Sines		Cotangents		Tangents		Angles

TRIGONOMETRIC FUNCTIONS.—(Continued)

Angles	Sines		Cosines		Tangents		Cotangents		Angles
	Nat.	Log.	Nat.	Log.	Nat.	Log.	Nat.	Log.	
36°00′	.5878	9.7692	.8090	9.9080	.7265	9.8613	1.3764	0.1387	54°00′
10	.5901	7710	.8073	9070	.7310	8639	1.3680	1361	50
20	.5925	7727	.8056	9061	.7355	8666	1.3597	1334	40
30	.5948	7744	.8039	9052	.7400	8692	1.3514	1308	30
40	.5972	7761	.8021	9042	.7445	8718	1.3432	1282	20
50	.5995	7778	.8004	9033	.7490	8745	1.3351	1255	10
37°00′	.6018	9.7795	.7986	9.9023	.7536	9.8771	1.3270	0.1229	53°00′
10	.6041	7811	.7969	9014	.7581	8797	1.3190	1203	50
20	.6065	7828	.7951	9004	.7627	8824	1.3111	1176	40
30	.6088	7844	.7934	8995	.7673	8850	1.3032	1150	30
40	.6111	7861	.7916	8985	.7720	8876	1.2954	1124	20
50	.6134	7877	.7898	8975	.7766	8902	1.2876	1098	10
38°00′	.6157	9.7893	.7880	9.8965	.7813	9.8928	1.2799	0.1072	52°00′
10	.6180	7910	.7862	8955	.7860	8954	1.2723	1046	50
20	.6202	7926	.7844	8945	.7907	8980	1.2647	1020	40
30	.6225	7941	.7826	8935	.7954	9006	1.2572	0994	30
40	.6248	7957	.7808	8925	.8002	9032	1.2497	0968	20
50	.6271	7973	.7790	8915	.8050	9058	1.2423	0942	10
39°00′	.6293	9.7989	.7771	9.8905	.8098	9.9084	1.2349	0.0916	51°00′
10	.6316	8004	.7753	8895	.8146	9110	1.2276	0890	50
20	.6338	8020	.7735	8884	.8195	9135	1.2203	0865	40
30	.6361	8035	.7716	8874	.8243	9161	1.2131	0839	30
40	.6383	8050	.7698	8864	.8292	9187	1.2059	0813	20
50	.6406	8066	.7679	8853	.8342	9212	1.1988	0788	10
40°00′	.6428	9.8081	.7660	9.8843	.8391	9.9238	1.1918	0.0762	50°00′
10	.6450	8096	.7642	8832	.8441	9264	1.1847	0736	50
20	.6472	8111	.7623	8821	.8491	9289	1.1778	0711	40
30	.6494	8125	.7604	8810	.8541	9315	1.1708	0685	30
40	.6517	8140	.7585	8800	.8591	9341	1.1640	0659	20
50	.6539	8155	.7566	8789	.8642	9366	1.1571	0634	10
41°00′	.6561	9.8169	.7547	9.8778	.8693	9.9392	1.1504	0.0608	49°00′
10	.6583	8184	.7528	8767	.8744	9417	1.1436	0583	50
20	.6604	8198	.7509	8756	.8796	9443	1.1369	0557	40
30	.6626	8213	.7490	8745	.8847	9468	1.1303	0532	30
40	.6648	8227	.7470	8733	.8899	9494	1.1237	0506	20
50	.6670	8241	.7451	8722	.8952	9519	1.1171	0481	10
42°00′	.6691	9.8255	.7431	9.8711	.9004	9.9544	1.1106	0.0456	48°00′
10	.6713	8269	.7412	8699	.9057	9570	1.1041	0430	50
20	.6734	8283	.7392	8688	.9110	9595	1.0977	0405	40
30	.6756	8297	.7373	8676	.9163	9621	1.0913	0379	30
40	.6777	8311	.7353	8665	.9217	9646	1.0850	0354	20
50	.6799	8324	.7333	8653	.9271	9671	1.0786	0329	10
43°00′	.6820	9.8338	.7314	9.8641	.9325	9.9697	1.0724	0.0303	47°00′
10	.6841	8351	.7294	8629	.9380	9722	1.0661	0278	50
20	.6862	8365	.7274	8618	.9435	9747	1.0599	0253	40
30	.6884	8378	.7254	8606	.9490	9772	1.0538	0228	30
40	.6905	8391	.7234	8594	.9545	9798	1.0477	0202	20
50	.6926	8405	.7214	8582	.9601	9823	1.0416	0177	10
44°00′	.6947	9.8418	.7193	9.8569	.9657	9.9848	1.0355	0.0152	46°00′
10	.6967	8431	.7173	8557	.9713	9874	1.0295	0126	50
20	.6988	8444	.7153	8545	.9770	9899	1.0235	0101	40
30	.7009	8457	.7133	8532	.9827	9924	1.0176	0076	30
40	.7030	8469	.7112	8520	.9884	9949	1.0117	0051	20
50	.7050	8482	.7092	8507	.9942	9975	1.0058	0025	10
45°00′	7071	9.8495	.7071	9.8495	1.0000	0.0000	1.0000	0.0000	45°00′
	Nat.	Log.	Nat.	Log.	Nat.	Log.	Nat.	Log.	
Angles	Cosines		Sines		Cotangents		Tangents		Angles

A Useful Table of Napierian or Natural Logarithms

Number	log$_e$	Number	log$_e$
1.0	0.0000	6.0	1.7918
1.1	0.0953	7.0	1.9459
1.2	0.1823	8.0	2.0794
1.5	0.4055	9.0	2.1972
1.7	0.5306	10.0	2.3026
2.0	0.6931	20.0	2.9957
2.2	0.7885	50.0	3.9120
2.5	0.9163	100.0	4.6052
2.7	0.9933	200.0	5.2983
2.8	1.0296	500.0	6.2146
3.0	1.0986	1,000.0	6.9078
3.5	1.2528	2,000.0	7.6010
4.0	1.3863	5,000.0	8.5172
4.5	1.5041	10,000.0	9.2104
5.0	1.6094	20,000.0	9.9035

INDEX

A

Applications of calculus, 326–373
aeronautics, airplane problem, 326
rate of paving landing field, 327
approximations by differentials, volume, value, error, percentage error, 370–373
bacteriology and chemistry, building up salt solution, 349
diluting salt solution, 349
number of bacteria in culture, 348
beams, resisting moment of, 346
simple, elastic curve and maximum deflection of, 347
relation of moment and shear in, 345
strongest rectangular, cut from round logs, 344
belt and pulley drives, relation of maximum power and centrifugal tension in, 351
relation of tight and slack tensions of, 352–354
bodies in motion, angular velocity and acceleration of rotating wheel, 357
distance of falling body, 356, 360
distance of moving body, 360
height to which ball will rise, 357
relation between angular and linear velocities, 358
relation between tangential and angular acceleration, 358
revolving flywheel, 361
velocity of falling body, 356

Applications of calculus, electricity, connecting cells of storage battery for maximum power, 336
current for maximum efficiency of transformer, 337
maximum current from storage battery, 337
relation of current and e.m.f., 338
excavation, minimum cost for, of pipe line, 344
expansion of metal, rate of, under heat, 366, 367
relation of increase of area and coefficient of expansion, 369
fan, relation of pressure and speed of, 365
gas and air, pressure of gas, 354
work done by expanding air, 355
illumination, placing light for maximum, 364
kinetic energy, of body, 351
maximum and minimum, forming box from square sheet of metal, 368
height of projectile, 331
most economical speed of steamer, 330
power output of impulse turbine, 331
surface of cylindrical tank of definite capacity, 367
projectiles, distance traveled by, 334
height of, 331
motion of, 332
velocity of bullet, 334, 358

423

A CATALOG OF SELECTED
DOVER BOOKS
IN ALL FIELDS OF INTEREST

A CATALOG OF SELECTED DOVER
BOOKS IN ALL FIELDS OF INTEREST

CONCERNING THE SPIRITUAL IN ART, Wassily Kandinsky. Pioneering work by father of abstract art. Thoughts on color theory, nature of art. Analysis of earlier masters. 12 illustrations. 80pp. of text. 5⅜ x 8½. 23411-8

ANIMALS: 1,419 Copyright-Free Illustrations of Mammals, Birds, Fish, Insects, etc., Jim Harter (ed.). Clear wood engravings present, in extremely lifelike poses, over 1,000 species of animals. One of the most extensive pictorial sourcebooks of its kind. Captions. Index. 284pp. 9 x 12. 23766-4

CELTIC ART: The Methods of Construction, George Bain. Simple geometric techniques for making Celtic interlacements, spirals, Kells-type initials, animals, humans, etc. Over 500 illustrations. 160pp. 9 x 12. (Available in U.S. only.) 22923-8

AN ATLAS OF ANATOMY FOR ARTISTS, Fritz Schider. Most thorough reference work on art anatomy in the world. Hundreds of illustrations, including selections from works by Vesalius, Leonardo, Goya, Ingres, Michelangelo, others. 593 illustrations. 192pp. 7⅛ x 10¼. 20241-0

CELTIC HAND STROKE-BY-STROKE (Irish Half-Uncial from "The Book of Kells"): An Arthur Baker Calligraphy Manual, Arthur Baker. Complete guide to creating each letter of the alphabet in distinctive Celtic manner. Covers hand position, strokes, pens, inks, paper, more. Illustrated. 48pp. 8¼ x 11. 24336-2

EASY ORIGAMI, John Montroll. Charming collection of 32 projects (hat, cup, pelican, piano, swan, many more) specially designed for the novice origami hobbyist. Clearly illustrated easy-to-follow instructions insure that even beginning papercrafters will achieve successful results. 48pp. 8¼ x 11. 27298-2

THE COMPLETE BOOK OF BIRDHOUSE CONSTRUCTION FOR WOODWORKERS, Scott D. Campbell. Detailed instructions, illustrations, tables. Also data on bird habitat and instinct patterns. Bibliography. 3 tables. 63 illustrations in 15 figures. 48pp. 5¼ x 8½. 24407-5

BLOOMINGDALE'S ILLUSTRATED 1886 CATALOG: Fashions, Dry Goods and Housewares, Bloomingdale Brothers. Famed merchants' extremely rare catalog depicting about 1,700 products: clothing, housewares, firearms, dry goods, jewelry, more. Invaluable for dating, identifying vintage items. Also, copyright-free graphics for artists, designers. Co-published with Henry Ford Museum & Greenfield Village. 160pp. 8¼ x 11. 25780-0

HISTORIC COSTUME IN PICTURES, Braun & Schneider. Over 1,450 costumed figures in clearly detailed engravings–from dawn of civilization to end of 19th century. Captions. Many folk costumes. 256pp. 8⅜ x 11¾. 23150-X

THE CLARINET AND CLARINET PLAYING, David Pino. Lively, comprehensive work features suggestions about technique, musicianship, and musical interpretation, as well as guidelines for teaching, making your own reeds, and preparing for public performance. Includes an intriguing look at clarinet history. "A godsend," *The Clarinet,* Journal of the International Clarinet Society. Appendixes. 7 illus. 320pp. 5⅜ x 8½. 40270-3

HOLLYWOOD GLAMOR PORTRAITS, John Kobal (ed.). 145 photos from 1926-49. Harlow, Gable, Bogart, Bacall; 94 stars in all. Full background on photographers, technical aspects. 160pp. 8⅜ x 11¼. 23352-9

THE ANNOTATED CASEY AT THE BAT: A Collection of Ballads about the Mighty Casey/Third, Revised Edition, Martin Gardner (ed.). Amusing sequels and parodies of one of America's best-loved poems: Casey's Revenge, Why Casey Whiffed, Casey's Sister at the Bat, others. 256pp. 5⅜ x 8½. 28598-7

THE RAVEN AND OTHER FAVORITE POEMS, Edgar Allan Poe. Over 40 of the author's most memorable poems: "The Bells," "Ulalume," "Israfel," "To Helen," "The Conqueror Worm," "Eldorado," "Annabel Lee," many more. Alphabetic lists of titles and first lines. 64pp. 5$\frac{3}{16}$ x 8¼. 26685-0

PERSONAL MEMOIRS OF U. S. GRANT, Ulysses Simpson Grant. Intelligent, deeply moving firsthand account of Civil War campaigns, considered by many the finest military memoirs ever written. Includes letters, historic photographs, maps and more. 528pp. 6⅛ x 9¼. 28587-1

ANCIENT EGYPTIAN MATERIALS AND INDUSTRIES, A. Lucas and J. Harris. Fascinating, comprehensive, thoroughly documented text describes this ancient civilization's vast resources and the processes that incorporated them in daily life, including the use of animal products, building materials, cosmetics, perfumes and incense, fibers, glazed ware, glass and its manufacture, materials used in the mummification process, and much more. 544pp. 6⅛ x 9¼. (Available in U.S. only.) 40446-3

RUSSIAN STORIES/RUSSKIE RASSKAZY: A Dual-Language Book, edited by Gleb Struve. Twelve tales by such masters as Chekhov, Tolstoy, Dostoevsky, Pushkin, others. Excellent word-for-word English translations on facing pages, plus teaching and study aids, Russian/English vocabulary, biographical/critical introductions, more. 416pp. 5⅜ x 8½. 26244-8

PHILADELPHIA THEN AND NOW: 60 Sites Photographed in the Past and Present, Kenneth Finkel and Susan Oyama. Rare photographs of City Hall, Logan Square, Independence Hall, Betsy Ross House, other landmarks juxtaposed with contemporary views. Captures changing face of historic city. Introduction. Captions. 128pp. 8¼ x 11. 25790-8

AIA ARCHITECTURAL GUIDE TO NASSAU AND SUFFOLK COUNTIES, LONG ISLAND, The American Institute of Architects, Long Island Chapter, and the Society for the Preservation of Long Island Antiquities. Comprehensive, well-researched and generously illustrated volume brings to life over three centuries of Long Island's great architectural heritage. More than 240 photographs with authoritative, extensively detailed captions. 176pp. 8¼ x 11. 26946-9

NORTH AMERICAN INDIAN LIFE: Customs and Traditions of 23 Tribes, Elsie Clews Parsons (ed.). 27 fictionalized essays by noted anthropologists examine religion, customs, government, additional facets of life among the Winnebago, Crow, Zuni, Eskimo, other tribes. 480pp. 6⅛ x 9¼. 27377-6

STICKLEY CRAFTSMAN FURNITURE CATALOGS, Gustav Stickley and L. & J. G. Stickley. Beautiful, functional furniture in two authentic catalogs from 1910. 594 illustrations, including 277 photos, show settles, rockers, armchairs, reclining chairs, bookcases, desks, tables. 183pp. 6½ x 9¼. 23838-5

AMERICAN LOCOMOTIVES IN HISTORIC PHOTOGRAPHS: 1858 to 1949, Ron Ziel (ed.). A rare collection of 126 meticulously detailed official photographs, called "builder portraits," of American locomotives that majestically chronicle the rise of steam locomotive power in America. Introduction. Detailed captions. xi+ 129pp. 9 x 12. 27393-8

AMERICA'S LIGHTHOUSES: An Illustrated History, Francis Ross Holland, Jr. Delightfully written, profusely illustrated fact-filled survey of over 200 American light-houses since 1716. History, anecdotes, technological advances, more. 240pp. 8 x 10¾. 25576-X

TOWARDS A NEW ARCHITECTURE, Le Corbusier. Pioneering manifesto by founder of "International School." Technical and aesthetic theories, views of industry, economics, relation of form to function, "mass-production split" and much more. Profusely illustrated. 320pp. 6⅛ x 9¼. (Available in U.S. only.) 25023-7

HOW THE OTHER HALF LIVES, Jacob Riis. Famous journalistic record, exposing poverty and degradation of New York slums around 1900, by major social reformer. 100 striking and influential photographs. 233pp. 10 x 7⅞. 22012-5

FRUIT KEY AND TWIG KEY TO TREES AND SHRUBS, William M. Harlow. One of the handiest and most widely used identification aids. Fruit key covers 120 deciduous and evergreen species; twig key 160 deciduous species. Easily used. Over 300 photographs. 126pp. 5⅜ x 8½. 20511-8

COMMON BIRD SONGS, Dr. Donald J. Borror. Songs of 60 most common U.S. birds: robins, sparrows, cardinals, bluejays, finches, more—arranged in order of increasing complexity. Up to 9 variations of songs of each species.
Cassette and manual 99911-4

ORCHIDS AS HOUSE PLANTS, Rebecca Tyson Northen. Grow cattleyas and many other kinds of orchids—in a window, in a case, or under artificial light. 63 illustrations. 148pp. 5⅜ x 8½. 23261-1

MONSTER MAZES, Dave Phillips. Masterful mazes at four levels of difficulty. Avoid deadly perils and evil creatures to find magical treasures. Solutions for all 32 exciting illustrated puzzles. 48pp. 8¼ x 11. 26005-4

MOZART'S DON GIOVANNI (DOVER OPERA LIBRETTO SERIES), Wolfgang Amadeus Mozart. Introduced and translated by Ellen H. Bleiler. Standard Italian libretto, with complete English translation. Convenient and thoroughly portable—an ideal companion for reading along with a recording or the performance itself. Introduction. List of characters. Plot summary. 121pp. 5¼ x 8½. 24944-1

TECHNICAL MANUAL AND DICTIONARY OF CLASSICAL BALLET, Gail Grant. Defines, explains, comments on steps, movements, poses and concepts. 15-page pictorial section. Basic book for student, viewer. 127pp. 5⅜ x 8½. 21843-0

FRANK LLOYD WRIGHT'S DANA HOUSE, Donald Hoffmann. Pictorial essay of residential masterpiece with over 160 interior and exterior photos, plans, elevations, sketches and studies. 128pp. 9¼ x 10¾. 29120-0

THE MALE AND FEMALE FIGURE IN MOTION: 60 Classic Photographic Sequences, Eadweard Muybridge. 60 true-action photographs of men and women walking, running, climbing, bending, turning, etc., reproduced from rare 19th-century masterpiece. vi + 121pp. 9 x 12. 24745-7

1001 QUESTIONS ANSWERED ABOUT THE SEASHORE, N. J. Berrill and Jacquelyn Berrill. Queries answered about dolphins, sea snails, sponges, starfish, fishes, shore birds, many others. Covers appearance, breeding, growth, feeding, much more. 305pp. 5¼ x 8¼. 23366-9

ATTRACTING BIRDS TO YOUR YARD, William J. Weber. Easy-to-follow guide offers advice on how to attract the greatest diversity of birds: birdhouses, feeders, water and waterers, much more. 96pp. 5³⁄₁₆ x 8¼. 28927-3

MEDICINAL AND OTHER USES OF NORTH AMERICAN PLANTS: A Historical Survey with Special Reference to the Eastern Indian Tribes, Charlotte Erichsen-Brown. Chronological historical citations document 500 years of usage of plants, trees, shrubs native to eastern Canada, northeastern U.S. Also complete identifying information. 343 illustrations. 544pp. 6½ x 9¼. 25951-X

STORYBOOK MAZES, Dave Phillips. 23 stories and mazes on two-page spreads: Wizard of Oz, Treasure Island, Robin Hood, etc. Solutions. 64pp. 8¼ x 11. 23628-5

AMERICAN NEGRO SONGS: 230 Folk Songs and Spirituals, Religious and Secular, John W. Work. This authoritative study traces the African influences of songs sung and played by black Americans at work, in church, and as entertainment. The author discusses the lyric significance of such songs as "Swing Low, Sweet Chariot," "John Henry," and others and offers the words and music for 230 songs. Bibliography. Index of Song Titles. 272pp. 6½ x 9¼. 40271-1

MOVIE-STAR PORTRAITS OF THE FORTIES, John Kobal (ed.). 163 glamor, studio photos of 106 stars of the 1940s: Rita Hayworth, Ava Gardner, Marlon Brando, Clark Gable, many more. 176pp. 8⅜ x 11¼. 23546-7

BENCHLEY LOST AND FOUND, Robert Benchley. Finest humor from early 30s, about pet peeves, child psychologists, post office and others. Mostly unavailable elsewhere. 73 illustrations by Peter Arno and others. 183pp. 5⅜ x 8½. 22410-4

YEKL and THE IMPORTED BRIDEGROOM AND OTHER STORIES OF YIDDISH NEW YORK, Abraham Cahan. Film Hester Street based on *Yekl* (1896). Novel, other stories among first about Jewish immigrants on N.Y.'s East Side. 240pp. 5⅜ x 8½. 22427-9

SELECTED POEMS, Walt Whitman. Generous sampling from *Leaves of Grass.* Twenty-four poems include "I Hear America Singing," "Song of the Open Road," "I Sing the Body Electric," "When Lilacs Last in the Dooryard Bloom'd," "O Captain! My Captain!"–all reprinted from an authoritative edition. Lists of titles and first lines. 128pp. 5³⁄₁₆ x 8¼. 26878-0

THE BEST TALES OF HOFFMANN, E. T. A. Hoffmann. 10 of Hoffmann's most important stories: "Nutcracker and the King of Mice," "The Golden Flowerpot," etc. 458pp. 5⅜ x 8½. 21793-0

FROM FETISH TO GOD IN ANCIENT EGYPT, E. A. Wallis Budge. Rich detailed survey of Egyptian conception of "God" and gods, magic, cult of animals, Osiris, more. Also, superb English translations of hymns and legends. 240 illustrations. 545pp. 5⅜ x 8½. 25803-3

FRENCH STORIES/CONTES FRANÇAIS: A Dual-Language Book, Wallace Fowlie. Ten stories by French masters, Voltaire to Camus: "Micromegas" by Voltaire; "The Atheist's Mass" by Balzac; "Minuet" by de Maupassant; "The Guest" by Camus, six more. Excellent English translations on facing pages. Also French-English vocabulary list, exercises, more. 352pp. 5⅜ x 8½. 26443-2

CHICAGO AT THE TURN OF THE CENTURY IN PHOTOGRAPHS: 122 Historic Views from the Collections of the Chicago Historical Society, Larry A. Viskochil. Rare large-format prints offer detailed views of City Hall, State Street, the Loop, Hull House, Union Station, many other landmarks, circa 1904-1913. Introduction. Captions. Maps. 144pp. 9⅜ x 12¼. 24656-6

OLD BROOKLYN IN EARLY PHOTOGRAPHS, 1865-1929, William Lee Younger. Luna Park, Gravesend race track, construction of Grand Army Plaza, moving of Hotel Brighton, etc. 157 previously unpublished photographs. 165pp. 8⅞ x 11¾. 23587-4

THE MYTHS OF THE NORTH AMERICAN INDIANS, Lewis Spence. Rich anthology of the myths and legends of the Algonquins, Iroquois, Pawnees and Sioux, prefaced by an extensive historical and ethnological commentary. 36 illustrations. 480pp. 5⅜ x 8½. 25967-6

AN ENCYCLOPEDIA OF BATTLES: Accounts of Over 1,560 Battles from 1479 B.C. to the Present, David Eggenberger. Essential details of every major battle in recorded history from the first battle of Megiddo in 1479 B.C. to Grenada in 1984. List of Battle Maps. New Appendix covering the years 1967-1984. Index. 99 illustrations. 544pp. 6½ x 9¼. 24913-1

SAILING ALONE AROUND THE WORLD, Captain Joshua Slocum. First man to sail around the world, alone, in small boat. One of great feats of seamanship told in delightful manner. 67 illustrations. 294pp. 5⅜ x 8½. 20326-3

ANARCHISM AND OTHER ESSAYS, Emma Goldman. Powerful, penetrating, prophetic essays on direct action, role of minorities, prison reform, puritan hypocrisy, violence, etc. 271pp. 5⅜ x 8½. 22484-8

MYTHS OF THE HINDUS AND BUDDHISTS, Ananda K. Coomaraswamy and Sister Nivedita. Great stories of the epics; deeds of Krishna, Shiva, taken from puranas, Vedas, folk tales; etc. 32 illustrations. 400pp. 5⅜ x 8½. 21759-0

THE TRAUMA OF BIRTH, Otto Rank. Rank's controversial thesis that anxiety neurosis is caused by profound psychological trauma which occurs at birth. 256pp. 5⅜ x 8½. 27974-X

A THEOLOGICO-POLITICAL TREATISE, Benedict Spinoza. Also contains unfinished Political Treatise. Great classic on religious liberty, theory of government on common consent. R. Elwes translation. Total of 421pp. 5⅜ x 8½. 20249-6

CATALOG OF DOVER BOOKS

THE STORY OF THE TITANIC AS TOLD BY ITS SURVIVORS, Jack Winocour (ed.). What it was really like. Panic, despair, shocking inefficiency, and a little heroism. More thrilling than any fictional account. 26 illustrations. 320pp. 5⅜ x 8½. 20610-6

FAIRY AND FOLK TALES OF THE IRISH PEASANTRY, William Butler Yeats (ed.). Treasury of 64 tales from the twilight world of Celtic myth and legend: "The Soul Cages," "The Kildare Pooka," "King O'Toole and his Goose," many more. Introduction and Notes by W. B. Yeats. 352pp. 5⅜ x 8½. 26941-8

BUDDHIST MAHAYANA TEXTS, E. B. Cowell and others (eds.). Superb, accurate translations of basic documents in Mahayana Buddhism, highly important in history of religions. The Buddha-karita of Asvaghosha, Larger Sukhavativyuha, more. 448pp. 5⅜ x 8½. 25552-2

ONE TWO THREE . . . INFINITY: Facts and Speculations of Science, George Gamow. Great physicist's fascinating, readable overview of contemporary science: number theory, relativity, fourth dimension, entropy, genes, atomic structure, much more. 128 illustrations. Index. 352pp. 5⅜ x 8½. 25664-2

EXPERIMENTATION AND MEASUREMENT, W. J. Youden. Introductory manual explains laws of measurement in simple terms and offers tips for achieving accuracy and minimizing errors. Mathematics of measurement, use of instruments, experimenting with machines. 1994 edition. Foreword. Preface. Introduction. Epilogue. Selected Readings. Glossary. Index. Tables and figures. 128pp. 5⅜ x 8½. 40451-X

DALÍ ON MODERN ART: The Cuckolds of Antiquated Modern Art, Salvador Dalí. Influential painter skewers modern art and its practitioners. Outrageous evaluations of Picasso, Cézanne, Turner, more. 15 renderings of paintings discussed. 44 calligraphic decorations by Dalí. 96pp. 5⅜ x 8½. (Available in U.S. only.) 29220-7

ANTIQUE PLAYING CARDS: A Pictorial History, Henry René D'Allemagne. Over 900 elaborate, decorative images from rare playing cards (14th–20th centuries): Bacchus, death, dancing dogs, hunting scenes, royal coats of arms, players cheating, much more. 96pp. 9¼ x 12¼. 29265-7

MAKING FURNITURE MASTERPIECES: 30 Projects with Measured Drawings, Franklin H. Gottshall. Step-by-step instructions, illustrations for constructing handsome, useful pieces, among them a Sheraton desk, Chippendale chair, Spanish desk, Queen Anne table and a William and Mary dressing mirror. 224pp. 8⅛ x 11¼. 29338-6

THE FOSSIL BOOK: A Record of Prehistoric Life, Patricia V. Rich et al. Profusely illustrated definitive guide covers everything from single-celled organisms and dinosaurs to birds and mammals and the interplay between climate and man. Over 1,500 illustrations. 760pp. 7½ x 10⅞. 29371-8

Paperbound unless otherwise indicated. Available at your book dealer, online at **www.doverpublications.com**, or by writing to Dept. GI, Dover Publications, Inc., 31 East 2nd Street, Mineola, NY 11501. For current price information or for free catalogues (please indicate field of interest), write to Dover Publications or log on to **www.doverpublications.com** and see every Dover book in print. Dover publishes more than 500 books each year on science, elementary and advanced mathematics, biology, music, art, literary history, social sciences, and other areas.